P9-CLC-622

Cosmological Constants

Cosmological Constants

Papers in Modern Cosmology

Edited by Jeremy Bernstein and
Gerald Feinberg

New York Columbia University Press 1986

Columbia University Press
New York Guildford, Surrey

Printed in the United States of America

Library of Congress Cataloging-in-Publication Data

Main entry under title:

Cosmological constants.

Includes bibliographies.
1. Cosmology—Addresses, essays, lectures.
2. Astrophysics—Addresses, essays, lectures.
I. Bernstein, Jeremy, 1929– . II. Feinberg, Gerald, 1933–
QB985.C66 1986 523.1 86-2220
ISBN 0-231-06376-8

To the memory of our teachers
Philipp Frank and Ernest Nagel

CONTENTS

ACKNOWLEDGMENT OF PERMISSIONS

The translation of the article "Cosmological Considerations on the General Theory of Relativity" by A. Einstein is reprinted from the book The Principle of Relativity, published by Methuen and Co., London, with permission of the publishers.

The paper "On Einstein's Theory of Gravitation and its Astronomical Consequences" by W. de Sitter is reprinted with the permission of the Royal Astronomical Society.

The translations of the papers "On the Curvature of Space" and "On the Possibility of a World with Constant Negative Curvature" by A. Friedmann are published with the permission of Springer-Verlag, who hold the copyright to the originals.

The translations of the papers "Remark on the Work of A. Friedmann 'On the Curvature of Space'" and "A Note on the Work of A. Friedmann 'On the Curvature of Space'", by A. Einstein are published with the permission of Springer-Verlag, who hold the copyright to the originals.

The paper "On the Foundations of Relativistic Cosmology" by H. P. Robertson is reprinted with the permission of the National Academy of Sciences.

The paper "A Relation Between Distance and Radial Velocity Among Extra-galactic Nebulae" by E. P. Hubble is reprinted with the permission of the National Academy of Sciences.

The paper "A New Determination of the Hubble Constant from Globular Clusters in M87" by A. Sandage is reprinted with the permission of the author.

The paper "A Homogeneous Universe of Constant Mass and Increasing Radius Accounting for the Radial Velocity of Extra-galactic Nebulae" by G. Lemaître reprinted with the permission of the Royal Astronomical Society.

The paper "An Upper Limit on the Neutrino Rest Mass" by R. Cowsik and J. McCelland is reprinted with the permission of the American Physical Society and the authors.

The paper "The Origin of Elements and the Separation of Galaxies" by G. Gamow is reprinted with the permission of the American Physical Society.

The paper "Remarks on the Evolution of the Expanding Universe" by R. Alpher and R. Herman is reprinted with the permission of the American Physical Society and the authors.

The paper "Cosmic Black-Body Radiation" by R. Dicke, P. Peebles, P. Roll and D. Wilkinson is reprinted with the permission of the authors.

The paper "A Measurement of Excess Antenna Temperature at 4080 Mc/s" by A. Penzias and R. Wilson is reprinted with the permission of the authors and of the AT&T Bell Laboratories, Holmdel, New Jersey, USA.

The paper "Measurement of the Spectrum of the Submillimeter Cosmic Background" by D. Woody, J. Mather, N. Nishioka and P. Richards is reprinted with the permission of the American Physical Society and the authors.

The paper "Physical Conditions in the Initial Stages of the Expanding Universe" by R. Alpher, J. Follin, Jr., and R. Herman is reprinted with the permission of the American Physical Society and the authors.

The paper "Primordial Helium Abundance and the Primordial Fireball, II" by P. Peebles is published with the permission of the author.

The translation of the paper "Violation of CP Invariance, C Asymmetry and Baryon Asymmetry of the Universe" by A. Sakharov is reprinted with permission from *Soviet Physics JETP Letters*, Vol. 5, N. 1, pp. 24-27 (1967) © 1967 American Institute of Physics.

The paper "Unified Gauge Theories and the Baryon Number of the Universe" by M. Yoshimura is reprinted with the permission of the American Physical Society and the author.

The paper "Baryon Number of the Universe" by S. Dimopoulos and L. Susskind is reprinted with the permission of the American Physical Society and the authors.

The paper "Cosmological Production of Baryons" by S. Weinberg is reprinted with the permission of the American Physical Society and the author.

The paper "Topology of Cosmological Domains and Strings" by T. Kibble is reprinted with the permission of the Institute of Physics and the author.

The paper "Cosmological Production of Superheavy Magnetic Monopoles" by J. Preskill is reprinted with the permission of the American Physical Society and the author.

The paper "Inflationary Universe: A Possible Solution to the Horizon and Flatness Problems" by A. Guth is reprinted with the permission of the American Physical Society and the author.

The paper "Cosmology for Grand Unified Theories with Radiatively Induced Symmetry Breaking" by A. Albrecht and P. Steinhardt is reprinted with the permission of the American Physical Society and the authors.

ACKNOWLEDGMENTS

We wish to thank a number of people who have helped us produce this book. Dr. Lowell Brown and Dr. Erick Weinberg read parts of the manuscript and gave us the benefit of their comments. Dr. Jacob Shaham and Dr. Steven Weinberg suggested several papers for inclusion. Susan Mescher word-processed the manuscript. James Danella redrew the illustrations for the various articles. Our editor, Edward Lugenbeel was highly supportive of our efforts. Finally, and most importantly, many of our colleagues kindly granted us permission to reprint their papers in this collection. This book would not have appeared without their contributions.

PREFACE

In this preface we would like to describe what we have done in this book and also what we have not done. It must be evident to every scientifically literate observer that in the last few years cosmology - the study of the universe at large, its history and its future - has become one of the most active fields in science. It is occupying the attention of scientists from several disciplines: astronomers and astro-physicists: particle theorists and mathematicians, as well as the odd chemist. The literature is proliferating so rapidly that even the specialists who devote their careers to the discipline cannot keep up with it. Much of this literature is destined to evaporate like last season's snow. However, a few papers will survive and become what we call "cosmological constants" - landmark papers. Some of these papers contain brilliant new ideas. Some contain observations of an absolutely crucial nature, and some have influenced the field because of their pedagogical clarity. This book contains a selection - only a selection - of such papers. We would have liked to include many more. But when we began adding up the words we realized that to have done so would have doubled or tripled the size of the book, making it both unwieldy and unaffordable. So we have been Draconian - even excluding a paper of our own to which we were rather attached. In this spirit we hope that friends and colleagues will forgive us our choices and our omissions.

The papers we present are grouped into four categories, which we believe are the aspects of cosmology that have had the most lasting value.

I. The theoretical and observational basis of the concept of an expanding universe.
II. The theory and evidence for the cosmological black body radiation.
III. The theoretical analysis of the formation of light elements in the early universe.
IV. Processes in the very early universe.

We have omitted as being not very relevant to current views, although they were quite influential for a time, such topics as the steady-state cosmology and Milne's kinematic relativity.

We have redone all the papers in a uniform format, essentially that used in the *Astrophysical Journal*. Some equations have been renumbered, and in a few of the early papers we have changed the notation to conform with present usage. We have corrected some typographical errors, but we have not tried to correct scientific errors in the original papers. However, in our introductions which begin each section, we comment on some such errors, as well as on the significance of the individual papers for later developments in cosmology.

HISTORICAL INTRODUCTION

It is said that man is the only animal that contemplates its own death. In view of this it is not surprising that one of the things that all races of mankind have had in common is an attempt to account for the birth and death of the universe as a whole - to make a cosmology. The flavor of the primitive cosmologies is, of course, culturally relativistic - often intermingling the supernatural and the familiar. Here, for example, is an early Chinese myth: "First there was the great cosmic egg. Inside the egg was chaos, and floating in chaos was P'an Ku, the Undeveloped, the divine Embryo. And P'an Ku burst out of the egg, four times larger than any man today, with a hammer and chisel in his hand with which he fashioned the world." The homey vision of universe as egg persisted well into the Middle Ages. The noted twelfth-century visionary Hildegard of Bingen - speaking of the universe - noted that she saw "a gigantic image, round and shadowy; like an egg, it was less large at the top, wider in the middle, and narrower again at the base." The illuminated manuscripts of the Middle Ages show the ovoid universe festooned like one of those Easter eggs Fabergé made for the Tsar.

The creation passages of the Brinandaranyka Upanishad start "In the beginning nothing at all existed here. This whole world was enveloped by Death - by Hunger. For what is death but hunger? And Death be thought himself! Would that I had a self! He roamed around, offering praise: and from him, as he offered praise water was born . . ." What could be more revealing about a civilization than the question, "For what is death but hunger?" "In the beginning of creation," begins the book of Genesis, "when God made heaven and earth, the earth was without form and void, with darkness over the face of the abyss, and a mighty wind that swept over the surface of the waters. God said, 'Let there be light and there was light . . . He called the light day and the darkness night . . ." According to the Big Bang cosmology - the proper subject of this book - while there was light and lots of other radiation at the origin of the universe, there was no darkness until many thousands of years later, when the universe had cooled down sufficiently that the radiation filling it was no longer in the visible part of the spectrum.

There were notable exceptions in the ancient world to the idea that the universe had to have a beginning, and hence a creator. One was Aristotle, who might be called the Fred Hoyle of ancient Greece. He, like Hoyle, believed in something like what has been named the "perfect cosmological principal." According to this the universe has neither a beginning nor an end but just is, unchanging throughout eternity. Lucretius, the Roman poet, gave arguments for the view that the universe was infinite in space. He asked a question that often occurs to those confronted for the first time with the notion of a finite universe, i.e., what lies beyond the end of space? From our perspective, this argument confuses finitude with boundedness, and is not conclusive.

In the post-Renaissance revival, speculative thought returned to the question of the age of the universe. By 1654, the Irish divine James Ussher provided, on the basis of a lifetime of study of biblical chronology, the exact date and time of the beginning. It was on October 26, 4004 B.C. at nine A.M. - presumably Greenwich mean time. This kind of literal biblical chronology - which seems absurd to us - was taken very seriously by no less a figure than Isaac Newton. Much of his long life - which overlapped the Bishop's - was devoted to biblical chronology and alchemy. However, Newton is responsible for the first cosmological statement that is recognizably modern. While Newton had no trouble in accepting a universe with a finite beginning, he managed to persuade himself that a universe with a finite spatial extent was potentially unstable. Gravitation, he had discovered, was both universal and attractive. Hence, he argued, if any clumping of the matter took place it would attract more matter and the effect could "run away"; i.e., all the matter in the universe might collect itself in one place - something that appeared to be in disagreement with experience.

Newton expressed his views on this matter in 1692 in a letter he wrote to Richard Bentley. "But," he wrote "if the matter was evenly disposed throughout an infinite space it would never convene into one mass; but some of it would convene into one mass and some into another, so as to make an infinite number of great masses scattered at great distances from one another throughout all that infinite space."

From our point of view this discussion is rather curious.[1] In the first place the finite universe need not collapse if, for example, it is rotating or if the matter has some non-zero average temperature. We will come back to this latter point when we discuss Einstein's motives for introducing the cosmological constant. Einstein, in fact, was thinking in terms of a finite universe which behaved, in some sense, like a Boltzmann gas with a temperature. In the second place, an infinite Newtonian universe with a uniform static distribution of matter - if that was what Newton was thinking of - does not exist. The gravitational field inside any evacuated spherical cavity should be zero, which means that if the matter is reintroduced, the sphere will collapse. In fact, if one computes or tries to compute the field inside the cavity, the result is divergent, from which one would conclude that this whole Newtonian model really makes no sense. This was appreciated by the nineteenth century when Neumann tried to deal with the dilemma by replacing Newton's gravitational potential by what we would call a Yukawa potential, thus introducing some kind of cosmological mass scale, a sort of precursor to Einstein's cosmological constant.

Finally, the collapse of matter into clumps under its own gravity is in fact what has happened in our universe. Since

[1]For an especially lucid discussion of these and other cosmological matters see Rindler (1979).

cosmology includes the study of what is in the universe, as well as its origin, we will make a small detour to describe some of the history of this question. Before the late seventeenth century the extent of the universe was not understood. In particular, the distance between the stars was unknown, except that it must be much greater than the size of the solar system, since stars, unlike planets, showed no visible parallax with instruments then available. However, Huygens, by ingenious arguments involving the relative apparent brightness of Sirius and the Sun, and reasonable assumptions about their actual brightness, obtained an estimate of the distance of Sirius which was correct as to order of magnitude. This result already implied a universe that extends far beyond the solar system.

Armed with this knowledge, Immanuel Kant, in the mid eighteenth century, in his book _Universal Natural History and Theory of the Heavens_, proposed a model of the universe which is quite modern in its outlook. (See the excerpt in Munitz [1957], which also gives selections from other early works on cosmology.) Kant first took up a suggestion by Thomas Wright that the Milky Way is a disk-shaped collection of stars which we view from inside, held together by gravity and slowly rotating. He went on to propose that some nebulae were similar collections of stars but viewed from the outside. Kant assumed that this collection of "island universes," as they came to be called, extends to infinity in space, although not homogeneously. He does not explicitly discuss the question of whether the whole system had an origin in the past, although he does imply that an evolution in the form of the island universes is still taking place. Most of these ideas are a part of contemporary cosmology, although by and large they were not accepted by most astronomers until the early twentieth century. At that time, through the work of Leavitt, Shapley, Hubble, and others, the scale of the universe was also established in something like the present form, although periodic revisions have occurred since then, usually in the direction of increasing the scale.

Meanwhile, the question of the age of the universe, or more specifically the age of the Earth, became a focus for scientific study in the nineteenth century. By this time a substantial amount of geological evidence had accumulated - the work of people like Hutton and the Scottish geologist Charles Lyell - that Bishop Ussher's creation date, taken literally, had to be wrong. Lyell summarized this work in his three volumes _The Principles of Geology_, published in the 1830s. Darwin took the first volume with him on his voyage on the _Beagle_. Lyell's _Principles_ summarized the so-called Doctrine of Uniformity according to which the present geology of the Earth could be accounted for by slow processes working uniformly over aeons. In 1866 Lord Kelvin - né Benjamin Thompson - dropped what he considered to be a spanner into the workings of the Doctrine. Kelvin reasoned that as the Earth is constantly losing heat, it should also continually be cooling down. Indeed, if it began as a molten ball and cooled down at something like its present rate, Kelvin felt the inside of the Earth would lose its heat in about 20 million years; a time which he felt was uncomfortably short for the Uniformitarian geologists. It has been pointed out, e.g., by Gould (1985), that until Kelvin reduced the Earth's putative age from his original 100 million years to his final 20 million years most of his scientific contemporaries were not unhappy with his figure. Compared to Bishop Ussher it was a liberation. Darwin was not happy with it, and eventually the geologists began to articulate the notion that perhaps Kelvin had left something out. He had - radioactivity - which continually supplies heat in a measure that is consistent with the Earth's real age; about 4.5 billion years. In 1904 Ernest Rutherford had the satisfaction of explaining this in a lecture attended by the then ancient Kelvin.

The stage is now nearly set for Albert Einstein and the first section of our collection, which begins with Einstein's 1917 paper on cosmology. But before turning to it, it is appropriate to say a few words about Ernst Mach, since what is called "Mach's principle" plays a large role in these early cosmological papers. Mach's dates are 1836-1916, so that he lived long enough to see his principle implemented in Einstein's work - something which might have given Mach satisfaction, but didn't. Mach made no pretense of being a *theoretical* physicist and was frank in saying that his mathematical education had been defective. He was, however, a man of vast culture, scientific and otherwise - a true polymath. He was also a strict positivist for whom direct experience was paramount. He was notorious, although far from alone, in his adamant belief that atoms did not "exist." He meant that they were, at best, a questionable speculation. In a debate with Boltzmann he once asked whether Boltzmann had ever "seen" an atom; a question that has a certain poignancy when one asks whether quarks "exist." Einstein was, of course, an atomist and this side of Mach's dada was of little concern to him. What had greatly influenced Einstein was Mach's great book *The Science of Mechanics*, Mach (1883) which Einstein had read as a student in Zurich around the turn of the century.

In this book Mach's concern was to go over the history of mechanics and to lay bare where investigators had gone wrong when they departed from strict empiricism. The book was written in a high polemic style and reaches its apex when Mach confronts Newton and the latter's view that space and time were absolutes. Newton, of course, realized that *uniform* motions were relative; a state of rest and a state of uniform motion were mechanically indistinguishable. But he was certain that accelerations were absolute, that it made sense to speak of an acceleration with respect to empty space. In a sense Newton was forced into this view because without it the equation $F = ma$ is a tautology. He even presented experimental "evidence" for it; i.e., the behavior of the surface of a rotating bucket of water in empty space. Mach thought that this "evidence" was completely fictitious, since it referred to an experiment that could never be carried out. As he wrote, "Newton's experiment with the rotating vessel of water simply informs us, that the relative rotation of the water with respect to the sides of the vessel produces no noticeable centrifugal forces, but that such forces are produced by its relative motion with respect to the mass of the Earth and the other celestial bodies. No one is competent to say how the experiment would turn out if the sides of the vessel increased in thickness and mass until they were several leagues thick." Einstein's version of Mach's principle might be summarized in a statement he apparently frequently made, namely, "space is not a thing." It is an empirical fact that bodies exhibit inertia when moving in physical space, i.e., against a background of masses like those of the fixed stars. There is no evidence that a body in truly empty space would exhibit inertia, and this intuition is what Einstein and his contemporaries tried to build into their cosmologies. We will come to closer grips with this matter in the next section. In any event Mach was not able to implement his own principle in a new physical theory. Moreover, he rejected Einstein's attempts, referring contemptuously to relativity as a "school or church." Einstein reacted rather strongly to Mach's criticisms, which were published only after Mach's death, and on one occasion remarked that Mach was a "good mechanician but a deplorable philosopher."

Thus the stage is set for the first paper in modern cosmology, Einstein's great and curious paper of 1917, and in the next section we present this paper and its sequels.

REFERENCES

Gould, S. 1985, The Flamingo's Smile (New York: W. W. Norton).

Mach, E. 1883, The Science of Mechanics (La Salle, Ill.: Open Court), 1980 edition.

Munitz, M.K. 1957, Theories of the Universe (Glencoe, Ill.: Free Press), p. 231.

Rindler, W. 1979, Essential Relativity (Berlin: Spinger-Verlag).

SECTION I

THE EXPANDING UNIVERSE

SECTION I

INTRODUCTION

The modern era of cosmology begins with the publication in 1917 of Einstein's paper "Cosmological Considerations on the General Theory of Relativity," reproduced here as paper 1. The year before he had published his paper on the general theory and gravitation, a subject that he had been working on for a decade. To us, who have grown up with the idea of an expanding universe with countless galaxies, some aspects of Einstein's paper seem very strange. It reminds one, in a way, of a man who chooses to stand on his hand with his own foot and looks around for an arcane solution to this excruciating dilemma which would consist of anything but simply removing his foot. The universe, as Einstein saw it, was basically static. It is true that the stars have small proper motions but there were, he was convinced, no large scale secular motions. If the universe was infinite, in both space and time, then in order to complete the equations of general relativity it was necessary to provide the boundary conditions at infinity. The boundary condition Einstein tried to impose appears a little odd because it is manifestly asymmetric between space and time. His idea was that a single particle spatially isolated from all matter should have no inertia - a version of Mach's principle. In general relativity theory this translated into the proposition that the spatial components of the metric tensor should vanish at infinity; i.e., that space, which in some sense does not exist without gravitation, should disappear. He was, however, as he tells us in his paper, unable to find any solutions to the theory satisfying both the boundary conditions and the condition that the stellar velocities are small. (The precise meaning of "small" is that all the components of the energy-momentum tensor are negligible except the time-time component.)

Thus Einstein gave up an infinite spatial universe and tried to find a solution for a finite universe filled with a static distribution of matter - static in the sense that there are no large proper motions of stars. But here he was foiled by the Newtonian dilemma; i.e., this universe is, as he argued, unstable. One of his arguments is especially interesting because it makes use of the kind of statistical mechanical reasoning of which he was such a master. He envisioned the finite universe of stars as a sort of Boltzmann gas in equilibrium at some finite temperature. If the number of stars per unit volume was to vanish at the boundary, then, Einstein argued, it must also vanish in the center of the distribution. This is because the ratio of the densities equals the ratio of the Boltzmann factors, $e^{-E/kT}$ which involves the difference of the gravitational potentials at the two reference points. As this difference does not vanish the result follows. But this result contradicts the fact that this stellar density averaged over the universe is sensibly constant. Hence, it would appear that general relativity, in Einstein's original formulation, is incompatible with a static universe. Since Einstein thought that astronomical observations required a static universe he had no choice but to modify the original theory. Enter the cosmological "constant" or gleider-member.

Since Einstein did not want to give up the general covariance of his equations, the only term he could add to them was a term proportional to the metric tensor, the constant of proportionality being, in fact, the cosmological constant. While this new term formally can be incorporated into the energy-momentum tensor, it must be regarded physically as something with a different origin. In particular, the energy-momentum tensor has essentially only one component reflecting the static universe, while the metric tensor is approximately Minkowskian - i.e., all its diagonal components have about unit absolute value. One can think of this as implying that a cosmological constant term in Einstein's equations includes a pressure term whose magnitude is similar to that of the energy density. Having modified the theory, Einstein then set out to find a solution to the new theory which would satisfy all his conditions. He begins by constructing a metric tensor. The conditions he uses are that the time component be independent of space; that the metric be diagonal; that the space be one of the constant curvature corresponding to the uniform distribution of matter. The solution he came up with is known as the Einstein "cylinder" world. If one picks an axis to represent the time direction and draws a cylinder with its center along this axis, then this is a kind of projective description of Einstein's space, with two of the spatial directions suppressed. It is a genuine static space, since displacements along the time axes are symmetries of it. Armed with the metric Einstein could then study the conditions for a solution to his equations. There are two conditions which relate the cosmological constant to the radius of curvature of the space and to the matter density. The possible presence of a cosmological constant in Einstein's field equations can be considered independently of Einstein's specific static space. However, it is still unknown whether there is a cosmological constant term, and if so, how it originated.

In his wonderful review article "Relativistic Cosmology," Robertson (1933) notes that the idea of the de Sitter universe - the next cosmological model after Einstein's original, and published the same year - was due to Ehrenfest. Paul Eherenfest and Wilhelm de Sitter (de Sitter was an astronomer) were colleagues in Holland. We have included as paper 2 in this collection de Sitter's English-language summary - also published in 1917 - of his cosmology. The modern reader of de Sitter's paper may be a little puzzled as to where in the paper is what we know as "de Sitter space." We are accustomed to seeing the de Sitter space metric written in its nonstatic form later derived by Robertson and others, in which the scale factor R(t) is an exponential in time. (Robertson makes the distinction between a "static universe" whose metric in some coordinate system has no coefficients that depend on time and a "stationary universe" whose essential properties are independent of time.) In fact, the recent introduction of the "inflationary universe" (see Part IV) makes use of just this metric. However, de Sitter uses coordinates in which the coefficients of the metric are independent of time. The thing that characterizes the de Sitter space in his paper (the equations designated with subscripts B) is that it contains no matter or, putting it more precisely, such forms of matter as the stars do not determine the average properties of the metric but are to be regarded as test particles in a fixed background metric. De Sitter finds a set of coordinates with the property that when the distance from the origin is taken to infinity the entire metric tensor collapses to zero, which is his covariant statement of Mach's principle. There is a nonvanishing cosmological constant which, if set equal to zero, causes the de Sitter space to go over into the Minkoswki space of special relativity.

In a "Postscriptum" to his paper - we do not reproduce it because of its length - de Sitter makes the first prediction of the cosmological red shift, noting that, because of the form of his metric, "the frequency of light vibrations diminishes with increasing distance from the origin of co-ordinates." He goes on to add, "The lines in the spectra of very distant stars or nebula must therefore be systematically displaced towards the red, giving rise to a spurious positive radial velocity." In fact, he goes on to discuss the very weak evidence that such an effect might have been observed. Two points occur to one in connection with this remarkable prediction. In the first place, why would de Sitter describe the radial velocity as "spurious"? From our point of view it seems no more spurious than any other expansion velocity. After all, the de Sitter universe is simply a special case of an expanding universe. Perhaps he had in mind the fact that the de Sitter space is "stationary." By "stationary" what is meant is that its "intrinsic" properties do not vary with time. Specifically, the three dimensional curvature, which is zero, and the four-dimensional curvature, which is positive, are both constant in time. Nonetheless, it is perfectly reasonable, we would claim, to say that the space is expanding - in fact inflating exponentially.

The second point we would like to make has to do with the notion that one sometimes comes across that general relativity 'should' have predicted the expanding universe but didn't - a failure of theory. This, it seems to us, is clearly wrong. Not only did de Sitter specifically make the prediction of a red shift, but in 1923, Hermann Weyl (1923) showed that this red shift increased uniformly with distance, something that is implicit in de Sitter's paper but which is not stated clearly. In fact this prediction was taken seriously by precisely the one person who mattered - Edwin Hubble. In the last paragraph of Hubble's 1929 paper, which we reproduce as paper 7, he says of his data, "The outstanding feature, however, is the possibility that the velocity-distance relation may represent the de Sitter effect and hence that numerical data may be introduced into discussions of the general curvature of space." Indeed, as has often been noted, Hubble's original data are so scanty and scattered - see his Figure 1 - that one is left to wonder if he could have drawn any conclusion from them without the guidance of the theoretical bias he inherited from general relativity. Nevertheless, de Sitter's metric could not be taken seriously as a model for the present world because the properties of space-time are independent of the matter it contains.

The real resolution of Einstein's dilemma came from an unlikely source. Aleksandr Aleksandrovich Friedmann was born in 1888 into a talented musical St. Petersburg family. After his graduation in 1910 from the University of St. Petersburg, where he had specialized in mathematics, he took up theoretical meteorology and in 1914 published a seminal paper on temperature inversion in the upper atmosphere. With the war Friedmann volunteered for an aviation detachment and by 1917, the year of Einstein's paper, Friedmann had become a section chief and later director of the first Russian factory in which aviation instruments were manufactured. But in 1920 Friedmann took a post in Petrograd (now Leningrad, previously St. Petersburg) where he taught physics and mathematics at the university. Friedmann was not a professional physicist but, like so many others, he got caught up in the wave of excitement that followed the 1919 confirmation of the bending of light by the sun's gravitational field by the amount Einstein's theory predicted. He not only taught himself the theory but began improving on it. His two great papers - of which more shortly - were published in 1922 and 1924, respectively. One can only imagine what he might gone on to accomplish, but the following year he contracted typhoid fever and at age 37 was dead.

In a certain sense Friedmann introduced a Copernican revolution into cosmology. Einstein and de Sitter had insisted that all the intrinsic properties of space-time be stationary. In order to achieve this in a fashion that was consistent with the rest of their constraints, they had been forced to introduce the cosmological constant. Friedmann simply dropped the requirements that the intrinsic properties be stationary. He began by showing that, given their assumptions, Einstein and de Sitter had exhausted all the possibilities for stationary universes. (The rigor of his argument was later criticized, and we have included a paper by H. P. Robertson [paper 6] in which Friedmann's arguments are redone with rigor and elegance using symmetry arguments that a modern reader will not find antique.) Equation 5 of Friedmann's first paper (paper 3) will look completely familiar to anyone who has studied modern cosmology. It is the general equation for a space of positive curvature which governs the time development of the expansion. Friedmann discusses this time development in detail in paper 3, whereas, in the companion paper 4 on a non-static space of negative curvature, he only indicates the possibility of expansion, without discussing details.

Friedmann, as well as Einstein and de Sitter, made the assumption that the pressure term was zero. The straightforward generalization to include a pressure term was made by the Abbé Georges Lemaitre. We have included an English language translation paper 9 of a 1931 paper by Lemaître because of its pegagogical excellence.

The most curious reaction to Friedmann's 1922 paper was that of Einstein himself. Rather than feeling liberated Einstein decided that Friedmann's paper was mathematically incorrect. We have included as paper 5 his short published statement to this effect. It is clear to a modern reader that Einstein has made a simple calculational mistake - something he acknowledged a year later. One is tempted to conjecture that he was misled by his theoretical prejudice against a nonstationary universe. While he did eventually agree that Friedmann's paper was mathematically correct, it would appear that in private he maintained that that is what it was - an exercise in mathematics. One of the things that Friedmann discovered was that under certain circumstances the size of the universe could be periodic, with alternating eras of expansion and contraction. It is remarkable that at the end of his first paper Friedmann gives an illustrative example of what the period might be and finds, setting the cosmological constant equal to zero, and taking what we would consider to be a reasonable value for the mass of the universe, a period of 10 billion years, which is about what we believe the present age of the universe to be.

If the metric in Friedmann's first paper looks vaguely familiar, the metric in his second one (paper 4), which deals with a universe of constant negative spatial curvature, seems completely unfamiliar - see equations D' and D". It is far from obvious, given the bizarre choice of coordinates, that this _is_ a space of constant curvature. We are grateful to our colleagues J. L. Anderson and D. W. Hobill who kindly ran this metric through the Stevens Institute CSYMA program which is specialized for general relativity computations, and found both the Einstein and Ricci tensors and did, indeed, verify that Friedmann's metric represents what he says it does. Actually, in writing the metric in this form, Friedmann returned to an old version of the metric for a 3-space of constant curvature, first given by Beltrami, (see the discussion in Levi-Civita 1927). We have not tried to find the transformation that takes this metric into the more familiar Robertson-Walker form, having been warned by a comment in Robertson's (1968) text on general relativity that even for two spatial dimensions "the transformation is hideous to write." (Incidentally, we have not included A. G. Walker's [1936] paper in this collection since it was preceded by the work of Robertson and was done in the context of a specific cosmology: that of Milne [1948] who thought that one could deduce the principles of cosmology from general kinematic considerations, without using the general theory of relativity. This view is not commonly accepted by most physicists.) Einstein had no specific published reaction to the second paper of Friedmann, but by 1931 he had decided that the non-vanishing cosmological constant was "theoretically unsatisfying." By that time, of course, Hubble's results had been published, and moreover, because of de Sitter's work, it was clear that the model with a cosmological constant did not satisfy what would seem to be the most straightforward formulation of Mach's principle. In de Sitter space there is a metric even though the average matter density vanishes. By 1950 when Einstein published the second edition of his book _The Meaning of Relativity_, with its new cosmological appendix, he noted in a footnote, "If Hubble's expansion had been discovered at the time of the creation of the general theory of relativity, the cosmologic member would never have been introduced." _Sic transit_ . . .

The result of the work on cosmological solutions to Einstein's equations is often summarized by the conclusion that if a 3-space has a uniform curvature, then there are only three distinct possibilities for it, corresponding to spaces of constant positive, zero, or negative curvature. When these three possibilities are discussed in the framework of Friedmann's equations, it is found that some of the positive curvature solutions correspond to a space that undergoes cyclic expansions and contractions, whereas the other solutions correspond to spaces which expand indefinitely. Of course, other possibilites exist if the assumption of constant 3-curvature is relaxed. A discussion of these possibilities is given in Ryan and Shepley (1975). Assuming that our space is one of the uniform curvature types, which is suggested by astronomical data concerning the isotropy of the universe over large regions, we would dearly like to know which of three kinds of universe we actually inhabit. Will it expand indefinitely or will it eventually contract back into the Big Crunch? The present theoretical prejudice is for a universe of zero curvature which expands indefinitely but at a constantly decreasing rate. The condition for this to be true is an equation that relates what is called the Hubble constant to the average energy density of the universe. The Hubble constant, which is a rate, is usually given in the baroque units of kilometers/second/megaparsec which, translated, is in units of $(9.78 \times 10^{11}\ \text{year})^{-1}$. In principle, the Hubble constant can be determined by measuring the red shift as a function of distance. That is the rub. An entire book could be written about the problem of determining the distance scale of the universe. Instead, we have included paper 8 by Allan Sandage, which indicates the methods and problems of such measurements. Sandage's result is, in the units above, about 75, which splits the difference between a lower bound of 50 and an upper bound of 100 - the usually quoted bounds. The immense expansion of the range of observational astronomy in this century can be seen from Figure 1 of Sandage's paper which, indicates, in a small box in the lower left corner, the distance range on which Hubble's original analysis was based.

Evidence that the Hubble expansion law extends to even greater distances, comparable to the size of the visible universe, and to correspondingly great red shifts has emerged in the last twenty years from a study of the spectral lines emitted by quasars. Most but not all astronomers are convinced that quasars are indeed at cosmological distances, and that the red shifts observed in their spectra are the result of the universal expansion. Some of the evidence for this, as well as a summary of the controversy about quasar distances, is given in Field, Arp and Bahcall (1973). But knowing the Hubble constant H is not enough. If the universe is truly flat, the Hubble constant must be related to the average energy density ρ_c and Newton's gravitational constant by the equation

$$\rho_c = \frac{H^2}{G}\frac{3}{8\pi} = \left(\frac{H}{\frac{75\ \mathrm{km}}{\mathrm{sMPc}}}\right)^2 \times 1.05 \times 10^{-29}\ \frac{\mathrm{g}}{\mathrm{cm}^3}$$

Thus we must also know the average energy density of the universe. Finding this by direct observation has also been extremely elusive, since most of the matter may well not be luminous. There is some evidence for the existence of nonluminous matter coming from observation of the motion of stars in galaxies and of galaxies in cluster, which seems to require more mass to bind them than is visible (see, e.g., Faber and Gallagher 1979). There are also many candidates for nonluminous matter from black holes to neutrinos if they are massive. The latter are especially interesting since the standard cosmological model predicts that neutrinos must be, at present, about as numerous as photons from the Big Bang. This means that they cannot be too massive because if they were they would make the universe more curved than we know it to be. This argument can be used to seek a "cosmological bound" on the mass of the neutrino. As paper 10 in this section we have included one of the early references on the details of finding this mass bound. It is an example of how cosmology can be used to inform us about elementary particles. More recent work on the subject of a universe whose energy density is dominated by massive neutrinos can be found in Bernstein and Feinberg (1981). Attempts have also been made to use information about the expansion of the universe to obtain bounds on other hypothetical subatomic particles (see, e.g., Pagels and Primack 1982).

REFERENCES

Bernstein, J., and Feinberg, G. 1981, Phys. Lett., **101B**, 39.

Einstein, A. 1931, S. B. preuss. Akad Wiss. **1931**, 235.

Einstein, A. 1950, The Meaning of Relativity (Princeton: Princeton University Press).

Faber, S. M., and Gallagher, J. S. 1979, Ann. Rev. Astron. Astrophys., **17**, 135.

Field, G. B., Arp. H., and Bahcall, J. 1973, The Redshift Controversy (Reading: E. A. Benjamin).

Levi-Civita, T. 1927, The Absolute Differential Calculus (London: Blackie & Son), p. 238.
Milne, E. A. 1948, Kinematic Relativity (Oxford: Clarendon Press).
Pagels, H., and Primack. J., 1982, Phys. Rev. Lett., **48**, 223.
Robertson, H. P. 1933, Rev. Mod. Phys., **5**, 62.
Robertson, H. P., and Noonan, T. W. 1968, Relativity and Cosmology (Philadelphia: W. B. Saunders), p. 328.
Ryan, M., and Shepley, L. C. 1975, Homogeneous Relativistic Cosmologies (Princeton: Princeton University Press).
Walker, A. G. 1936, Proc. Lond. Math. Soc., **42**, 70.
Weyl, H. 1923, Phys. Zeit., **24**, 230.

COSMOLOGICAL CONSIDERATIONS ON THE GENERAL THEORY OF RELATIVITY

A. Einstein

It is well known that Poisson's equation

$$\nabla^2\phi = 4\pi\kappa\rho \tag{1}$$

in combination with the equations of motion of a material point is not as yet a perfect substitute for Newton's theory of action at a distance. There is still to be taken into account the condition that at spatial infinity the potential ϕ tends toward a fixed limiting value. There is an analogous state of things in the theory of gravitation in general relativity. Here, too, we must supplement the differential equations by limiting conditions at spatial infinity, if we really have to regard the universe as being of infinite spatial extent.

In my treatment of the planetary problem I chose these limiting conditions in the form of the following assumption: it is possible to select a system of reference so that at spatial infinity all the gravitational potentials $g_{\mu\nu}$ become constant. But it is by no means evident a priori that we may lay down the same limiting conditions when we wish to take larger portions of the physical universe into consideration. In the following pages the reflections will be given which, up to the present, I have made on this fundamentally important question.

1. THE NEWTONIAN THEORY

It is well known that Newton's limiting condition of the constant limit for ϕ at spatial infinity leads to the view that the density of matter becomes zero at infinity. For we imagine that there may be a place in universal space round about which the gravitational field of matter, viewed on a large scale, possesses spherical symmetry. It then follows from Poisson's equation that, in order that ϕ may tend to a limit at infinity, the mean density ρ must decrease toward zero more rapidly than $1/r^2$ as the distance r from the center increases.[1] In this sense, therefore, the universe according to Newton is finite, although it may possess an infinitely great total mass.

[1] ρ is the mean density of matter, calculated for a region which is large as compared with the distance between neighboring fixed stars, but small in comparison with the dimensions of the whole stellar system.

From this it follows in the first place that the radiation emitted by the heavenly bodies will, in part, leave the Newtonian system of the universe, passing radially outwards, to become ineffective and lost in the infinite. May not entire heavenly bodies fare likewise? It is hardly possible to give a negative answer to this question. For it follows from the assumption of a finite limit for ϕ at spatial infinity that a heavenly body with finite kinetic energy is able to reach spatial infinity by overcoming the Newtonian forces of attraction. By statistical mechanics this case must occur from time to time, as long as the total energy of the stellar system - transferred to one single star - is great enough to send that star on its journey to infinity, whence it never can return.

We might try to avoid this peculiar difficulty by assuming a very high value for the limiting potential at infinity. That would be a possible way, if the value of the gravitational potential were not itself necessarily conditioned by the heavenly bodies. The truth is that we are compelled to regard the occurrence of any great differences of potential of the gravitational field as contradicting the facts. These differences must really be of so low an order of magnitude that the stellar velocities generated by them do not exceed the velocities actually observed.

If we apply Boltzmann's law of distribution for gas molecules to the stars, by comparing the stellar system with a gas in thermal equilibrium, we find that the Newtonian stellar system cannot exist at all. For there is a finite ratio of densities corresponding to the finite difference of potential between the center and spatial infinity. A vanishing of the density at infinity thus implies a vanishing of the density at the center.

It seems hardly possible to surmount these difficulties on the basis of the Newtonian theory. We may ask ourselves the question whether they can be removed by a modification of the Newtonian theory. First of all we will indicate a method which does not in itself claim to be taken seriously; it merely serves as a foil for what is to follow. In place of Poisson's equation we write

$$\nabla^2\phi - \lambda\phi = 4\pi\kappa\rho \qquad (2)$$

where λ denotes a universal constant. If ρ_0 be the uniform density of a distribution of mass, then

$$\phi = -\frac{4\pi\kappa}{\lambda}\rho_0 \qquad (3)$$

is a solution of equation (2). This solution would correspond to the case in which the matter of the fixed stars was distributed uniformly through space, if the density ρ_0 is equal to the actual mean density of the matter in the universe. The solution then corresponds to an infinite extension of the central space, filled uniformly with matter. If, without making any change in the mean density, we imagine matter to be non-uniformly distributed locally, there will be, over and above the ϕ with the constant value of equation (3), an additional ϕ, which in the neighborhood of denser masses will so much the more resemble the Newtonian field as $\lambda\phi$ is smaller in comparison with $4\pi\kappa\rho$.

A universe so constituted would have, with respect to its gravitational field, no center. A decrease of density in spatial infinity would not have to be assumed, but both the mean potential and mean density would remain constant to infinity. The conflict with statistical mechanics which we found in the case of the Newtonian theory is not repeated. With a definite but extremely small density, matter is in equilibrium, without any internal material forces (pressures) being required to maintain equilibrium.

2. THE BOUNDARY CONDITIONS ACCORDING TO THE GENERAL THEORY OF RELATIVITY

In the present paragraph I shall conduct the reader over the road that I have myself travelled, rather a rough and winding road, because otherwise I cannot hope that he will take much interest in the result at the end of the journey. The conclusion I shall arrive at is that the field equations of gravitation which I have championed hitherto still need a slight modification, so that on the basis of the general theory of relativity those fundamental difficulties may be avoided which have been set forth in §1 as confronting the Newtonian theory. This modification corresponds perfectly to the transition from Poisson's equation (1) to equation (2) of §1. We finally infer that boundary conditions in spatial infinity fall away altogether, because the universal continuum in respect of its spatial dimensions is to be viewed as a self-contained continuum of finite spatial (three-dimensional) volume.

The opinion which I entertained until recently, as to the limiting conditions to be laid down in spatial infinity, took its stand on the following considerations. In a consistent theory of relativity there can be no inertia relatively to "space," but only an inertia of masses relatively to one another. If, therefore, I have a mass at a sufficient distance from all other masses in the universe, its inertia must fall to zero. We will try to formulate this condition mathematically.

According to the general theory of relativity the negative momentum is given by the first three components, the energy by the last component of the covariant tensor multiplied by $\sqrt{-g}$

$$m \sqrt{-g}\, g_{\mu\alpha} \frac{dx_\alpha}{ds} \tag{4}$$

where, as always, we set

$$ds^2 = g_{\mu\nu} dx_\mu dx_\nu \tag{5}$$

In the particularly perspicuous case of the possibility of choosing the system of co-ordinates so that the gravitational field at every point is spatially isotropic, we have more simply

$$ds^2 = -A(dx_1^2 + dx_2^2 + dx_3^2) + Bdx_4^2 .$$

If, moreover, at the same time

$$\sqrt{-g} = 1 = (A^3B)^{1/2}$$

we obtain from (4), to a first approximation for small velocities,

$$m \frac{A}{\sqrt{B}} \frac{dx_1}{dx_4}, \quad m \frac{A}{\sqrt{B}} \frac{dx_2}{dx_4}, \quad m \frac{A}{\sqrt{B}} \frac{dx_3}{dx_4}$$

for the components of momentum, and for the energy (in the static case)

$$m \sqrt{B} .$$

From the expressions for the momentum, it follows that $m \frac{A}{\sqrt{B}}$ plays the part of the rest mass. As m is a constant peculiar to the point of mass, independently of its position, this expression, if we retain the condition $\sqrt{-g} = 1$ at spatial infinity, can vanish only when A diminishes to zero, while B increases to infinity. It seems, therefore, that such a degeneration of the co-efficients $g_{\mu\nu}$ is required by the postulate of relativity of all inertia. This requirement implies that the potential energy $m \sqrt{B}$ becomes infinitely great at infinity. Thus a point of mass can never leave the system; and a more detailed investigation shows that the same thing applies to light-rays. A system of the universe with such behavior of the gravitational potentials at infinity would not therefore run the risk of wasting away which was mooted just now in connection with the Newtonian theory.

I wish to point out that the simplifying assumptions as to the gravitational potentials on which this reasoning is based, have been introduced merely for the sake of lucidity. It is possible to find general formulations for the behavior of the $g_{\mu\nu}$ at infinity which express the essentials of the question without further restrictive assumptions.

At this stage, with the kind assistance of the mathematician J. Grommer, I investigated centrally symmetrical, static gravitational fields, degenerating at infinity in the way mentioned. The gravitational potentials $g_{\mu\nu}$ were applied, and from them the energy-tensor $T_{\mu\nu}$ of matter was calculated on the basis of the field equations of gravitation. But here it proved that for the system of the fixed stars no boundary conditions of the kind can come into question at all, as was also rightly emphasized by the astronomer de Sitter recently.

For the contravariant energy-tensor $T^{\mu\nu}$ of ponderable matter is given by

$$T^{\mu\nu} = \rho \frac{dx_\mu}{ds} \frac{dx_\nu}{ds} ,$$

where ρ is the density of matter in natural measure. With an appropriate choice of the system of co-ordinates the stellar velocities are very small in comparison with that of light. We may, therefore, substitute $\sqrt{g_{44}}\, dx_4$ for ds. This shows us that all components of $T^{\mu\nu}$ must be very small in comparison with the last component T^{44}. But it was quite impossible to reconcile this condition with the chosen boundary conditions. In the retrospect this result does not appear astonishing. The fact of the small velocities of the stars allows the conclusion that wherever there are fixed stars, the gravitational potential (in our case $\sqrt{B}$) can never be much greater than here on earth. This follows from statistical reasoning, exactly as in the case of the Newtonian theory. At any rate, our calculations have convinced me that such conditions of degeneration for the $g_{\mu\nu}$ in spatial infinity may not be postulated.

After the failure of this attempt, two possibilities next present themselves.

a) We may require, as in the problem of the planets, that, with a suitable choice of the system of reference, the $g_{\mu\nu}$ in spatial infinity approximate to the values

$$\begin{matrix} -1 & 0 & 0 & 0 \\ 0 & -1 & 0 & 0 \\ 0 & 0 & -1 & 0 \\ 0 & 0 & 0 & 1 \end{matrix}$$

b) We may refrain entirely from laying down boundary conditions for spatial infinity claiming general validity; but at the spatial limit of the domain under consideration we have to give the $g_{\mu\nu}$ separately in each individual case, as hitherto we were accustomed to give the initial conditions for time separately.

The possibility (b) holds out no hope of solving the problem, but amounts to giving it up. This is an incontestable position, which is taken up at the present time by de Sitter (1916). But I must confess that such a complete resignation in this fundamental question is for me a difficult thing. I should not make up my mind to it until every effort to make headway toward a satisfactory view had proved to be vain.

Possibility (a) is unsatisfactory in more respects than one. In the first place those boundary conditions pre-suppose a definite choice of the system of reference, which is contrary to the spirit of the relativity principle. Secondly, if we adopt this view, we fail to comply with the requirement of the relativity of inertia. For the inertia of a material point of mass m (in natural measure) depends upon the $g_{\mu\nu}$; but these differ but little from their postulated values, as given above, for spatial infinity. Thus inertia would indeed be influenced, but would not be conditioned by matter (present in finite space). If only one single point of mass were present, according to this view, it would possess inertia, and in fact an inertia almost as great as when it is surrounded by the other masses of the actual universe. Finally, those statistical objections must be raised against this view which were mentioned in respect of the Newtonian theory.

From what has now been said it will be seen that I have not succeeded in formulating boundary conditions for spatial infinity. Nevertheless, there is still a possible way out, without resigning as suggested under (b). For if it were possible to regard the universe as a continuum which is finite (closed) with respect to its spatial dimensions, we should have no need at all of any such boundary conditions. We shall proceed to show that both the general postulate of relativity and the fact of the small stellar velocities are compatible with the hypothesis of a spatially finite universe; though certainly, in order to carry through this idea, we need a generalizing modification of the field equations of gravitation.

3. THE SPATIALLY FINITE UNIVERSE WITH A UNIFORM DISTRIBUTION OF MATTER

According to the general theory of relativity the metrical character (curvature) of the four-dimensional space-time continuum is defined at every point by the matter at that point and the state of that matter. Therefore, on account of the lack of uniformity in the distribution of matter, the metrical structure of this continuum must necessarily be extremely complicated. But if we are concerned with the structure only on a large scale, we may represent matter to ourselves as being uniformly distributed over enormous spaces, so that its density is a variable function which varies extremely slowly. Thus our procedure will somewhat resemble that of the geodesists who, by means of an ellipsoid, approximate to the shape of the earth's surface, which on a small scale is extremely complicated.

The most important fact that we draw from experience as to the distribution of matter is that the relative velocities of the stars are very small as compared with the velocity of light. So I think that for the present we may base our reasoning upon the following approximative assumption. There is a system of reference relatively to which matter may be looked upon as being permanently at rest. With respect to this system, therefore, the contravariant energy-tensor $T^{\mu\nu}$ of matter is, by reason of (5), of the simple form

$$\begin{matrix} 0 & 0 & 0 & 0 \\ 0 & 0 & 0 & 0 \\ 0 & 0 & 0 & 0 \\ 0 & 0 & 0 & \rho \end{matrix} \qquad (6)$$

The scalar ρ of the (mean) density of distribution may be a priori a function of the space co-ordinates. But if we assume the universe to be spatially finite, we are prompted to the hypothesis that ρ is to be independent of locality. On this hypothesis we base the following considerations.

As concerns the gravitational field, it follows from the equation of motion of the material point

$$\frac{d^2 x_\nu}{ds^2} + \{\alpha\beta, \nu\} \frac{dx_\alpha}{ds} \frac{dx_\beta}{ds} = 0$$

that a material point in a static gravitational field can remain at rest only when g_{44} is independent of locality. Since, further, we presuppose independence of the time co-ordinate x_4 for all magnitudes, we may demand for the required solution that, for all x_ν,

$$g_{44} = 1 \qquad (7)$$

Further, as always with static problems, we shall have to set

$$g_{14} = g_{24} = g_{34} = 0 \qquad (8)$$

It remains now to determine those components of the gravitational potential which define the purely spatial-geometrical relations of our continuum ($g_{11}, g_{12}, \dots g_{33}$). From our assumption as to the uniformity of distribution of the masses generating the field, it follows that the curvature of the required space must be constant. With this distribution of mass, therefore, the required finite continuum of the x_1, x_2, x_3, with constant x_4, will be a spherical space.

We arrive at such a space, for example, in the following way. We start from a Euclidean space of four dimensions, ξ_1, ξ_2, ξ_3, ξ_4, with a linear element $d\sigma$; let, therefore,

$$d\sigma^2 = d\xi_1^2 + d\xi_2^2 + d\xi_3^2 + d\xi_4^2 \tag{9}$$

In this space we consider the hyper-surface

$$R^2 = \xi_1^2 + \xi_2^2 + \xi_3^2 + \xi_4^2 , \tag{10}$$

where R denotes a constant. The points of this hyper-surface form a three-dimensional continuum, a spherical space of radius of curvature R.

The four-dimensional Euclidean space with which we started serves only for a convenient definition of our hyper-surface. Only those points of the hyper-surface are of interest to us which have metrical properties in agreement with those of physical space with a uniform distribution of matter. For the description of this three-dimensional continuum we may employ the co-ordinates ξ_1, ξ_2, ξ_3 (the projection upon the hyper-plane $\xi_4 = 0$) since, by reason of (10), ξ_4 can be expressed in terms of ξ_1, ξ_2, ξ_3. Eliminating ξ_4 from (9), we obtain for the linear element of the spherical space the expression

$$\left.\begin{aligned} d\sigma^2 &= \gamma_{\mu\nu} d\xi_\mu d\xi_\nu \\ \gamma_{\mu\nu} &= \delta_{\mu\nu} + \frac{\xi_\mu \xi_\nu}{R^2 - \rho^2} \end{aligned}\right\} \tag{11}$$

where $\delta_{\mu\nu} = 1$, if $\mu = \nu$; $\delta_{\mu\nu} = 0$, if $\mu \neq \nu$, and $\rho^2 = \xi_1^2 + \xi_2^2 + \xi_3^2$. The co-ordinates chosen are convenient when it is a question of examining the environment of one of the two points $\xi_1 = \xi_2 = \xi_3 = 0$.

Now the linear element of the required four-dimensional space-time universe is also given us. For the potential $g_{\mu\nu}$, both indices of which differ from 4, we have set

$$g_{\mu\nu} = - \left(\delta_{\mu\nu} + \frac{x_\mu x_\nu}{R^2 - (x_1^2 + x_2^2 + x_3^2)}\right) \tag{12}$$

which equation, in combination with (7) and (8), perfectly defines the behavior of measuring-rods, clocks, and light-rays.

4. ON AN ADDITIONAL TERM FOR THE FIELD EQUATIONS OF GRAVITATION

My proposed field equations of gravitation for any chosen system of coordinates run as follows: -

$$
\begin{aligned}
R_{\mu\nu} &= -\kappa\left(T_{\mu\nu} - \frac{1}{2} g_{\mu\nu} T\right), \\
R_{\mu\nu} &= -\frac{\partial}{\partial x_\alpha}\{\mu\nu,\alpha\} + \{\mu\alpha,\beta\}\{\nu\beta,\alpha\} \\
&\quad + \frac{\partial^2 \log\sqrt{-g}}{\partial x_\mu \partial x_\nu} - \{\mu\nu,\alpha\}\frac{\partial \log\sqrt{-g}}{\partial x_\alpha}
\end{aligned}
\qquad (13)
$$

The system of equations (13) is by no means satisfied when we insert for the $g_{\mu\nu}$ the values given in (7), (8), and (12), and for the (contravariant) energy-tensor of matter the values indicated in (6). It will be shown in the next paragraph how this calculation may conveniently be made. So that, if it were certain that the field equations (13) which I have hitherto employed were the only ones compatible with the postulate of general relativity, we should probably have to conclude that the theory of relativity does not admit the hypothesis of a spatially finite universe.

However, the system of equations (14) allows a readily suggested extension which is compatible with the relativity postulate, and is perfectly analogous to the extension of Poisson's equation given by equation (2). For on the left-hand side of field equation (13) we may add the fundamental tensor $g_{\mu\nu}$, multiplied by a universal constant, $-\lambda$, at present unknown, without destroying the general covariance. In place of field equation (13) we write

$$R_{\mu\nu} - \lambda g_{\mu\nu} = -\kappa\left(T_{\mu\nu} - \frac{1}{2} g_{\mu\nu} T\right) \qquad (13a)$$

This field equation, with λ sufficiently small, is in any case also compatible with the facts of experience derived from the solar system. It also satisfies laws of conservation of momentum and energy, because we arrive at (13a) in place of (13) by introducing into Hamilton's principle, instead of the scalar of Riemann's tensor, this scalar increased by a universal constant; and Hamilton's principle, of course, guarantees the validity of laws of conservation. It will be shown in §5 that field equation (13a) is compatible with our conjectures on field and matter.

5. CALCULATION AND RESULT

Since all points of our continuum are on an equal footing it is sufficient to carry through the calculation for one point, e.g. for one of the two points with the co-ordinates

$$x_1 = x_2 = x_3 = x_4 = 0 .$$

Then for the $g_{\mu\nu}$ in (13a) we have to insert the values

$$\begin{matrix} -1 & 0 & 0 & 0 \\ 0 & -1 & 0 & 0 \\ 0 & 0 & -1 & 0 \\ 0 & 0 & 0 & 1 \end{matrix}$$

wherever they appear differentiated only once or not at all. We thus obtain in the first place

$$R_{\mu\nu} = \frac{\partial}{\partial x_1} [\mu\nu, 1] + \frac{\partial}{\partial x_2} [\mu\nu, 2] + \frac{\partial}{\partial x_3} [\mu\nu, 3] + \frac{\partial^2 \log\sqrt{-g}}{\partial x_\mu \partial x_\nu} .$$

From this we readily discover, taking (7), (8), and (13) into account, that all equations (13a) are satisfied if the two relations

$$-\frac{2}{R^2} + \lambda = -\frac{\kappa\rho}{2}, \quad -\lambda = -\frac{\kappa\rho}{2},$$

or

$$\lambda = \frac{\kappa\rho}{2} = \frac{1}{R^2} \tag{14}$$

are fulfilled.

Thus the newly introduced universal constant λ defines both the mean density of distribution ρ which can remain in equilibrium and also the radius R and the volume $2\pi^2 R^3$ of spherical space. The total mass M of the universe, according to our view, is finite, and is in fact

$$M = \rho \cdot 2\pi^2 R^3 = 4\pi^2 \frac{R}{\kappa} = \pi^2 (32/\kappa^3 \rho)^{1/2} \tag{15}$$

Thus the theoretical view of the actual universe, if it is in correspondence with our reasoning, is the following.

The curvature of space is variable in time and place, according to the distribution of matter, but we may roughly approximate to it by means of a spherical space. At any rate, this view is logically consistent, and from the standpoint of the general theory of relativity lies nearest at hand; whether, from the standpoint of present astronomical knowledge, it is tenable, will not here be discussed. In order to arrive at this consistent view, we admittedly had to introduce an extension of the field equations of gravitation which is not justified by our actual knowledge of gravitation. It is to be emphasized, however, that a positive curvature of space is given by our results, even if the supplementary term is not introduced. That term is necessary only for the purpose of making possible a quasi-static distribution of matter, as required by the fact of the small velocities of the stars.

REFERENCE

de Sitter, W. 1916, Akad. van Wetensch te Amsterdam, 8 Nov. 1916.

ON EINSTEIN'S THEORY OF GRAVITATION AND ITS ASTRONOMICAL CONSEQUENCES (Third Paper)[1]

W. de Sitter

I. In Einstein's theory of general relativity there is no essential difference between gravitation and inertia. The combined effect of the two is described by the fundamental tensor $g_{\mu\nu}$, and how much of it is to be called inertia and how much gravitation is entirely arbitrary. We might abolish one of the two words, and call the whole by one name only. Nevertheless it is convenient to continue to make a difference. Part of the $g_{\mu\nu}$ can be directly traced to the effect of known material bodies, and the common usage is to call this part "gravitation," and the rest "inertia." Then, if we take as a system of reference three rectangular cartesian space coordinates and the time multiplied by c (the velocity of light in vacuo), we know that, in that portion of the four-dimensional time-space which is accessible to our observations, the $g_{\mu\nu}$ of pure inertia are, within certain limits of uncertainty,

$$\begin{matrix} -1 & 0 & 0 & 0 \\ 0 & -1 & 0 & 0 \\ 0 & 0 & -1 & 0 \\ 0 & 0 & 0 & +1 \end{matrix} \qquad (1)$$

In our immediate neighborhood, within the solar system, the limits of uncertainty are very narrow: say the eighth decimal place. As we get further away in space, or in time, or in both, the limits become wider: at a distance of a million light-years we can perhaps only guarantee the second decimal place.[2] How the $g_{\mu\nu}$ are in those portions of space and time

[1]See first paper, M.N.R.A.S., vol. lxxvi. p. 69; second paper, M.N.R.A.S., vol. lxxvii. p. 155. The present paper gives an account of the questions treated in the following communications: A. Einstein (1917); W. de Sitter (1917a); W. de Sitter (1917b).

The notations used are the same as in first and second papers. We may recall that $\delta_{\mu\mu} = 1$, $\delta_{\mu\nu} = 0$ for $\mu \neq \nu$, and that Σ is a sum from 1 to 4, and Σ' from 1 to 3.

[2]There are two criteria by which we can judge the value of the fundamental tensor at great distances from us. The frequency of light-vibrations is proportional to $\sqrt{g_{44}}$. Consequently, objects in whose spectra we are able to identify definite spectral lines must be situated in a portion of space where g_{44} is still of the order of unity. The motion of material particles, on the other hand, depends on all $g_{\mu\nu}$. We know that the relative velocities of the fixed stars are small. From this we conclude that also the accelerations are

to which our observations have not yet penetrated, we do not know, and how they are at infinity (of space or of time) we shall never know. All assumption regarding the values of the $g_{\mu\nu}$ at infinity are therefore extrapolations, which we are free to choose in accordance with theoretical or philosophical requirements.

The extrapolation which most naturally offers itself, and which is also tacitly made in Newton's theory of inertia, is that the $g_{\mu\nu}$ retain the values (1) for all distances and times up to infinity. It has been pointed out in the second paper[3] that in this theory inertia is not relative. The values (1) are not invariant: the boundary-values of the $g_{\mu\nu}$ at infinity are different in different systems of co-ordinates. Einstein and others have therefore tried to find another extrapolation, by which the $g_{\mu\nu}$, while in our neighborhood retaining the values (1) with the approximation demanded by the observations, would at infinity degenerate to a set of values which would be the same for all systems of reference.

The $g_{\mu\nu}$ are determined by the field-equations, which in Einstein's theory of 1915 are

$$R_{\mu\nu} = -\kappa T_{\mu\nu} + \frac{1}{2}\kappa g_{\mu\nu} T , \tag{2}$$

or

$$R_{\mu\nu} - \frac{1}{2} g_{\mu\nu} \bar{R} = -\kappa T_{\mu\nu} , \tag{2'}$$

and

$$\bar{R} = \kappa T .$$

Once the system of reference of space- and time-variables has been chosen, these equations determine the $g_{\mu\nu}$ apart from constants of integration, or boundary-conditions at infinity.

small. Let the velocities be of the order α, and let g_{44} be of the order γ, and $g_{ij} + \delta_{ij}$ of the order $\beta (i,j = 1,2,3)$. Then the acceleration contains terms of the order γ, γ^2, $\beta \cdot \gamma$, $\alpha^2 \cdot \gamma$, $\alpha^2 \cdot \beta$, etc., but none of the order β. Thus here also we can only be sure of the smallness of γ, and not of β. Within the solar system the case is different, for there we have not only a statistical knowledge of the velocities, but we know the accelerations themselves; and our observations are so exact as to carry us to quantities of the second order. Consequently, we can be sure of g_{ij} to the first order, and of g_{44} to the second, the first order corresponding to about 10^{-8}.

[3] M.N.R.A.S., Vol. lxxvii, pp. 181-183.

Only the deviations of the actual $g_{\mu\nu}$ from these values at infinity are thus due to the effect of matter, through the mechanisms of the equations (2) or (2'). If at infinity all $g_{\mu\nu}$ were zero, then we could truly say that the whole of inertia, as well as gravitation, is thus produced. This is the reasoning which has led to the postulate that at infinity all $g_{\mu\nu}$ shall be zero. I have called this the mathematical postulate of relativity of inertia.

If all matter were destroyed, with the exception of one material particle, then would this particle have inertia or not? The school of Mach requires the answer No. If, however, by "all matter" is meant all matter known to us, stars, nebulae, clusters, etc., then the observations very decidedly give the answer Yes. The followers of Mach[4] are therefore compelled to assume the existence of still more matter. This matter, however, fulfills no other purpose than to enable us to suppose it not to exist, and to assert that in that case there would be no inertia. This point of view, which denies the logical possibility of the existence of a world without matter, I call the material postulate of relativity of inertia. The hypothetical matter introduced in accordance with it I call world-matter. Einstein originally supposed that the desired effect could be brought about by very large masses at very large distances. He has, however, now convinced himself that this is not possible. In the solution which he now proposes, the world-matter is not accumulated at the boundary of the universe, but distributed over the whole world, which is finite, though unlimited. Its density (in natural measure) is constant, when sufficiently large units of space are used to measure it. Locally its distribution may be very unhomogeneous. In fact, there is no essential difference between the nature of ordinary gravitating matter and the world-matter. Ordinary matter, the sun, stars, etc., are only condensed world-matter, and it is possible, though not necessary, to assume all world-matter to be so condensed. In this theory "inertia" is produced by the whole of the world-matter, and "gravitation" by its local deviations from homogeneity.

In Einstein's new solution the three-dimensional world is not infinite, but spherical.[5] Thus no boundary conditions at infinity are required. From the point of view of the theory of relativity it seems at first sight to be incorrect to say: the world is finite, since by a transformation of co-ordinates it can be made infinite, euclidean, or hyperbolical. Such transformations, however, leave the invariant $\bar{R}$ unaltered, and consequently also after the introduction of euclidean or hyperbolical co-ordinates the world remains finite and spherical in natural measure. The length of the semi-axis of x_1 in natural measure is

[4] Mach himself still thought that the fixed stars would be sufficient. This, however, is not so.

[5] Or elliptical, see article 2.

$$L_1 = \int_0^\infty \sqrt{-g_{11}}\, dx_1 .$$

If this is to be finite, it is necessary that g_{11} shall become zero for $x_1 = \infty$; and inversely, if g_{11} becomes zero of sufficiently high order for $x_1 = \infty$, then L_1 is finite. It is thus evident that the condition that the $g_{\mu\nu}$ shall be zero at infinity is equivalent to the finiteness of the world in natural measure.

It is found, however, that the $g_{\mu\nu}$ of this finite world do not satisfy the equations (2). Einstein is thus compelled to add a new term to these equations, which then become

$$R_{\mu\nu} - \lambda g_{\mu\nu} = - \kappa T_{\mu\nu} + \frac{1}{2} g_{\mu\nu} T , \tag{3}$$

or

$$R_{\mu\nu} - \frac{1}{2} g_{\mu\nu} (\bar{R} - 2\lambda) = - \kappa T_{\mu\nu} ; \tag{3'}$$

from which we find easily

$$\bar{R} - 4\lambda = kT . \tag{4}$$

If we put

$$R'_{\mu\nu} = R_{\mu\nu} - \lambda g_{\mu\nu} ,$$

we have

$$\bar{R}' = \bar{R} - 4\lambda .$$

Therefore the equations (3) and (3') are found if in (2) or (2') we replace $R_{\mu\nu}$ and $\bar{R}$ by $R'_{\mu\nu}$ and $\bar{R}'$. Consequently the equations (3) can be derived from the generalized principle of Hamilton,[6] if we now take

$$H_3 = \int \sqrt{-g}\, (\bar{R} - 4\lambda) d\tau .$$

[6] See first paper, M.N.R.A.S., lxxvi, p. 707.

All the conservative properties which follow from the principle of Hamilton thus remain true after the introduction of λ.

The curvature of the four-dimensional time-space is proportional to $\bar{R}$. In the new theory we have $\bar{R} = \kappa T + 4\lambda$: thus if there were no matter ($T = 0$), this curvature would not be zero.

Einstein's solution of the equations (3) implies the existence of a "world matter" which fills the whole universe, as has already been mentioned. It is, however, also possible to satisfy the equations without this hypothetical world-matter. Then, of course, the "material postulate of relativity of inertia" is not satisfied, but the "mathematical postulate," which makes no mention of matter, but only requires the $g_{\mu\nu}$ to be zero at infinity, is satisfied. This is brought about by the introduction of the term with λ, and not by the world-matter, which, from this point of view, is not essential.

If we neglect all pressures and other internal forces, and if we suppose all matter to be at rest, then the tensor $T_{\mu\nu}$ becomes

$$T_{44} = g_{44}\rho, \qquad \text{all other } T_{\mu\nu} = 0, \tag{5}$$

ρ being the density in natural measure. We can put

$$\rho = \rho_0 + \rho_1 , \tag{6}$$

where ρ_0 is the average density of the world-matter. If ρ_0 is positive, then ρ_1 may be positive or negative; but in the latter case the numerical value must not exceed ρ_0.

If we wish to neglect gravitation, we must neglect ρ_1, and take ρ_0 constant. The equations (3) then become[7]

$$\begin{aligned} R_{ij} - \left(\lambda + \frac{1}{2}\kappa\rho_0\right) g_{ij} &= 0 . \\ R_{44} - \left(\lambda + \frac{1}{2}\kappa\rho_0\right) g_{44} &= -\kappa\rho_0 g_{44} . \end{aligned} \tag{7}$$

These can be satisfied by the $g_{\mu\nu}$ implied by the line element

$$ds^2 = - dr'^2 - R^2 \sin^2 \frac{r'}{R} \left[d\psi^2 + \sin^2\psi d\theta^2 \right] + c^2 dt^2 , \tag{8A}$$

[7]The equations will be further developed in article 5.

if

$$\kappa\rho_0 = 2\lambda, \qquad \lambda = \frac{1}{R^2}. \tag{9A}$$

This is Einstein's new solution.
The equations are also satisfied by

$$ds^2 = -dr'^2 - R^2 \sin^2\frac{r'}{R}\left[d\psi^2 + \sin^2\psi d\theta^2\right] + \cos^2\frac{r'}{R}c^2dt^2, \tag{8B}$$

if

$$\rho_0 = 0, \qquad \lambda = \frac{3}{R^2}; \tag{9B}$$

and, of course, also by

$$ds^2 = -dr'^2 - r'^2\left[d\psi^2 + \sin^2\psi d\theta^2\right] + c^2dt^2, \tag{8C}$$

with

$$\rho_0 = 0, \qquad \lambda = 0. \tag{9C}$$

This last solution (C) gives the $g_{\mu\nu}$ of the old theory of relativity, or of Newton's theory of inertia. In it three-dimensional space is euclidean, in (A) and (B) it has a constant positive curvature. In (A) there is world-matter; in (B) and (C) we have $\rho_0 = 0$: the hypothetical world-matter does not exist.

2. If in (8A) and (8B) we put

$$r' = R\chi, \tag{10}$$

the three-dimensional line-element becomes

$$d\sigma^2 = R^2\{d\chi^2 + \sin^2\chi[d\psi^2 + \sin^2\psi d\theta^2]\}. \tag{11}$$

This is the line-element of a three dimensional space with a constant positive curvature, which is

$$\varepsilon = \frac{1}{R^2} .$$

There are two possible forms of space with constant positive curvature, viz. the spherical space, or space of Riemann and the elliptical space, which has been investigated by Newcomb. In the spherical space all straight lines starting from a point intersect again in the "antipodal" point, whose distance from the first point measured along any of these lines is πR. In the elliptical space any two straight lines cannot have more than one point in common. In both forms of space the straight line is closed: its total length is $2\pi R$ in the spherical space, and πR in the elliptical space. In the spherical space the largest possible distance between two points is πR, and there is only one point, the "antipodal point," at that distance from a given point. In the elliptical space the largest possible distance is $(1/2)\pi R$, and all points at that distance from a given point lie on a straight line - the "polar line" of the point. Both spaces are finite. The total volume of the spherical space is $2\pi^2 R^3$, and of the elliptical $\pi^2 R^3$.

Einstein only mentions the spherical space, which by the two-dimensional analogy of the sphere is easier to represent to our imagination. The elliptical space is, however, really the simpler case, and it is preferable to adopt this for the physical world.[8] Also the spherical space would give rise to difficulties, which will be pointed out below.

We can, instead of the co-ordinates r', ψ, θ, introduce other co-ordinates by which the elliptical, or spherical, space is projected on an euclidean or on a hyperbolical space. By the transformation

$$r = R \tan \chi \qquad (12)$$

the whole of the elliptical space is projected on the whole of

[8]This is also the opinion of Einstein (communicated to the writer by letter).

the euclidean space.[9] The projection of the spherical space

[9]By the transformation

$$\mathbf{r}_1 = R \sin \chi$$

the elliptical space is made to correspond with the inside of the sphere $\mathbf{r}_1 < R$ in the euclidean space. The representation of the spherical space fills this sphere twice. If we put

$$x_1 = \mathbf{r}_1 \sin \phi \sin \theta ,$$

$$y_1 = \mathbf{r}_1 \sin \phi \cos \theta ,$$

$$z_1 = \mathbf{r}_1 \cos \phi ,$$

the co-ordinates x_1, y_1, z_1 are those used by Einstein in his paper of 1917 February. In these coordinates the three-dimensional line-element is

$$d\sigma^2 = \sum_i{}' dx_i^2 + \sum_i{}'\sum_j{}' \frac{x_i x_j dx_i dx_j}{R^2 - \mathbf{r}_1^2} .$$

If we add

$$u_1 = R \cos \chi ,$$

then x_1, y_1, z_1 are the coordinates used by Weierstrass.

Riemann used the co-ordinates found by the transformation

$$\mathbf{r}_2 = 2R \tan (\chi/2) .$$

The line-element then becomes

$$d\sigma^2 = \frac{d\mathbf{r}_2^2 + \mathbf{r}_2^2[d\phi^2 + \sin^2\phi d\theta^2]}{(1 + \mathbf{r}_2^2/4R^2)^2} .$$

By this transformation (which was also used in the paper by the writer of 1917 March) the whole of the spherical space corresponds to the whole of the euclidean space. The elliptical space corresponds to the inside of the sphere $\mathbf{r}_2 < 2R$.

The transformation used in the text leads to the co-ordinates of Beltrami.

fills the euclidean space twice, the projections of antipodal points being the same.

The four-dimensional line-element in these co-ordinates is, for the two systems,

$$ds^2 = -\frac{dr^2}{(1+\varepsilon r^2)^2} - \frac{r^2[d\psi^2+\sin^2\psi d\theta^2]}{1+\varepsilon r^2} + c^2dt^2 , \qquad (13A)$$

$$ds^2 = -\frac{dr^2}{(1+\varepsilon r^2)^2} - \frac{r^2[d\psi^2+\sin^2\psi d\theta^2]}{1+\varepsilon r^2} + \frac{c^2dt^2}{1+\varepsilon r^2} . \qquad (13B)$$

If now we put

$$x_1 = r \sin\psi \sin\theta$$

$$x_2 = r \sin\psi \cos\theta$$

$$x_3 = r \cos\psi$$

$$x_4 = ct$$

then the $g_{\mu\nu}$ for these coordinates are

$$g_{ij} = -\frac{\delta_{ij}}{1+\varepsilon r^2} + \frac{\varepsilon x_i x_j}{(1+\varepsilon r^2)^2} ,$$

A. $g_{44} = 1.$

B. $g_{44} = \dfrac{1}{1+\varepsilon r^2} .$

The $g_{\mu\nu}$ for $r = 0$ have the values (1) in both systems A and B. For $r = \infty$ they degenerate to

$$\begin{matrix} 0 & 0 & 0 & 0 \\ 0 & 0 & 0 & 0 \\ 0 & 0 & 0 & 0 \\ 0 & 0 & 0 & 1 \end{matrix} \qquad (1A)$$

$$\begin{matrix} 0 & 0 & 0 & 0 \\ 0 & 0 & 0 & 0 \\ 0 & 0 & 0 & 0 \\ 0 & 0 & 0 & 0 \end{matrix} \qquad (1B)$$

The set (1A) is invariant for all transformations for which (at infinity) $t' = t$; the set (1B) is invariant for all transformations.[10] It thus appears that the system A only satisfies the mathematical postulate of relativity if the latter is applied to three-dimensional space only. In other words, if we conceived the three-dimensional space (x_1, x_2, x_3) with its world-matter as movable in an absolute space, its movements can never be detected by observations: all motions of material bodies are relative to the space (x_1, x_2, x_3) with the world-matter, not to the absolute space. The world-matter thus takes the place of the absolute space in Newton's theory, or of the "inertial system." It is nothing else but this inertial system materialized. It should be pointed out this relativity of inertia is in the system A only realized by making the time practically absolute. It is true that the fundamental equations of the theory, the field-equations (3) and the equations of motion, i.e. the differential equations of the geodetic line, remain invariant for all transformations. But only such transformations for which at infinity $t' = t$ can be carried out without altering the values (1A). In the system B, on the other hand, there is complete invariance for all transformations involving the four variables.

The system B is the four-dimensional analogy of the three-dimensional space of the system A. If we put

$$ds^2 = - R^2\{d\omega^2 + \sin^2\omega(d\zeta^2 + \sin^2\zeta[d\psi^2 + \sin^2\psi d\theta^2])\} , \tag{14}$$

the $g_{\mu\nu}$ implied by this line-element satisfy the equations (3), with the conditions (9B). In order to avoid imaginary angles, we can put

$$\omega = i\omega', \qquad \zeta = i\zeta' .$$

[10] With the restriction that none of the coefficients dx_i/dx'_j becomes infinite at infinity.

Then the line-element becomes[11]

[11]If we take

$r = R \sinh \omega' \sinh \zeta'$, $t = R \sinh \omega' \cosh \zeta'$,

$x = r \sin \psi \sin \theta$,

$y = r \sin \psi \cos \theta$, $u = R \cosh \omega'$,

$z = r \cos \psi$,

we have

$$ds^2 = -dx^2 - dy^2 - dz^2 + dt^2 - du^2 ,$$

and

$$R^2 - x^2 - y^2 - z^2 + t^2 - u^2 = 0 . \qquad (a)$$

The latter equation represents an hyperboloid (one-bladed) in the five-dimensional space (x, y, z, t, u). The projection of a point x, y, z, t, u, of this hyperboloid from the point $x = y = z = t = u = 0$ on the four dimensional space $u = R$ has the co-ordinates (ξ, η, ζ, τ), where

$\xi = \rho \sin \psi \sin \theta$,

$\eta = \rho \sin \psi \cos \theta$,

$\zeta = \rho \cos \psi$.

This projection is limited by the "hyperbola"

$$R^2 + \xi^2 + \eta^2 + \zeta^2 - \tau^2 = 0 , \quad \text{or} \quad 1 + \varepsilon(\rho^2 - \tau^2) = 0 , \quad (b)$$

which is the projection of the points at infinity on the hyperboloid (a). The part of $u = R$ which is outside the hyperbola (b) is the projection of the (two-bladed) hyperboloid which is conjugated to (a). It will be seen from (16) that on the limiting "hyperbola" (b) all $g_{\mu\nu}$ become infinite.

$$ds^2 = R^2\{d\omega'^2 - \sinh^2\omega'(d\zeta'^2 + \sinh^2\zeta'[d\psi^2 + \sin^2\psi d\theta^2])\} \quad . \tag{15}$$

If now we put

$$\rho = R \tanh \omega' \sinh \zeta' \ ,$$

$$\tau = R \tanh \omega' \cosh \zeta' \ ,$$

then we have

$$ds^2 = \frac{-(1-\varepsilon\tau^2)d\rho^2 - 2\varepsilon\rho\tau d\rho d\tau + (1+\varepsilon\rho^2)d\tau^2}{[1+\varepsilon(\rho^2-\tau^2)]^2} - \frac{\rho^2[d\psi^2 + \sin^2\psi d\tau^2]}{1+\varepsilon(\rho^2-\tau^2)} \ . \tag{16}$$

Finally, by the transformation

$$R \sin\frac{r'}{R} = \frac{\rho}{\sqrt{1+\varepsilon(\rho^2-\tau^2)}} \ , \qquad R \sinh\frac{ct}{R} = \frac{\tau}{\sqrt{1-\varepsilon\tau^2}} \ ,$$

we find the formula (8B).

In the three-dimensional space, whose line-element is (11), we can transfer the origin to a point $(\chi_1, \psi_1, \theta_1)$, and the line-element expressed in co-ordinates referred to this new origin will again have the same form (11). Exactly in the same way we can in (14) transfer the origin to a point $(\omega_1, \zeta_1, \psi_1, \theta_1)$, corresponding to $(\chi_1, \psi_1, \theta_1, ct_1)$ in (8B). The line-element in the co-ordinates referred to this new origin will again have the same form (14), and this can again be transformed to new variables χ', ψ', θ', ct', and will then again have the form (8B). Of course ct' will generally be different from ct.

In both systems A and B it is always possible, at every point of the four-dimensional time-space, to find systems of reference in which the $g_{\mu\nu}$ depend only on one space-variable (the "radius-vector"), and not on the "time." In the system A the "time" of these systems of reference is the same always and everywhere, in B it is not. In B there is no universal time; there is no essential difference between the "time" and the other three co-ordinates. None of them has any real physical meaning. In A on the other hand, the time is essentially different from the space-variables.

3. In order further to compare the two systems, we will consider the course of rays of light. In A, if we use the co-ordinates r, ψ, θ, ct, the velocity of light is constant, and the rays of light, which are geodetic lines in the four-dimensional time-space, are also geodetic in the three-dimensional space r, ψ, θ. On triangles formed by such lines the ordinary formulae of spherical trigonometry are applicable. Thus, if we suppose the sun to be at rest in the origin of co-ordinates, and if the distance sun-earth be called a, then the parallax[12] p of a star whose distance from the sun is r', is given by

$$\tan p = \sin \frac{a}{R} \cot \frac{r'}{R} ;$$

or, since the square of a/R can be neglected,

$$p = \frac{a}{R} \cot \frac{r'}{R} . \tag{17}$$

The same result is found in the reference system r, ψ, θ, ct. By the transformation (12) all straight lines remain straight in the projection. We can moreover, easily verify that the rays of light must be straight lines in the system r, ψ, θ, ct. The velocity of light in this system is

$$v = \frac{c(1 + \varepsilon r^2)}{\sqrt{(1 + \varepsilon r^2 \sin^2 V)}} ,$$

where V is the angle between the radius-vector and the tangent to the ray of light. The equation of the ray of light then becomes[13]

$$\sin V = \frac{k}{r} ,$$

k being a constant. This is the equation of a straight line. The parallax is thus determined by the ordinary formulas of euclidean geometry, and we have

$$p = \frac{a}{r} = \frac{a}{R} \cot \frac{r'}{R} ,$$

[12] The parallax is 90° $-A$, if A is the angle at the earth, the angle at the sun being 90°. In spherical geometry, of course, 90° $-A$ is not equal to the angle at the star, as it is in euclidean geometry.

[13] See first paper, M.N.R.A.S., lxxvi, p. 717.

which is the same as (17).

The parallax vanishes for $r' = \frac{1}{2}\pi R$, i.e. for the largest distance which can occur in elliptical space. If we adopted the spherical space, so that still larger distances could occur, p would become negative, and for $r = \pi R$ we would have $p = -90°$.

In the system B the rays of light are not geodetic lines in the three-dimensional space (r', ψ, θ), nor in (r, ψ, θ). In (r, ψ, θ) the velocity of light is $v = c \cos\chi$. If now we introduce a new variable h by the condition

$$\frac{dr'}{dh} = \cos\chi ,$$

of which the integral is

$$\sinh\frac{h}{R} = \tan\frac{r'}{R} = \frac{r}{R} , \tag{18}$$

then the velocity of light in the radial direction will be constant. The line-element becomes[14]

$$ds^2 = \frac{-dh^2 - R^2\sinh^2\frac{h}{R}[d\psi^2 + \sin^2\psi d\theta^2] + c^2dt^2}{\cosh^2\frac{h}{R}} . \tag{19B}$$

The three-dimensional space of this system of reference is the space with constant negative curvature, or hyperbolical space, or space of Lobatschewski. It is evident from (18) that the whole of elliptical space corresponds to the whole of the hyperbolical space; the representation of the spherical space would fill the hyperbolical space twice.

In the system of reference h, ψ, θ, ct the velocity of light is constant [in all directions, though the transformation (18) was found from the condition that it should be constant in the radial direction], and the rays of light are

[14]The transformation (18) can, of course, also be applied in the system A. The the line-element becomes

$$ds^2 = \frac{-dh^2 - R^2\sinh^2\frac{h}{R}[d\psi^2+\sin^2\psi d\theta^2]}{\cosh^2\frac{h}{R}} + c^2dt^2 . \tag{19A}$$

In (19A) all g^{ij} become zero for $h = \infty$, but g_{44} remains 1; in (19B) g_{44} also becomes zero.

straight (i.e. geodetic) lines in the three-dimensional hyperbolical space (h, ψ, θ). This hyperbolical space thus in the system B plays the same part as the elliptical space in the system A (and the euclidean space in the system C), so far as the propagation of light is concerned. If the motion of material particles (mechanics) also is considered, then the analogy breaks down, owing to the numerator $\cosh^2 h/R$.

The light-rays being straight lines, we can for the derivation of the parallax use the formulas of trigonometry in hyperbolical geometry. We thus find

$$\tan p = \sinh \frac{a}{R} \coth \frac{h}{R} ,$$

or

$$p = \frac{a}{R} \coth \frac{h}{R} .$$

It follows that in the system B the parallax of a star can never be zero. For $h = \infty$ we have $p = \frac{a}{R}$. By the transformation (18) we have

$$p = \frac{a}{R \sin \chi} = \frac{a}{r} \left(1 + r^2/R^2\right)^{1/2} . \qquad (20')$$

Thus p reaches its minimum value a/R for $\chi = \frac{1}{2}\pi$. For larger values of χ, which can only occur in the spherical space, p would increase again, and for $r' = \pi R$ we would have $p = 90°$. In fact, if the spherical space is projected by (18) on the hyperbolical space, the projections of antipodal points coincide: a star at the antipodal point of the sun would be projected in the sun.

It may be interesting to derive the formula (20') from the course of rays of light in the system (r, ψ, θ). In this system the velocity of light is

$$v = c \left(\frac{1 + \varepsilon r^2}{1 + \varepsilon r^2 \sin^2 V}\right)^{1/2} .$$

The equation of the ray of light becomes

$$\sin V = \frac{a}{r(1+\varepsilon r^2)} .$$

The parallax is determined by the equation[15]

$$\frac{dp}{dr} = -\frac{\tan V}{r},$$

from which, if we neglect a^2/r^2, we find

$$p = \frac{a}{r}\left(1 + \varepsilon r^2\right)^{1/2},$$

which is the same as (20').

4. The equations of motion of a material particle in the field of pure inertia are the differential equations of the geodetic line, viz.

$$\frac{d^2x_i}{c^2dt^2} = -\sum_p \sum_q \left[\left\{{p\,q \atop i}\right\} - \left\{{p\,q \atop 4}\right\}\dot{x}_i\right]\dot{x}_p\dot{x}_q \; ; \qquad (21)$$

or, if we restrict ourselves to such systems of reference in which the $g_{\mu\nu}$ do not depend on $x_4 = ct$,

[15]Strictly speaking r here is the distance from the star to the earth, instead of to the sun. The square of a^2/r^2 being neglected, these two distances may be interchanged. We thus have, in the notation of the first paper,

$$p = V - x .$$

Now we have (see first paper, p. 718)

$$\frac{dx}{dV} = 1 + \frac{\tan V}{r}\frac{dr}{dV},$$

or

$$\frac{dx}{dr} = \frac{dV}{dr} + \frac{\tan V}{r},$$

from which the equation for p follows immediately.

$$\frac{d^2 x_i}{c^2 dt^2} = - \left\{ {44 \atop i} \right\} - \sum_p{}' \sum_q{}' \left\{ {p\,q \atop i} \right\} \dot{x}_p \dot{x}_q + 2 \sum_p{}' \left\{ {p\,4 \atop 4} \right\} \dot{x}_i \dot{x}_p \,. \tag{21'}$$

In the system C, which represents Newton's theory of inertia, if we take rectangular cartesian space-co-ordinates, the g_{ij} are given by (1), and all the brackets are zero. Consequently

$$\frac{d^2 x_i}{c^2 dt^2} = 0 \,.$$

The orbit of a particle under the influence of inertia without gravitation is thus a straight line in euclidean space, and the velocity is constant.

In the system A we have for the co-ordinates r', ψ, θ, ct,

$$\left\{ {2\,2 \atop 1} \right\} = - R \sin\chi \cos\chi \,, \quad \left\{ {3\,3 \atop 1} \right\} = - R \sin\chi \cos\chi \sin^2\psi \,,$$

$$\left\{ {1\,2 \atop 2} \right\} = \left\{ {1\,3 \atop 3} \right\} = \frac{1}{R} \cot\chi \,, \quad \left\{ {3\,3 \atop 2} \right\} = - \sin\psi \cos\psi \,, \quad \left\{ {2\,3 \atop 3} \right\} = \cot\psi \,.$$

The other brackets are zero. We find now:

$$\frac{d^2 r'}{c^2 dt^2} = R \sin\chi \cos\chi \left[\left(\frac{d\psi}{cdt} \right)^2 + \sin^2\psi \left(\frac{d\theta}{cdt} \right)^2 \right] \,,$$

$$\frac{d^2 \theta}{c^2 dt^2} = - \frac{2}{R} \cot\chi \, \frac{dr'}{cdt} \cdot \frac{d\theta}{cdt} - 2 \cot\psi \, \frac{d\psi}{cdt} \cdot \frac{d\theta}{cdt} \,,$$

$$\frac{d^2 \psi}{c^2 dt^2} = - \frac{2}{R} \cot\chi \, \frac{dr'}{cdt} \cdot \frac{d\psi}{cdt} + \sin\psi \cos\psi \left(\frac{d\theta}{cdt} \right)^2 \,.$$

We can take $\psi = 90°$, $\frac{d\psi}{dt} = 0$. Then we find the integrals of areas and of living force:

$$R^2 \sin^2\chi\left(\frac{d\theta}{dt}\right) = c$$

$$R^2 \sin^2\chi\left(\frac{d\theta}{dt}\right)^2 + \left(\frac{dr'}{dt}\right)^2 = k \,. \tag{22}$$

Eliminating dt, we find the differential equation of the orbit

$$\left(\frac{dr'}{d\theta}\right)^2 + R^2 \sin^2 \chi = \frac{k}{c^2} R^4 \sin^4 \chi \,. \tag{23}$$

The integral is

$$\tan\chi \cos(\theta - \theta_0) = \frac{c}{kR^2 - c} \,.$$

This is the equation of a straight (i.e. geodetic) line in the spherical or elliptical space. By the second of (22) the velocity is constant. Thus in the system A a material particle under the action of inertia alone describes a straight line in elliptical space with a constant velocity.

In the case of the system B we will use the co-ordinates r, ψ, θ, ct. Then we have:

$$\left\{{1\,1 \atop 1}\right\} = -\frac{2\varepsilon r}{1 + \varepsilon r^2}, \left\{{2\,2 \atop 1}\right\} = -r, \left\{{3\,3 \atop 1}\right\} = -r\sin^2\psi, \left\{{4\,4 \atop 1}\right\} = -\varepsilon r,$$

$$\left\{{1\,2 \atop 2}\right\} = \left\{{1\,3 \atop 3}\right\} = \frac{1}{r}, \left\{{3\,3 \atop 2}\right\} = -\sin\psi\cos\psi,$$

$$\left\{{2\,3 \atop 3}\right\} = \cot\psi, \left\{{1\,4 \atop 4}\right\} = -\frac{\varepsilon r}{1 + \varepsilon r^2}$$

The others are zero. We find now:

$$\frac{d^2 r}{c^2 dt^2} = \varepsilon r + r\left[\left(\frac{d\psi}{cdt}\right)^2 + \sin^2\psi\left(\frac{d\theta}{cdt}\right)^2\right] ,$$

$$\frac{d^2\theta}{c^2 dt^2} = -\frac{2}{r}\frac{dr}{cdt}\frac{d\theta}{cdt} - 2\cot\psi \frac{d\psi}{cdt}\frac{d\theta}{cdt} ,$$

$$\frac{d^2\psi}{c^2 dt^2} = -\frac{2}{r}\frac{dr}{cdt}\frac{d\psi}{cdt} + \sin\psi\cos\psi\left(\frac{d\theta}{cdt}\right)^2 .$$

We can again take $\psi = 90°$, $\frac{d\psi}{dt} = 0$. The integrals of areas and living force are

$$r^2 \frac{d\theta}{dt} = c \; .$$

$$\left(\frac{dr}{dt}\right)^2 + r^2\left(\frac{d\theta}{dt}\right)^2 = \varepsilon r^2 + k \; . \tag{25}$$

The differential equation of the orbit is

$$\left(\frac{dr}{d\theta}\right)^2 + r^2 = \frac{\varepsilon r^2 + k}{c^2} \cdot r^4 \; . \tag{26}$$

The integration is easily effected by putting $y = \frac{1}{2r^2}$. We find

$$r^2[1 + e \cos 2(\theta - \theta_0)] = \frac{2c^2}{k} \; , \tag{27}$$

where

$$e = \frac{(4\varepsilon c^2 + k^2)^{1/2}}{k} \; .$$

This becomes a straight line in elliptical space[16] only if $e = 1$ or $c = 0$, i.e. $d\theta/dt = 0$. The orbit is thus only straight if it passes through the origin.

We can complete the integration by introducing an auxilliary angle u. We find the formuals

[16]In the co-ordinates h, ψ, θ (hyperbolical space) the equation (27) becomes

$$R^2\tanh^2 \frac{h}{R} \, [k + 2\varepsilon c^2 + ke \cos 2(\theta - \theta_0)] = 2c^2 \; .$$

which is a straight line when

$$e = 1 + \frac{2\varepsilon c^2}{k} \; .$$

$$r^2 = \frac{1}{2} R^2 k(e \cosh 2u - 1) ,$$

$$r^2 \cos 2(\theta - \theta_0) = \frac{1}{2} R^2 k(e - \cosh 2u) ,$$

$$r^2 \sin 2(\theta - \theta_0) = \frac{1}{2} R^2 k \left(e^2 - 1\right)^{1/2} \sinh 2u , \qquad (28)$$

$$\tan (\theta - \theta_0) = [(e + 1)/(e - 1)]^{1/2} \tanh u ,$$

$$u = \frac{t}{R} + u_0 .$$

We have

$$\frac{dr}{dt} = \cos^2 \chi \frac{dr}{dt} .$$

Consequently, the integrals (25) expressed in the co-ordinates r', ψ, θ of elliptical space become[17]

$$R^2 \tan^2 \chi \frac{d\theta}{dt} = c .$$

$$(25')$$

$$\left(\frac{dr'}{dt}\right)^2 + R^2 \sin^2\chi \left(\frac{d\theta}{dt}\right)^2 = \sin^2\chi \cos^2\chi + (k + \varepsilon c^2)\cos^4\chi .$$

In the system B, therefore, a material particle under the influence of inertia alone does not describe a straight line with constant velocity. The orbit can only be straight if it passes through the origin, but even then the velocity is not

[17]In the co-ordinates of hyperbolical space we have

$$R^2 \sinh^2 \frac{h}{R} \frac{d\theta}{dt} = c,$$

$$\left(\frac{dh}{dt}\right)^2 + R^2 \sinh^2 \frac{h}{R} \left(\frac{d\theta}{dt}\right)^2 = \tan h^2 \frac{h}{R} + (k+\varepsilon c^2)\operatorname{sech}^2 \frac{h}{R} .$$

The law of areas is therefore true in hyperbolical space in the system B, as it is elliptical space in the system A, and in the euclidean space in C.

constant. For small values of the χ the equations (25') do, however, not differ from (22). Those parts of the orbit which come within the reach of our observations therefore are sensibly straight if we adopt a sufficiently large value of R.

The velocity becomes zero for $r' = 1/2\ \pi R$. Thus a material particle which is on the polar line of the origin can have no velocity. It also has no energy, for the energy of a material particle is

$$m \sum_p g_{p4} \frac{dx_p}{ds},$$

which also vanishes for $r = (1/2)\pi R$. Also the velocity of light is zero on the polar line.

All these results sound very strange and paradoxical. They are, of course, all due to the fact that g_{44} becomes zero for $r' = (1/2)\pi R$. We can say that the polar line for the four-dimensional time-space is reduced to the three-dimensional space: there is no time, and consequently no motion.

It may be pointed out that the time taken by light to reach the distance $(1/2)\pi R$ from the origin (or from any other point) is

$$T = \frac{R}{c} \int_0^{\pi/2} \sec \chi\, d\chi = \infty .$$

A fortiori the time needed by a material particle for the same journey is also infinite. This also follows from the equations (28), for the distance $r' = (1/2)\pi R$ corresponds to $r = \infty$, and consequently to $u = \pm \infty$, or $t = \pm \infty$. A particle which has not always been on the polar line can therefore only reach it after an infinite time, i.e. it can never reach it at all. We can thus say that all the paradoxical phenomena (or rather negations of phenomena) which have been enumerated above can only happen after the end or before the beginning of eternity.[18]

Of course such things as "velocity" and "energy" are relative to the system of co-ordinates. They are not tensors, and consequently different in different systems of references. It may well be that the system, r', ψ, θ, ct is not the most simple or the most convenient to describe the phenomena. When described in other co-ordinates the same results may present themselves in a different form. But the fact remains that the extrapolation according to the hypothesis B is more different from what we are used to in our neighborhood than that according to the hypotheses A or C.

[18]In the systems of reference in which the radius-vector is measured by r (projection on euclidean space) and h (projection on hyperbolical space) they are also relegated to infinity of space.

The system A satisfies the "material postulate of relativity of inertia," but it restricts the admissible transformations to those for which at infinity t' = t, and thus introduces a quasi-absolute time, as has been explained in article 2. In B and C the time is entirely relative, and completely equivalent to the other three co-ordinates. In A there is a world-matter, with which the whole world is filled. and this can be in a state of equilibrium without any any internal stresses or pressures if it is entirely homogeneous and at rest. In B there may, or may not, be matter, but if there is more than one material particle these cannot be at rest, and if the whole world were filled homogeneously with matter this could not be at rest without internal pressure or stress; for if it were, we would have the system A, with $g_{44} = 1$ for all values of the four co-ordinates. The system B satisfies the "mathematical postulate" of relativity of inertia, which does not appear to admit of a simple physical interpretation.

In the system C we have no relativity of inertia at all. It cannot be denied that the introduction of the constant λ, which distinguishes the systems A and B from C, is somewhat artificial, and detracts from the simplicity and elegance of the orignal theory of 1915, one of whose great charms was that it embraced so much without introducing any new empirical constant.

REFERENCES

de Sitter, W. 1917a, Proc. Akad. Amsterdam, March 31, **xix**, 1217.
de Sitter, W. 1917b, Proc. Akad. Amsterdam, June 30, **xx**, (not yet published in English).
Einstein, A. 1917, Sitzungsber. Berlin, February 8, 142.
Newcomb, Crelles Journal, **lxxxiii**, 293.
Riemann, Werke, 272.

ON THE CURVATURE OF SPACE

A. Friedmann
Petersburg

Received: June 29, 1922

§1. In their well known works on cosmological questions Einstein (1917) and de Sitter (1917) have arrived at two types of world structure; Einstein discovered the so-called "cylindrical world", with a time independent curvature; the spatial[1] radius of curvature depending on the total mass in the space; de Sitter has developed a spherical world in which not only space but also, in a certain sense, space-time has a constant curvature (Klein 1918). To this end, Einstein, and also de Sitter, have made assumptions about the matter tensor; namely that the matter is incoherent and nearly at rest; i.e., that the velocity of the matter is sufficiently small compared to the velocity of light.

The purpose of this note is firstly to show that the cylindrical and spherical worlds are special cases of more general assumptions, and secondly to demonstrate the possibility of a world in which the curvature of space is independent of the three spatial coordinates but does depend on time; i.e., the fourth coordinate; the rest of the properties of this new type of world are analogous to the Einstein's cylinder world.

§2. The assumptions on which our treatment rests fall into two classes. To the first class belong assumptions which are common to those of Einstein and de Sitter. They refer to the equations that the gravitational potentials satisfy; and to the state and motion of the matter. The second class of assumptions have to do with, so to say, the geometrical character of the world; from our hypotheses there follow as special cases Einstein's cylinder world and also the spherical world of de Sitter.

The assumptions of the first class are the following:

1) The gravitational potentials satisfy the Einstein equations with a cosmological constant [Gleide-member] that one can also set equal to zero:

$$R_{ik} - 1/2 g_{ik}\bar{R} + \lambda g_{ik} = -\kappa T_{ik} \qquad (i,k = 1,2,3,4) \qquad \text{(A)}$$

Here the g_{ik} are the gravitational potentials, T_{ik} the matter tensor, κ a constant, $\bar{R} = g^{ik} R_{ik}$; R_{ik} is determined by the equation

[1]By 'space' we mean here a three dimensional manifold; by 'world' we mean a four dimensional manifold.

$$R_{ik} = \frac{\partial^2(\ln\sqrt{g})}{\partial x_i \partial x_k} - \frac{\partial(\ln\sqrt{g})}{\partial x_\sigma}\left\{{i\atop \sigma}{k}\right\}$$

$$-\partial/\partial x_\sigma\left\{{i\atop \sigma}{k}\right\} + \left\{{i\atop \sigma}{\alpha}\right\}\left\{{k\atop \alpha}{\sigma}\right\} \qquad \text{(B)}$$

in which the $x_i (i = 1,2,3,4)$ are the world-coordinates and $\{{i\atop \ell}{k}\}$ the Christoffel symbols of the second kind.[2]

2) The matter is incoherent and at relative rest; or put less strictly, the relative velocity of the matter is vanishingly small compared to that of light. As a consequence of this assumption the matter tensor satisfies the equations:

$$T_{ik} = 0 \text{ for } i \text{ and } k \text{ not } = 4$$

$$T_{44} = c^2 \rho\, g_{44} \qquad \text{(C)}$$

Here ρ is the matter density and c the velocity of light; furthermore the world coordinates are split into three space coordinates x_1, x_2, x_3 and a time coordinate x_4.

3) The assumption of the second class are the following:

I. Corresponding to each distribution of coordinates is a space of constant curvature, which however can change with time (x_4). The interval ds,[3] defined by $ds^2 = g_{ik}dx_i dx_k$, can, by the introduction of suitable coordinates, be brought into the form:

$$ds^2 = R^2(dx_1^{\,2} + \sin^2 x_1 dx_2^{\,2} + \sin^2 x_1 \sin^2 x_2 dx_3^{\,2})$$

$$+ 2g_{14}dx_1dx_4 + 2g_{24}dx_2dx_4 + 2g_{34}dx_3dx_4 + g_{44}dx_4^{\,2}$$

Here R is only dependent on x_4; as R is proportional to the radius of curvature of space, that will be proportional to time also.

II. In the expression for ds^2 we can by an appropriate choice of the time coordinate make g_{14}, g_{24}, g_{34} vanish; or,

[2]The signs of R_{ik} and $\bar{R}$ given here differ from the usual ones.

[3]See for example, Eddington, 1921.

putting it briefly, space can be made orthogonal to time. We cannot offer any philosophical or physical justification for these assumptions; they simplify the calculations. We must now demonstrate that Einstein's and de Sitter's world emerge as special cases.

As a consequence of assumptions 1 and 2, ds^2 can be brought into the form

$$ds^2 = R^2(dx_1^2 + \sin^2 x_1 dx_2^2 + \sin^2 x_1 \sin^2 x_2 dx_3^2) + M^2 dx_4^2 \qquad (D)$$

where R is a function of x_4 and, in general, M is a function of all four coordinates. The Einstein world metric arises from making the replacement in (D) of R^2 by $-R^2/c^2$ and setting M equal to 1, whereby R becomes the constant (independent of x_4) radius of curvature. The metric of de Sitter arises when in (D) one replaces R^2 by $-R^2/c^2$ and M by $\cos x_1$:[4]

$$d\tau^2 = -R^2/c^2(dx_1^2 + \sin^2 x_1 dx_2^2 + \sin^2 x_1 \sin^2 x_2 dx_3^2) + dx_4^2 \qquad (D_1)$$

$$d\tau^2 = -R^2/c^2(dx_1^2 + \sin^2 x_1 dx_2^2 + \sin^2 x_1 \sin^2 x_2 dx_3^2) + \cos^2 x_1 dx_4^2 \qquad (D_2)$$

We must now define the limits of the various world coordinates; that is, which points in the four dimensional manifold we take as distinct. Without giving a detailed justification we bound the world coordinates in the following intervals: x_1 in the interval $(0,\pi)$; x_2 in the interval $(0,2\pi)$. With respect to the time coordinate we make for now no special assumption and return to this matter later.

§2. From assumptions (C) and (D) it follows, when one takes in equation (A), i = 1,2,3 and k=4:

$$R'(x_4)\partial M/\partial x_1 = R'(x_4)\partial M/\partial x_2 = R'(x_4)\partial M/\partial x_3 = 0$$

[4] If we replace ds by $d\tau$ which has dimensions of time then the constant κ has dimensions length/mass and is in c.g.s. units equal to 1.8×10^{-27} cm/gm. See Laue 1921.

which suggests two cases: (1) $R'(x_4) = 0$, R is independent of x_4. We call this a stationary world. (2) $R'(x_4) \neq 0$, M depends only on x_4; we call this a non-stationary world.

We treat first the stationary world and write equations (A) for $i,k = 1,2,3$ and $i \neq k$ leading to the following system

$$\frac{\partial^2 M}{\partial x_1 \partial x_2} - \operatorname{ctn}(x_1)\, \partial M/\partial x_2 = 0$$

$$\frac{\partial^2 M}{\partial x_1 \partial x_3} - \operatorname{ctn}(x_1)\, \partial M/\partial x_3 = 0$$

$$\frac{\partial^2 M}{\partial x_2 \partial x_3} - \operatorname{ctn}(x_2)\, \partial M/\partial x_3 = 0$$

The integration of these equations leads to the following expression for M

$$M = A(x_3,x_4)\sin x_1 \sin x_2 + B(x_2,x_4)\sin x_1 + C(x_1,x_2) \qquad (1)$$

where A, B, and C are arbitrary functions of their arguments. If we solve equation (A) for R_{ik} and use the extra equations to eliminate the unknown[5] density ρ after a rather long but elementary calculation we find the following possibilities for M:

$$M = M_0 = \text{const.} \qquad (2)$$

$$M = (A_0 x_4 + B_0) \cos x_1 \qquad (3)$$

with M_0, A_0, B_0 constants.

If M is a constant then the stationary world is a cylinder world. Here it is advantageous to use the gravitational potentials derived form the formula (D_1) and thus compute the density and the constant λ. We derive the well-known results of Einstein

$$\lambda = c^2/R^2 \ , \ \rho = 2/\kappa R^2 \ , \ \bar{M} = 4\pi^2 R/\kappa$$

where $\bar{M}$ is the total mass in the space.

[5]The density ρ is for us an unknown function of the world coordinates x_1,x_2,x_3,x_4.

In the second possible case, when M is given by equation 3 we can by a useful transformation[6] of x_4 arrive at the spherical world of de Sitter, in which $M = \cos x_1$. With the help of (D_2) we obtain the results of de Sitter

$$\lambda = 3c^2/R^2 \ , \ \rho = 0 \ , \ \bar{M} = 0$$

We may then make the following statement: The stationary world is either an Einstein cylinder world or a de Sitter spherical world.

2) We will now treat the non-stationary world. M is only a function of x_4; by a judicious choice of x_4 one can without loss of generality take M = 1; joining on to our previous representations of ds^2, we have analogously to (D_1) and (D_2)

$$d\tau^2 = -R^2(x_4)/c^2 \left(dx_1^2 + \sin^2 x_1 dx_2^2 + \sin^2 x_1 \sin^2 x_2 dx_3^2\right) + dx_4^2 \qquad (D_3)$$

Our task is now to determine R and ρ from equations (A). It is clear that equations (A) with distinct indices do not give anything. Equation (A) for i=k=1,2,3 yields the relation

$$R'^2/R^2 + 2RR''/R^2 + c^2/R^2 - \lambda = 0 \qquad (4)$$

Equation (A) with i=k=4 leads to

$$3R'^2/R^2 + 3c^2/R^2 - \lambda = \kappa c^2 \rho \qquad (5)$$

with $R' = dR/dx_4$ and $R'' = d^2R/dx_4^2$.

Since R' is not equal to zero, if we replace x_4 by t, the integration of equation (4) leads to the following equation:

$$1/c^2 (dR/dt)^2 = (A - R + \lambda R^3/3c^2)/R \qquad (6)$$

[6]This transformation is given by the formula $d\bar{x}_4 = (A_0 x_4 + B_0)^{1/2} dx_4$.

with A an arbitrary constant. From this equation we can derive R in terms of an elliptic integral; that is R can be found from the equation

$$t = 1/c \int_a^R x^{1/2}(A-x + \lambda x^3/3c^2)^{-1/2} dx + B \tag{7}$$

where B and a are constants. We must take into account the possibility of a negative sign under the square root. From eq. (5) ρ is determined to be

$$\rho = 3A/\kappa R^3 \tag{8}$$

and thus A is related to the rest mass of space by

$$A = \kappa \bar{M}/6\pi^2 \tag{9}$$

As $\bar{M}$ is positive so is A.

3. The treatment of the non-stationary world is based on equations (6) and (7). The quantity λ is not determined. We assume that it can take on any value. In terms of this we can find the value of x at which the argument of the square root in eq. (7) changes sign. As we are restricting our treatment to positive curvature x will lie in the interval $(0, \infty)$ and in this interval there is a value of x for which the argument of the radical is repectively 0 or ∞. A value which makes the argument zero is $x = 0$. The values which make it infinite are given by the positive roots of the equation

$$A - x + \lambda x^3/3c^2 = 0$$

We let $y = \lambda/3c^2$ and study in the (x,y) plane the third degree curve

$$yx^3 - x + A = 0$$

The parameter A varies over the interval $(0, \infty)$. The curve crosses the x-axis at $x = A$, $y = 0$ and has a maximum at the point

$$x = 3A/2$$

$$y = 4/(27A^2) .$$

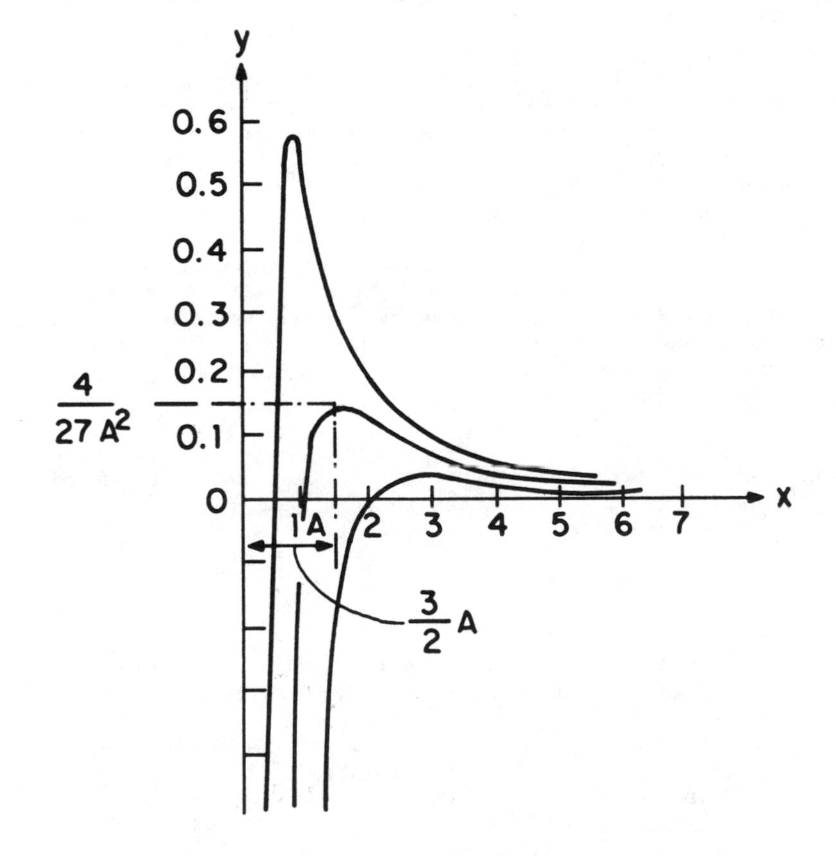
y
0.6
0.5
0.4
0.3
0.2
0.1
0
$\frac{4}{27A^2}$
1 A
2
3
4
5
6
7
x
$\frac{3}{2}A$

From the figure it is clear that for negative λ the equation

$$A - x + \lambda x^3/3c^2 = 0$$

has a positive root x_0 as a function of λ and A:

$$x_0 = \theta(\lambda, A)$$

and that θ is an increasing function of λ and A. If λ is in the interval $(0, 4c^2/9A^2)$ the equation has two positive roots $x_0 = \theta(\lambda, A)$ and $x_0' = \phi(\lambda, A)$, where x_0 is in the interval (A, 3A/2) and x_0' is the interval (3A/2, A); $\theta(\lambda, A)$ is an increasing function of λ and A, while ϕ is a decreasing function of its arguments. If λ is greater than $4c^2/9A^2$ the equation has no positive root.

In treating formula (7) we note the following; for $t = t_0$ the square root in eq. 7 can be either positive or negative. We can, by choosing, if necessary, $-t$ rather than t, always arrange it so that at $t = t_0$ it is positive; i.e., we arrange things at $t = t_0$ so that the curvature radius is an increasing function of time.

4) We treat first the case $\lambda > 4c^2/9A^2$; i.e., the case in which the equation $A-x + \lambda x^3/3c^2 = 0$ has no positive root. The equation (7) can be written in this case as

$$t - t_0 = 1/c \int_{R_o}^{R} x^{1/2}(A-x + \lambda x^3)^{-1/2} dx \tag{11}$$

from which one can see that the square root is positive. Thus R is an increasing function of t. The positive number R_0 is arbitrary.

Since R cannot be negative there must be, as one decreases the time, a time when R vanishes. The time since this beginning of the world, t', is given by[7]

$$t' = 1/c \int_0^{R_0} x^{1/2}(A-x + \lambda x^3/3c^2)^{-1/2} dx \tag{12}$$

We call the world we have just described a "monotone world of the first class."

[7]The time since the creation of the world is the time interval between which $R = 0$ and $R = R_0$; this time might be infinite.

The time since the creation of a monotone world of the first class thought of as a function of R_0, A, λ has the following properties.

1) It increases with R_0. 2) It decreases with increasing A; i.e., with increasing mass. 3) It decreases when λ increases. If $A > 2R_0/3$ the time since creation is finite for arbitrary λ. If $A \leqslant 2R_0/3$ there is always a chance of $\lambda = \lambda_1 = 4c^2/9A^2$ for which the time since the creation is unlimited.

5. Now we consider λ in the interval $(0, 4x^2/9A^2)$. Then R_0 can lie in the intervals $(0,x_0)$, (x_0,x_0') or (x_0',∞). If R_0 lies in the interval (x_0,x_0') the square root in formula (7) is imaginary; a space with such an R_0 is impossible.

The case that R_0 lies in the interval $(0,x_0)$ we will treat in the next section. Here we treat the third case: $R_0 > x_0'$ or $R_0 > \theta(\lambda,A)$. In an analogous way in which we dealt with the last case we can show that R is a function that increases with time. The smallest value R can have is $x_0' = \phi(\lambda,A)$. Here we take the age of the universe to be the time elapsed from when $R = x_0'$ and $R = R_0$. This time is

$$t' = 1/c \int_{x_0'}^{R_0} x^{1/2}(A - x + \lambda x^3/3c^2)^{-1/2}dx \qquad (13)$$

This world we call the monotone world of the second kind.

6. We now treat the case in which λ lies between $(-\infty,0)$. Here if $R_0 > x_0 = \theta(\lambda,A)$ the square root is imaginary, leading to an impossible world. If $R_0 < x_0$ the treatment is the same as before. We take λ in the interval $(-\infty, 4c^2/9A^2)$ and $R_0 < x_0$. Using well known arguments[8] one can show that R is a periodic function of its arguments. The period increases with λ and becomes infinite if λ takes on the value $\lambda_1 = 4c^2/9A^2$.

For small λ the period is given by

$$t_\pi = \pi A/c \qquad (15)$$

We can look at the periodic world in two ways: we can call two events coincident if their space coordinates coincide and their time coordinates differ by an integer number of periods. In this case, the radius of curvature oscillates between 0 and R_0 and the lifetime of the universe is finite. On the other hand if we allow the time to vary between $-\infty$ and $+\infty$ - that is if we treat two events as identical only if all four of their coordinates are the same - then we can speak of a genuine periodicity of the spatial curvature.

[8]See for example Weierstrass 1866; and Horn 1902. We must modify the treatment of these authors for our case but the periodicity is easy to establish.

7) Our knowledge is insufficient for a numerical comparison to decide which world is ours. It is possible that the causality problem and the centrifugal force problem will illuminate these questions. Finally, we may remark that the cosmological constant in our formulas is undetermined since it is an arbitrary constant in our problem. Perhaps electrodynamics will help to define it. If we set $\lambda = 0$ and $M = 5\times10^{21}$ solar masses we get a world period of about ten billion years. These figures are clearly only illustrative of our calculations.

REFERENCES

de Sitter, W. 1917, M.N.R.A.S., **78**, 3.
Eddington, A. 1921, Space, Time and Gravitation, Part 2, (Paris).
Einstein, A. 1917, (Sitzungsberichte Berl Akad).
Horn, 1902, Zs. f. Math. und Physik, **47**, 400.
Klein, 1918, (Götting. Nachr.).
Laue, 1921, Die Relativitätstheorie, Bd.II, S. 185, Braunschweig.
Weierstrass, 1866, Monatsber. d. Konigl. Akad. d. Wissensch.

ON THE POSSIBILITY OF A WORLD WITH CONSTANT NEGATIVE CURVATURE

A. Friedmann
Petersburg

Received January 7, 1924

§1.1 In our note "On the Curvature of Space" (Friedmann 1922) we have treated solutions to Einstein's 'world' equations which have as a common feature that they lead to a space of constant positive curvature; we discussed all possible cases which lead to such a solution. According to the equations the possibility of having a world of positive curvature depends on the finiteness of space. For that reason it is interesting to see if these same equations can lead to a world with constant negative curvature and thus, so to speak, free our discourse from this 'finiteness'.

In the note that follows we therefore consider the possibility that the Einstein equations coexist with a space of negative curvature. As in the previous paper we have here two cases to consider; namely

1) The case of a stationary world in which the curvature remains constant in time.
2) The case of a non-stationary world in which the curvature does not vary in space but does change with time. Between a stationary world with positive curvature and one with negative curvature there is an essential difference. A stationary world with negative curvature cannot have a positive matter density; it must be either zero or negative. The physical possibility of such a stationary world (one with a non-negative matter density) is then the analog of a de Sitter space rather than an Einstein space.[1]

§1.2 According to our general assumptions we can then make a division into the same two classes as used in our previous note; thus we can make use of our earlier calculations. The assumptions of the first class mean that we can take over equations (A), (B), (C) of our first note. However, the assumptions of the second class (in the terminology of our previous note) is the first place where a difference appears. Allowing x_4 to remain the time coordinate in our choice of world coordinates, we can in the case of negative curvature, making use of the assumptions of class two, write the expression for the interval ds in the form

[1]My friend Professor Dr. Tamarkine called my attention to the special treatment required for a world with a negative mass density.

$$ds^2 = \frac{R^2(dx_1^2 + dx_2^2 + dx_3^2)}{x_3^2} + M^2\,dx_4^2 \qquad \text{(D')}$$

where R is a function of time and M is a function of all four world coordinates. The constant negative curvature of our world space is therefore proportional to $-1/R^2$.[2]

Recalling that ds^2 is an indefinite form we can, by changing the definition, rewrite the formula D' as

$$dt^2 = \frac{-R^2\,(dx_1^2 + dx_2^2 + dx_3^2)}{x_3^2\,c^2} + M^2\,dx_4^2 \qquad \text{(D")}$$

It is clear, then, that the spatial curvature of our world is negative and proportional to $-1/R^2$.

The problem then becomes finding two functions R and M which satisfy the Einstein equations given in (A), (B), (C) of our previous note.

Using (A) we set $i = 1, 2, 3$, $k = 4$ in eq. 1 and derive the following equations.

$$R'(x_4)\,\partial M/\partial x_1 = R'(x_4)\,\partial M/\partial x_2 = R'(x_4)\,\partial M/\partial x_3 = 0$$

These equations show that negative curvature worlds are of two types

1st type: Stationary world, $R' = 0$, R is constant in time.
2nd type: Non stationary world, $R' \neq 0$, M depends only on time.

We treat first of all the case of the stationary world; the case of the non-stationary world is very similar to the case of a non-stationary world with positive curvature; for that reason this second case can be dealt with relatively briefly.

§2.1 The equations (A) of the previous note with indices i, k = 1, 2, 3 lead to

$$\frac{\partial^2 M}{\partial x_1 \partial x_2} = 0$$

[2]In connection with the line element ds^2 see, for example, B. Bianchi (no date available).

$$\partial^2 M/\partial x_2 \partial x_3 + (1/x_3)\, \partial M/\partial x_2 = 0$$

$$\partial^2 M/\partial x_1 \partial x_3 + (1/x_3)\, \partial M/\partial x_1 = 0$$

The integration of these equations gives

$$M = \frac{P(x_1, x_4) + Q(x_2, x_4)}{x_3} + L(x_3, x_4) \qquad (1)$$

where, for the present P, Q and L are arbitrary functions of their arguments.

To determine P, Q, and L we again use equations (A) with $i, k = 1, 2, 3$. A calculation gives

$$-(1/M)\,(\partial^2 M/\partial x_2^2 + \partial^2 M/\partial x_3^2) = (1 - \lambda R^2)/x_3^2$$

$$-(1/M)\,(\partial^2 M/\partial x_1^2 + \partial^2 M/\partial x_3^2) = (1 - \lambda R^2)/x_3^3 \qquad (2)$$

$$-(1/M)(\partial^2 M/\partial x_1^2 + \partial^2 M/\partial x_2^2) + (2/x_3)(1/M)\partial M/\partial x_3 = (1-\lambda R^2)/x_3^2$$

If we subtract the first equation from the second we find

$$\partial^2 P/\partial x_1^2 = \partial^2 Q/\partial x_2^2$$

From this it follows that

$$P = n(x_4)\, x_1^2 + a_1(x_4)\, x_1 + b_1(x_4)$$

$$Q = n(x_4)\, x_2^2 + a_2(x_4)\, x_2 + b_2(x_4) \qquad (3)$$

We note that equations (1) and (3) can be used to write the last of the equations (2) in the form

$$-(3-\lambda R^2)(P+Q)/x_3^2 = 4n/x_3 + (1-\lambda R_3^2)/x_3^2 - (2/x_3)\partial L/\partial x_3 \quad (4)$$

Since the right hand side of this equation does not depend on x_1 or x_2, if any of the functions, n, a_1, a_2 are different from zero the coefficient of P+Q on the left hand side must vanish. We shall treat as a separate case the situation in which all three of these functions vanish.

If n, a_1 and a_2 don't all vanish there is a relation between λ and the curvature of space; namely

$$\lambda R^2 = 3 \quad (5)$$

With the help of eq. 5 we can simplify eq. 2 so that it becomes an equation for determining the function L, namely

$$\partial L/\partial x_3 + L/x_3 = 2n \quad (6)$$

§2.2 In what follows we must differentiate two cases: 1) $n \neq 0$, 2) $n = 0$. In the first case we can see from eqs. (D'), (1) and (3) that by a suitable change of variables $\bar{x}_4 = \phi(x_4)$ we can choose $n = 1$ without any loss of generality. Having done this we have for the solution to eq. (6)

$$L = \frac{L_0(x_4)}{x_3} + x_3 \quad (7)$$

To determine ρ we set $i = k = 4$ in the equations of our previous note and then a simple calculation shows that ρ vanishes. Thus this case corresponds to a vanishing matter density and to an interval

$$ds^2 = \frac{R^2(dx_1^2 + dx_2^2 + dx_3^2)}{x_3^2} + \frac{(x_1^2 + x_2^2 + a_1(x_4)x_1 + a_2(x_4)x_2 + a_3(x_4) + x_3^2)^2\, dx_4^2}{x_3^2} \quad (D_1)$$

Going to the second case ($n = 0$) we find for L the expression

$$L = \frac{L_0(x_4)}{x_3} \tag{8}$$

In this case as well, ρ vanishes. The second case is therefore a case of vanishing matter density and an interval

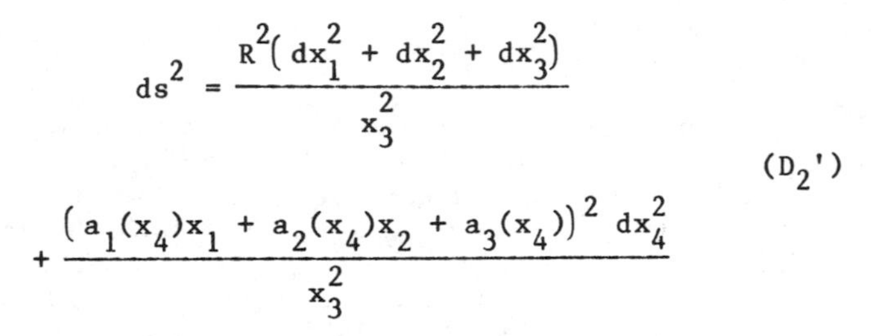

$$ds^2 = \frac{R^2(dx_1^2 + dx_2^2 + dx_3^2)}{x_3^2} + \frac{(a_1(x_4)x_1 + a_2(x_4)x_2 + a_3(x_4))^2\, dx_4^2}{x_3^2} \tag{D_2'}$$

We treat, finally, the case in which all three coefficients n, a_1, a_2 vanish so that M does not depend on x_1 and x_2. Upon integrating eq. 2 we may again consider two cases:

1.) $\lambda R^2 = 3$ $\qquad M = \frac{M_0(x_4)}{x_3}$

2.) $\lambda R^2 = 1$ $\qquad M = M(x_4)$

where M_0 and M are arbitrary functions of their arguments.

The first case is a special case of the interval given by formula (D_2'); an easy calculation shows that here the density of matter vanishes.

The second case[3] leads, as one can readily convince oneself, to a matter density which differs from zero. However, to determine whether it is positive or negative we must use the form of the interval as it is given by the indefinite quadratic form (D"). Using the gravitational potential corresponding to the form (D") we find that M is a function of x_4 alone; it follows that without loss of generality we can take M = 1 [for this purpose one replaces x_4 by $\bar{x}_4 = \phi(x_4)$]. Using this representation we can determine the density ρ. One finds

$$\lambda = -c^2/R^2 \qquad \rho = -2/\kappa R^2$$

$$d\tau^2 = -R^2/c^2 \; \frac{(dx_1^2 + dx_2^2 + dx_3^2)}{x_3^2} + dx_4^2 \tag{D_3''}$$

[3]The possibility of this case was called to my attention by Dr. W. Fock.

This case, therefore, also corresponds to a negative value for ρ.

Thus putting everything together we can say that a stationary world with constant negative curvature is only possible with either a vanishing or negative matter density; this world is expressed by the interval formulae given by (D_1'), (D_2') and (D_3").

§2.3 We turn now to the case of non-stationary world. We note, in the first place, that here M is a function of x_4 alone; the now often used considerations show that we can set M equal to unity. Under these assumptions we show without difficulty that Equations (A) for i = 1, 2, 3; k = 4 and for i, k = 1, 2, 3 are satisfied. To determine $R(x_4)$ we set i = k = 1, 2, 3 and find that R satisfies a second order differential equation namely:

$$R'^2/R^2 + 2R\,R''/R^2 + 1/R^2 - \lambda = 0 \qquad (9)$$

This equation is completely analogous to one we have previously derived [eq. (4) of our cited note]; the latter goes precisely into eq. (9) when one set c = 1. We can therefore apply the whole discussion appropriate to eq. (4) of that note to the equation we have just written down. We will not repeat that discussion but only compute the matter density for the non-stationary world.

For the non-stationary world we write the interval in the form (D"), and then find for R the differential equation

$$R'^2/R^2 + 2RR''/R^2 - c^2/R^2 - \lambda = 0$$

The integration of this equation leads to the expression

$$R'^2/c^2 = (A + R + \lambda R^3/3c^2)/R$$

where A is an arbitrary constant. We can then show that the density ρ is given by

$$\rho = 3A/\kappa R^3 \qquad (10)$$

Formula (10) shows that for positive A the matter density is positive.

Thus follows the possibility of a non-stationary world with constant negative spatial curvature and with a positive matter density.

§3.1 We turn to a discussion of the physical meaning of the results of the last paragraph. We have shown that the Einstein equations have as a solution a world with constant negative spatial curvature. The fact demonstrates that the world equations alone are not enough to allow a decision as to whether or not the world is finite. The knowledge we have about the spatial curvature gives as yet no direct hint about the finiteness or infiniteness. To definitely decide about its finiteness one needs additional conditions. Indeed, we would call a space finite if the distance between any two distinct points cannot be larger than some finite number. Therefore, before we can treat the problem of the finiteness of space we must know which points in the space are to be regarded as distinct. Consider for example a spherical surface in a three dimensional Euclidean space and choose points on the same latitude separated by 360°; should we regard these as different points on a many sheeted spherical surface in Euclidean space? The distance between two points on a surface of a sphere ought not to exceed a finite number; whereas if we consider the sphere as a many sheeted surface we can make the distance arbitrarily large. This example illuminates the difficulty in treating the finiteness of space; i.e., we must specify which points are the same and which are different.

§3.2 As a criterion for the distinctness of points we may take what is known as the principle of the "Phantomenangst." We mean by this the axiom that between any two distinct points only one straight (geodesic) line can be drawn. Thus if we accept this principle, two points through which more than one geodesic can be drawn are not distinct. From this principle, for example, it follows that the two end points of a diameter of a sphere should not be regarded as distinct. It is clear that this criterion allows the possibility of 'ghosts'; objects and their own images occuring at the same point.

This formulation of the sameness and distinctness of points implies that a space of positive curvature is finite. However it does not allow us to settle the question of the finiteness of a space of negative curvature. This is the foundation of our view that Einstein's world equations, without additional assumptions, do not allow us to decide the question of the finiteness of our world.

REFERENCES

Bianchi, B. (no date known), Lessons in Differential Geometry, 1, 345.

Friedmann, A. 1922, Zs. f. Phys., **10**, 377.

REMARK ON THE WORK OF A. FRIEDMANN (FRIEDMANN 1922) "ON THE CURVATURE OF SPACE"

A. Einstein
Berlin

Received September 18, 1922

The work cited contains a result concerning a non-stationary world which seems suspect to me. Indeed, those solutions do not appear compatible with the field equations (A). From the field equations it follows necessarily that the divergence of the matter tensor T_{ik} vanishes. This along with the anzatzes (C) and (D) leads to the condition

$$\partial\rho/\partial x_4 = 0$$

which together with (8) implies that the world-radius R is constant in time. The significance of the work therefore is to demonstrate this constancy.

REFERENCES

Friedmann, A. 1922, Zs. f. Phys., **10**, 377.

A NOTE ON THE WORK OF A. FRIEDMANN "ON THE CURVATURE OF SPACE"

A. Einstein
Berlin

Received May 31, 1923

I have in an earlier note (Einstein 1922) criticized the cited work (Friedmann 1922). My objection rested however - as Mr. Krutkoff in person and a letter from Mr. Friedmann convinced me - on a calculational error. I am convinced that Mr. Friedmann's results are both correct and clarifying. They show that in addition to the static solutions to the field equations there are time varying solutions with a spatially symmetric structure.

REFERENCES

Einstein, A. 1922, Zs. f. Phys., **11**, 326.
Friedmann, A. 1922, Ebenda, **10**, 377.

ON THE FOUNDATIONS OF RELATIVISTIC COSMOLOGY

H. P. Robertson
Department of Physics, Princeton University

Received October 11, 1929

The general theory of relativity attributes the particular metrical properties of the space-time universe, considered as a 4-dimensional Riemannian manifold, directly to the distribution of matter within it, and has naturally led to speculations concerning the structure of the universe as a whole, in which the local irregularities caused by the agglomeration of matter into stars and stellar systems are disregarded. Chief among the resulting relativistic cosmologies are those based on the cylindrical world of Einstein (1917) and the spherical world of de Sitter (1917); the line elements on which these interpretations are based have not, however, been derived from the intrinsic properties of homogeneity and isotropy attributable _a priori_ to such an idealized universe, but rather are presented as defining manifolds which do possess the desired uniformity. It is the purpose of the present note to formulate explicitly an assumption embodying the uniformity demanded by such a cosmology and to deduce all line elements satisfying it (Friedmann 1922, 1924; Tolman 1929).[1] We shall find that the only possible _stationary_ cosmologies - i.e., the intrinsic properties of which are independent of time - are in fact those of Einstein and de Sitter, and that they arise from particular cases of a class of solutions whose general member defines a non-stationary cosmology.

We first introduce coordinates x^α (Greek indices = 0,1,2,3; we write $x^0 = t$) in which the line element assumes the form

[1]A. Friedmann (1922, 1924) and, more recently, R.C. Tolman (1929) have also attacked the problem of deriving the most general line element suitable for relativistic cosmology. But Friedmann's reduction to a normal form by means of his "assumptions of the second class" is unsatisfactory and, it seems to the present author, is not possible on his assumptions alone. Tolman, on the other hand, has restricted the form of the line element _a priori_ and without taking full advantage of the isotropy which he mentions, and cannot deal with non-stationary possibilities (cf. loc. cit., p. 304). Both introduce untenable assumptions on the matter-energy tensor (cf. footnote 6) and require that Einstein's field equations be satisfied instead of making full use of the intrinsic uniformity of such a space, as we do here. The non-stationary solution found by Friedmann is contained in (2) below as the case in which the proper pressure p vanishes.

$$ds^2 = dt^2 + g_{ij}\, dx^i\, dx^j \text{ (Latin indices = 1,2,3)} , \qquad (1)$$

where the $g_{ij}(t, x^1, x^2, x^3)$ define a negative definite form, in such a way that the matter in the universe has on the whole the time-like geodesics x^i = const. as world lines (Hilbert 1924; Eisenhart 1926).[2] The coordinate t can then be interpreted as a mean time which serves to define proper time and simultaneity for the entire universe; in any portion t will be a mean proper time for all matter contained therein and the 3-space t = const. will be interpreted as the 3-dimensional physical space at time t.

Having described the actual universe in terms of these coordinates and having obtained thereby a natural and significant separation of space-time into space and time, we are now in position to state our only assumption, the assumption which is to express the intrinsic uniformity of the universe when local irregularities are disregarded. We demand that to any stationary observer ("test body") in this idealized universe all (spacial) directions about him shall be fully equivalent in the sense that he shall be unable to distinguish between them by any intrinsic property of space-time, and he shall similarly be unable to detect any difference between his observations and those of any contemporary observer; in the absence of irregularities the world offers no landmarks which enable us to distinguish between simultaneous events or spacial directions. Our assumption may be formulated mathematically:

> I. Space-time shall be spacially homogeneous and isotropic in the sense that it shall admit a transformation which sends an arbitrary configuration in any of the 3-spaces t = const., consisting of (a) an arbitrary point, (b) an arbitrary (spacial) direction through the chosen point and (c) an arbitrary

[2]Professor Eisenhart has remarked in a conversation with the author that for the purposes of the present paper we need not be so specific in introducing a coordinate system; we need only require that the time coordinate x^0 be chosen in such a way that the difference of x^0 between any two world points on the (geodesic) world line of any of the particles of which the material content of the universe is composed measure the proper time interval between the two events in question - or rather that x^0 measure proper time along a mean world line in each portion of the universe in the sense of our treatment above. Although in this more elegant treatment the remainder of the coordinate net may then be filled in at will, assumption I will lead to exactly the same result as in the above, viz., the line element (2) below, and the form of the resulting matter-energy (5) will tell us that the matter in this idealized universe is at rest with respect to the spacial coordinates x^i.

2-flat of (spacial) directions through the chosen point and direction, into any other such configuration in the same 3-space in such a way that all intrinsic properties of space-time are left unaltered by the transformation. That is, any such configuration shall be fully equivalent to any other in the same 3-space in the sense that it shall be impossible to distinguish between them by any intrinsic property of space-time.[3]

Manifolds satisfying this condition alone will be suitable for a cosmology representing the ideal background of the actual universe - provided, of course, that they satisfy Einstein's field equations for some suitable choice of the matter-energy tensor - but they are not of necessity stationary. If we wish to require in addition that their intrinsic properties be independent of the time t, we may amend the above assumption to state that any configuration, as there described, in one of the 3-spaces t = const. is fully equivalent to any such in any 3-space of the family. That the transformations involved send spacial directions into spacial directions implies, as is readily verified, that the 3-spaces t = const. are transformed among themselves; with this in mind we may alternatively - and shall in the following - characterize stationary space-times as ones which satisfy in addition to I the assumption:

II. There exists a transformation of space-time transforming the 3-spaces t = const. among themselves and sending an arbitary member of the family into any other without affecting the intrinsic properties of space-time. That is, any two of the 3-spaces t = const. shall be fully equivalent in the sense that it shall be imposible to distinguish between them by any intrinsic property of space-time.[4]

[3]We could in fact impose merely the requirement of spacial isotropy, i.e., that any such configuration be transformable into any other containing the same point (element a), without altering the results, as in either case space-time admits a group G_6 of motions (cf. below). Thus here, analogously to Schur's theorem, spacial isotropy implies spacial homogeneity. Furthermore, as follows from the sequel, it is only necessary to require that I be true for a given 3-space, say t = 0.

[4]It is to be noted that this use of the word "stationary" is at variance with that of Friedmann, who defines (loc. cit., p. 380) a stationary manifold to be one, the coefficients of whose line element are independent of t. If we wish to reserve the word "static" to describe this circumstance, it is evident that a static manifold is stationary but the converse is not necessarily true; this would also seem at variance with Tolman's use of the word "static" (cf. loc. cit., p. 304).

In order to determine all manifolds satisfying I or I and II we first express these assumptions in the language of the theory of continuous groups. We note that I is satisfied if, and only if, it be possible to find a transformation which sends an arbitrary configuration into some standard one, so the transformations contemplated in I must consequently constitute a 6-parameter transformation group - corresponding to the 3, 2, 1 parameters required to specify the elements (a), (b), (c), respectively, of the configuration - which is intransitive and has as minimum invariant varieties the 3-spaces t = const. That two configurations are equivalent under one of these transformations implies that they, together with the unique time axes at the points of the configurations, may be taken as defining the origin and direction of axes of two coordinate systems with reference to which the line element of space-time assumes identically the same form. This 6-parameter transformation group is consequently a group G_6 of motions of space-time into itself,[5] and assumption I may now be stated:

> I'. Space-time shall admit an intransitive group G_6 of motions which has as minimum invariant varieties the 3-spaces t = const.

Similarly, II may be stated:

> II'. In order that space-time be stationary it shall possess a group G_1 of motions in which ξ^0 is non-vanishing and depends at most on the time t, where ξ^0 is the time component of the infinitesimal transformation of the group.

Now Fubini has shown that a 4-space which admits an intransitive group G_6 of motions has as minimum invariant varieties a family of geodesically parallel 3-spaces of constant curvature which are mapped conformally on each other by their orthogonal trajectories (Fubini 1904; Bianchi, 1918). In virtue of assumption I our line element (1) must consequently assume the form

$$ds^2 = dt^2 - e^{2f} h_{ij} dx^i dx^j \qquad (2)$$

where f is an arbitrary real function of the time t and the coefficients h_{ij} are functions of the spacial variables x^1, x^2, x^3 alone such that the differential form

[5]For an account of the theory of groups of motions consult Eisenhart, loc. cit., Chap. VI, in particular pp. 233ff.

$$ds^{*2} = h_{ij}\, dx^i\, dx^j \tag{3}$$

is positive definite and defines a 3-space of constant curvature, say κ.

In order that this line element (2), which has been derived from the single assumption I of spacial uniformity, be suitable for relativistic cosmology it must satisfy Einstein's field equations

$$R_{\alpha\beta} - \frac{g_{\alpha\beta}}{2}(R - 2\lambda) = -\, 8\pi T_{\alpha\beta} \; .$$

for some appropriate choice of the matter energy tensor $T_{\alpha\beta}$. But we find on inserting in (4) the values of the Ricci tensor $R_{\alpha\beta}$ computed for (2) that

$$T_{\alpha\beta} = (\rho + p)\, \delta^{\alpha}_{0}\, \delta^{\beta}_{0} - g^{\alpha\beta}\, p \; , \tag{5}$$

where

$$\begin{aligned} 8\pi\rho(t) &= \lambda + 3f'^2 + 3\kappa e^{-2f} \\ 8\pi p(t) &= \lambda - 2f'' - 3f'^2 - \kappa e^{-2f} \; , \end{aligned} \tag{6}$$

and this allows us to consider (2) as defining an ideal universe containing a uniform distribution of matter of proper density ρ and proper pressure p at rest with respect to the spacial coordinates x^i (Tolman, 1929).[6]

Before proceeding to a discussion of the general line element (2) we apply assumption II' in order to determine which cases of it lead to stationary cosmologies. The components ξ^{α} of the motion G_1 must satisfy the equations of Killing (Eisenhart 1926)

$$\xi_{\alpha;\beta} + \xi_{\beta;\alpha} = 0 \; , \tag{7}$$

[6]But it is to be noted that whereas in our coordinates the world lines x^i = const. are actually geodesics, so that rest is a possible state for the material content of the universe, this is not the case in Tolman's or Friedmann's coordinates. The condition that this be permissible in coordinates in which the g_{oi} vanish is that g_{00} depend at most on t, and is not satisfied in their discussion of the de Sitter universe.

where $\xi_{\alpha;\beta}$ is the covariant derivative of ξ_α with respect to x^β; in the case contemplated in II' they are

$$\frac{\partial \xi^0}{\partial t} = 0, \quad h_{ij}\frac{\partial \xi^i}{\partial t} = 0 \tag{8a}$$

$$2\xi^0 f' h_{ij} + \xi^k \frac{\partial h_{ij}}{\partial x^k} + h_{kj}\frac{\partial \xi^k}{\partial x^i} + h_{ik}\frac{\partial \xi^k}{\partial x^j} = 0 \,. \tag{8b}$$

Consequently ξ^0, which can depend at most on t, is a constant, say unity, and the ξ^k are independent of t. Since $f^i(t)$ is then the only function in (8b) which could depend on t it must be a constant, say 1/a, and we may take f = t/a without loss of generality. Furthermore, the conditions of integrability of (8b) are[7]

$$\kappa f' h_{ij} = 0, \quad \text{i.e.} \quad \kappa f' = 0 \,, \tag{9}$$

from which it follows that either f' = 0, in which case we may take f = 0 and (2) defines Einstein's cosmology, or $\kappa = 0$ and f = t/a, from which direct computation shows that (2) has constant curvature $- 1/a^2$ and defines de Sitter's cosmology (Robertson 1928; Weyl 1928; Lemaitre 1925).[8] The only stationary cosmologies satisfying our requirements are those of Einstein and de Sitter.

We now waive assumption II and return to the cosmology defined by the general line element (2). This (conformal-Minkowskian) space-time is of class one, i.e., it can be

[7]Eisenhart (1926) p. 237; or more readily, from eq. (69.8), p. 232, on noting that (8b) defines an infinitesimal conformal transformation of the 3-space with line element ds^* defined by (3).

[8]For a discussion of the de Sitter universe in terms of these coordinates see H.P. Robertson (1928), and also a forthcoming paper by H. Weyl (1928) in the same journal. (I have since discovered that these coordinates have also been employed by G. Lemaitre (1925), and wish to take this opportunity to correct the omission of reference to this work in my previous paper.) It is to be noted, however, that this is not the only form of (2) which describes the de Sitter universe, as the line element for which $e^f = a\sqrt{\kappa}\cosh(t/a)$ is also of constant curvature $-1/a^2$; although both ρ and p are constant in this case it does not represent a stationary cosmology in which t is interpreted as time. The stationary form can be obtained from this form by replacing t by t + a log(2R/a) and allowing $R \to \infty$.

immersed in a 5-flat (Eisenhart 1926);[9] we shall actually express it in this form, which will be of value in discussing the macroscopic properties of the universe. We take $\kappa > 0$ and write $\kappa = 1/R^2$; our spacial coordinates can then be so chosen that (2) assumes the form

$$ds^2 = dt^2 - e^{2f(t)}\left(\frac{dr^2}{1 - r^2/R^2} + r^2 d\theta^2 + r^2 \sin^2\theta \, d\phi^2\right) . \quad (10)$$

On defining

$$z_0 = \int (1 + R^2 f'^2 e^{2f})^{1/2} dt \qquad \begin{array}{l} z_1 = e^f r \sin\theta \cos\phi \\ z_2 = e^f r \sin\theta \cos\phi \end{array} \quad (11)$$

$$z_4 = Re^f (1 - k^2/R^2)^{1/2} \qquad Z_3 = e^f r \cos\theta$$

(10) may be written

$$ds^2 = dz_0^2 - \left(dz_1^2 + dz_2^2 + dz_3^2 + dz_4^2\right) \quad (12)$$

and the universe is represented by the general hypersurface of revolution about the z_0 or time axis

$$z_1^2 + z_2^2 + z_3^2 + z_4^2 = R^2 e^{2f} \quad (13)$$

where f(t) is expressed as a function $f[t(z_0)]$ of z_0 by inverting $z_0(t)$ as defined in (11).

Einstein's universe, for which f is constant, is represented as a cylinder with axis along the z_0 or time direction, and de Sitter's (in the form given in footnote 8 for $\kappa = 0$) by a pseudo-sphere. Unfortunately the stationary form of de Sitter's universe is here represented as a pseudo-sphere lying, as do all manifolds (2) for which $\kappa = 0$, in the infinite regions of the variables, z_0, z_4. To obtain a representation of these in the finite domain we may take z_1, z_2, z_3 as before and define

$$z_0 = \frac{e^f}{2a}\left(a^2 + r^2 + \alpha(t)\right), \quad z_4 = \frac{e^f}{2a}\left(a^2 - r^2 - \alpha(t)\right) , \quad (14)$$

[9]These results follow, after some calculation, from the general conditions; for these consult Eisenhart (1926), pp. 92, 197-198.

where

$$\alpha(t) = e^{-f} \int \frac{e^{-f}}{f'} dt .$$

The universe is then represented by

$$- z_0^2 + z_1^2 + z_2^2 + z_3^2 + z_4^2 = - e^{2f} \alpha \qquad (15)$$

where the function of t on the right is expressed as a function of $z_0 + z_4$ by the inversion of $z_0 + z_4 = a e^{f(t)}$. The stationary form of de Sitter's universe is, as is known, represented as a pseudo-sphere. The discussion of the remaining case, that in which $\kappa < 0$, can be obtained from the above for $\kappa > 0$ by replacing R, z_4 by iR, iz_4.

Finally, we consider briefly the question of Doppler effect in light from distant objects. We have found that the material content of our idealized universe is at rest, so the Doppler effect will be obtained by computing the difference in time of arrival at the observer between two flashes of light emitted at times t_0, $t_0 + \Delta t_0$ from a point source at $P_0(x_0^i)$ at rest with respect to the spacial coordinates x^i. The spacial projection of geodesics in the space-time with line element ds^2 (2) are geodesic in the 3-space with element ds^{*2} (3), so light which leaves P_0 at time t_0 arrives at the observer at time t defined implicity by

$$\int_{t_0}^{t} e^{-f(t)} dt = \int_{P_0}^{0} ds^* , \qquad (16)$$

where the integral on the right is the geodesic distance between P_0 and the observer, measured in the 3-space (3) - and is $R \sin^{-1} r_0/R$ in the coordinates employed in (10). Light leaving P_0 in the interval t_0, $t_0 + \Delta t_0$ will consequently arrive at the observer in the interval t, $t + \Delta t$ where $\Delta t = \Delta t_0 \exp[f(t)-f(t_0)]$; the resulting Doppler shift $\Delta\lambda/\lambda = \Delta t/\Delta t_0 - 1$ will be attributed to a velocity of recession

$$v = c \tanh [f(t) - f(t_0)] , \qquad (17)$$

where t is defined by (16) and c is the velocity of light. Our choice of coordinates in the actual universe has been such that the above considerations apply to the residual Doppler effect, after averaging to eliminate the effect due to accidental "proper" motions. Of the two stationary cosmologies, only that of de Sitter will show such a residual effect, as in Einstein's f is constant.

To summarize briefly our conclusions, we have described the actual world in terms of coordinates which effect a natural separation of it into space and time and have determined its ideal background by a single assumption (I) which is but the concrete expression of uniformity implied by the very concept of a system of relativistic cosmology; in requiring that this ideal background can be fitted into the actual universe in the way described we have expressed in another way the assumption made by other writers on the subject, above all by H. Weyl (1923),[10] that the world lines of all matter in the universe form a coherent pencil of geodesics. We have shown that our idealized space-time can be represented as a hypersurface of revolution in a 5-dimensional flat space, and that the additional requirement (II) that it be stationary leads to the cylindrical world of Einstein and the spherical world of de Sitter as the only possibilities. The material content of this idealized background of the actual universe, being at rest with respect to the spacial coordinates, leads to a unique Doppler shift which will appear in the actual universe as a residual effect.

REFERENCES

Bianchi, L. 1918, Teoria dei Gruppi Continuii, (Pisa), p. 544
de Sitter, W. 1917, M.N.R.A.S., **78**, 3.
Einstein, A. 1917, Sitzungber. Berl. Akad., p. 142.
Eisenhart, L. P. 1926, Riemannian Geometry, (Princeton: Princeton University Press).
Friedmann, A. 1922, Z. Physik, **10**, 377.
Friedmann, A. 1924, Z. Physik, **21**, 326.
Fubini, G. 1904, Annali di Matematica (iii), **9**, 64.
Hilbert, D. 1924, Math. Ann., **92**, 15.
Lemaitre, G. 1925, Jour. of Math. and Phys., **4**, 188.
Robertson, H. P. 1928, Phil Mag, **5**, 385.
Tolman, R. C. 1929, Proc. of the Nat. Acad. of Sc., **15**, 297.
Weyl, H. 1923, Phys. Zeitschr., **29**, 230.
Weyl, H. 1930, Phil Mag., **9**, 936.

[10]See also papers cited in footnote 8.

A RELATION BETWEEN DISTANCE AND RADIAL VELOCITY AMONG EXTRA-GALACTIC NEBULAE

Edwin Hubble
Mount Wilson Observatory, Carnegie Institution of Washington

Received January 17, 1929

Determinations of the motion of the sun with respect to the extragalactic nebulae have involved a K term of several hundred kilometers which appears to be variable. Explanations of this paradox have been sought in a correlation between apparent radial velocities and distances, but so far the results have not been convincing. The present paper is a re-examination of the question, based on only those nebular distances which are believed to be fairly reliable.

Distances of extra-galactic nebulae depend ultimately upon the application of absolute-luminosity criteria to involved stars whose types can be recognized. These include, among others, Cepheid variables, novae, and blue stars involved in emission nebulosity. Numerical values depend upon the zero point of the period-luminosity relation among Cepheids, the other criteria merely check the order of the distances. This method is restricted to the few nebulae which are well resolved by existing instruments. A study of these nebulae, together with those in which any stars at all can be recognized, indicates the probability of an approximately uniform upper limit to the absolute luminosity of stars, in the late-type spiral and irregular nebulae at least, of the order of M (photographic) = -6.3 (Hubble, 1926). The apparent luminosities of the brightest stars in such nebulae are thus criteria which, although rough and to be applied with caution, furnish reasonable estimates of the distances of all extra-galactic systems in which even a few stars can be detected.

Finally, the nebulae themselves appear to be of a definite order of absolute luminosity, exhibiting a range of four or five magnitudes about an average value M (visual) = -15.2 (Hubble, 1926). The application of this statistical average to individual cases can rarely be used to advantage, but where considerable numbers are involved, and especially in the various clusters of nebulae, mean apparent luminosities of the nebulae themselves offer reliable estimates of the mean distances.

Radial velocities of 46 extra-galactic nebulae are now available, but individual distances are estimated for only 24. For one other, NGC 3521, an estimate could probably be made, but no photographs are available at Mount Wilson. The data are given in table 1. The first seven distances are the most reliable, depending, except for M 32 the companion of M 31, upon extensive investigations of many stars involved. The next thirteen distances, depending upon the criterion of a uniform upper limit of stellar luminosity, are subject to considerable probable errors but are believed to be the most reasonable values at present available. The last four objects appear to be in the Virgo Cluster. The distance assigned to the cluster, 2×10^6 parsecs, is derived from the distribution of nebular luminosities, together with luminosities of stars in some of the later-type spirals, and differs somewhat from the Harvard estimate of ten million light years (Harvard College Observatory Circular, 1926).

TABLE 1

NEBULAE WHOSE DISTANCES HAVE BEEN ESTIMATED FROM STARS INVOLVED OR FROM MEAN LUMINOSITIES IN A CLUSTER

OBJECT	m_s	r	v	m_t	M_t
S. Mag.	--	0.032	+ 170	1.5	-16.0
L. Mag.	--	0.034	+ 290	0.5	-17.2
NGC 6822	--	0.214	- 130	9.0	-12.7
NGC 598	--	0.263	- 70	7.0	-15.1
NGC 221	--	0.275	- 185	8.8	-13.4
NGC 224	--	0.275	- 220	5.0	-17.2
NGC 5457	17.0	0.45	+ 200	9.9	-13.3
NGC 4736	17.3	0.5	+ 290	8.4	-15.1
NGC 5194	17.3	0.5	+ 270	7.4	-16.1
NGC 4449	17.8	0.63	+ 200	9.5	-14.5
NGC 4214	18.3	0.8	+ 300	11.3	-13.2
NGC 3031	18.5	0.9	- 30	8.3	-16.4
NGC 3627	18.5	0.9	+ 650	9.1	-15.7
NGC 4826	18.5	0.9	+ 150	9.0	-15.7
NGC 5236	18.5	0.9	+ 500	10.4	-14.4
NGC 1068	18.7	1.0	+ 920	9.1	-15.9
NGC 5055	19.0	1.1	+ 450	9.6	-15.6
NGC 7331	19.0	1.1	+ 500	10.4	-14.8
NGC 4258	19.5	1.4	+ 500	8.7	-17.0
NGC 4151	20.0	1.7	+ 960	12.0	-14.2
NGC 4382	--	2.0	+ 500	10.0	-16.5
NGC 4472	--	2.0	+ 850	8.8	-17.7
NGC 4486	--	2.0	+ 800	9.7	-16.8
NGC 4649	--	2.0	+ 1090	9.5	-17.0
Mean					-15.5

m_s = photographic magnitude of brightest stars involved.
r = distance in units of 10^6 parsecs. The first two are Shapley's values.
v = measured velocities in km/sec. NGC 6822, 221, 224 and 5457 are recent determinations by Humason.
m_t = Holetscheck's visual magnitude as corrected by Hopmann. The first three objects were not measured by Holetschek, and the values of m_t represent estimates by the author based upon such data as are available.
M_t = total visual absolute magnitude computed from m_t and r.

The data in the table indicate a linear correlation between distances and velocities, whether the latter are used directly or corrected for solar motion, according to the older solutions. This suggests a new solution for the solar motion in which the distances are introduced as coefficients of the K term, i.e., the velocities are assumed to vary directly with the distances, and hence K represents the velocity at unit distance due to this effect. The equations of condition then take the form

$$rK + X \cos\alpha \cos\delta + Y \sin\alpha \cos\delta + Z \sin\delta = v .$$

Two solutions have been made, one using the 24 nebulae individually, the other combining them into 9 groups according to proximity in direction and in distance. The results are

	24 OBJECTS	9 GROUPS
X	− 65 ± 50	+ 3 ± 70
Y	+226 ± 95	+230 ± 120
Z	−195 ± 40	−133 ± 70
K	+465 ± 50	+513 ± 60 km/sec per 10^6 parsecs
A	286°	269°
D	+40°	+33°
V_0	306 km/sec	247 km/sec

For such scanty material, so poorly distributed, the results are fairly definite. Differences between the two solutions are due largely to the four Virgo nebulae, which, being the most distant objects and all sharing the peculiar motion of the cluster, unduly influence the value of K and hence of V_0. New data on more distant objects will be required to reduce the effect of such peculiar motion. Meanwhile round numbers, intermediate between the two solutions, will represent the probable order of the values. For instance, let A = 277°, D = +36° (Gal. long. = 32°, lat. = +18°), V_0 = 280 km/sec, K = +500 km/sec per million parsecs. Mr. Stromberg has very kindly checked the general order of these values by independent solutions for different groupings of the data.

A constant term, introduced into the equations, was found to be small and negative. This seems to dispose of the necessity for the old constant K term. Solutions of this sort have been published by Lundmark (1925), who replaced the old K by $k + lr + mr^2$. His favored solution gave k = 513, as against the former value of the order of 700, and hence offered little advantage.

The residuals for the two solutions given above average 150 and 110 km/sec and should represent the average peculiar motions of the individual nebulae and of the groups, respectively. In order to exhibit the results in a graphical form, the solar motion has been eliminated from the observed velocities and the remainders, the distance terms plus the residuals, have been plotted against the distances. The run of the residuals is about as smooth as can be expected, and in general the form of the solutions appears to be adequate.

The 22 nebulae for which distances are not available can be treated in two ways. First, the mean distance of the group derived from the mean apparent magnitudes can be compared with the mean of the velocities but corrected for solar motion. The result, 745 km/sec for a distance of 1.4×10^6 parsecs, falls between the two previous solutions and indicates a value for K of 530 as against the proposed value, 500 km/sec.

TABLE 2

NEBULAE WHOSE DISTANCES ARE ESTIMATED FROM RADIAL VELOCITIES

OBJECT	v	v_s	r	m_t	M_t
NGC 278	+ 650	- 110	1.52	12.0	-13.9
NGC 404	- 25	- 65	--	11.1	--
NGC 584	+1800	+ 75	3.45	10.9	-16.8
NGC 936	+1300	+ 115	2.37	11.1	-15.7
NGC 1023	+ 300	- 10	0.62	10.2	-13.8
NGC 1700	+ 800	+ 220	1.16	12.5	-12.8
NGC 2681	+ 700	- 10	1.42	10.7	-15.0
NGC 2683	+ 400	+ 65	0.67	9.9	-14.3
NGC 2841	+ 600	- 20	1.24	9.4	-16.1
NGC 3034	+ 290	- 105	0.79	9.0	-15.5
NGC 3115	+ 600	+ 105	1.00	9.5	-15.5
NGC 3368	+ 940	+ 70	1.74	10.0	-16.2
NGC 3379	+ 810	+ 65	1.49	9.4	-16.4
NGC 3489	+ 600	+ 50	1.10	11.2	-14.0
NGC 3521	+ 730	+ 95	1.27	10.1	-15.4
NGC 3623	+ 800	+ 35	1.53	9.9	-16.0
NGC 4111	+ 800	- 95	1.79	10.1	-16.1
NGC 4526	+ 580	- 20	1.20	11.1	-14.3
NGC 4565	+1100	- 75	2.35	11.0	-15.9
NGC 4594	+1140	+ 25	2.23	9.1	-17.6
NGC 5005	+ 900	- 130	2.06	11.1	-15.5
NGC 5866	+ 650	-215	1.73	11.7	-14.5
Mean				10.5	-15.3

Secondly, the scatter of the individual nebulae can be examined by assuming the relation between distances and velocities as previously determined. Distances can then be calculated from the velocities corrected for solar motion, and absolute magnitudes can be derived from the apparent magnitudes. The results are given in table 2 and may be compared with the distribution of absolute magnitudes among the nebulae in table 1, whose distances are derived from other criteria.

NGC 404 can be excluded since the observed velocity is so small that the peculiar motion must be large in comparison with the distance effect. The object is not necessarily an exception, however, since a distance can be assigned for which the peculiar motion and the absolute magnitude are both within the range previously determined. The two mean magnitudes, -15.3 and -15.5, the ranges, 4.9 and 5.0 mag., and the frequency distributions are closely similar for these two entirely independent sets of data; and even the slight difference in mean magnitudes can be attributed to the selected, very bright, nebulae in the Virgo Cluster. This entirely unforced agreement supports the validity of the velocity-distance relation in a very evident matter. Finally, it is worth recording that the frequency distribution of absolute magnitudes in the two tables combined is comparable with those found in the various clusters of nebulae.

The results establish a roughly linear relation between velocities and distances among nebulae for which velocities have been previously published, and the relation appears to dominate the distribution of velocities. In order to investigate the matter on a much larger scale, Mr. Humason at Mount Wilson has initiated a program of determining velocities of the most distant nebulae that can be observed with confidence. These, naturally, are the brightest nebulae in clusters of nebulae. The first definite result (Humason, 1929), v = +3779 km/sec for NGC 7619, is thoroughly consistent with the present conclusions. Corrected for the solar motion, this velocity is +3910, which, with K = 500, corresponds to a distance of 7.8×10^6 parsecs. Since the apparent magnitude is 11.8, the absolute magnitude at such a distance is -17.65, which is of the right order for the brightest nebulae in a cluster. A preliminary distance, derived independently from the cluster of which this nebula appears to be a member, is of the order of 7×10^6 parsecs.

New data to be expected in the near future may modify the significance of the present investigation or, if confirmatory, will lead to a solution having many times the weight. For this reason it is thought premature to discuss in detail the obvious consequences of the present results. For example, if the solar motion with respect to the clusters represents the rotation of the galactic system, this motion could be subtracted from the results for the nebulae and the remainder would represent the motion of the galactic system with respect to the extra-galactic nebulae.

The outstanding feature, however, is the possibility that the velocity-distance relation may represent the de Sitter effect, and hence that numerical data may be introduced into discussions of the general curvature of space. In the de Sitter cosmology, displacements of the spectra arise from two sources, an apparent slowing down of atomic vibrations and a general tendency of material particles to scatter. The latter involves an acceleration and hence introduces the element of time. The relative importance of these two effects should de-

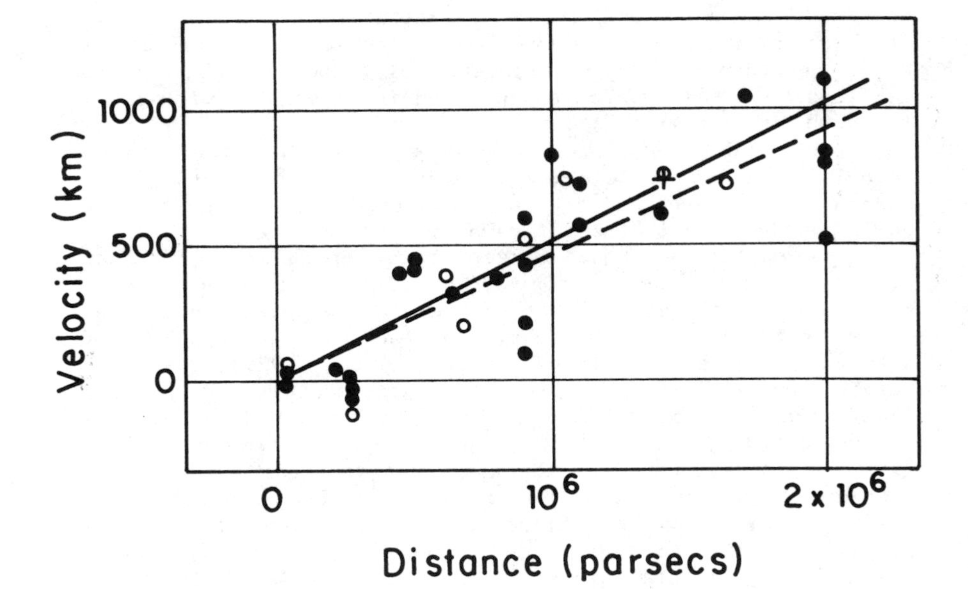

Velocity-Distance Relation among Extragalactic Nebulae.

Radial velocities, corrected for solar motion, are plotted against distances estimated from involved stars and mean luminosities of nebulae in a cluster. The black discs and full line represent the solution for solar motion using the nebulae individually; the circles and broken line represent the solution combining the nebulae into groups; the cross represents the mean velocity corresponding to the mean distance of 22 nebulae whose distances could not be estimated individually.

termine the form of the relation between distances and observed velocities; and in this connection it may be emphasized that the linear relation found in the present discussion is a first approximation representing a restricted range in distance.

REFERENCES

Harvard Coll. Obs. Cir. 1926, 294.
Hubble, E. 1926, _Ap.J._, **64**, 321.
Humason, M. 1929, _Proc. of the Nat. Acad. of Sc._, **15**, 167.
Lundmark, K. 1925, _M.N.R.A.S._, **85**, 865.

A NEW DETERMINATION OF THE HUBBLE CONSTANT FROM GLOBULAR CLUSTERS IN M87

Allan Sandage
Mount Wilson and Palomar Observatories, Carnegie Institution of Washington, California Institute of Technology

Received April 29, 1968

ABSTRACT

An apparent blue distance modulus of $(m - M)_{AB} = 31.1$ is derived for the Virgo Cluster from Racine's measurement of B(first) = 21.3 for the brightest globular cluster in M87, using a calibration of M_B(first) = −9.83 for cluster B282 in M31. This assumes that the brightest globular clusters in the two galaxies have the same absolute magnitude. The assumption provides an upper limit to H in the present context. For this distance modulus, the absolute luminosity of NGC 4472 - the first-ranked E galaxy in the Virgo Cluster - is $M_B = -21.68$.

A new redshift-apparent magnitude relation (the Hubble diagram) for the first-ranked cluster members (Fig. 1) shows that M_B(first) has a remarkably small dispersion of ±0.3 mag in a sample of forty-one clusters. Using M_B from NGC 4472 and reading the Hubble diagram at $cz = 10^4$ km sec^{-1}, which is beyond the local anisotropy in the velocity field, we find that $H = 75.3_{-15}^{+19}$ km sec^{-1} Mpc^{-1}, using the precepts in the text. Systematic errors are discussed. It seems possible that H could be as small as 50 km sec^{-1} Mpc^{-1} ($H^{-1} = 19 \times 10^9$ years) with the present data.

The local anisotropy in the velocity field at Virgo Cluster is estimated to be $H_\infty/H_{VC} = 1.17 \pm 0.09$.

I. INTRODUCTION

The value of the Hubble constant is not well known at present. Work by Holmberg (1958), van den Bergh (1960), Sersic (1960), Sandage (1958, 1962), and others has shown that H probably lies in the range $125 > H > 50$ km sec^{-1} Mpc^{-1}, but a more precise value is not available.

The determination of H is difficult. Two requirements exist: (1) Accurate distances to galaxies of known redshifts are needed. (2) The redshifts must be large compared to the random virial velocities. These boundary conditions restrict the range of moduli within which H can be measured to $32 \gtrsim m - M \gtrsim 30$. At $m - M \simeq 30$, the cosmological redshift is about 750 km sec^{-1}, which is only a few times the mean random motion. Beyond $m - M \simeq 32$, precision distance indicators fade below plate limit.

A further complication arises from the local anisotropy of the velocity field (de Vaucouleurs 1958, 1966; Kristian 1967) for redshifts less than ~4000 km sec^{-1}. Measured radial velocities differ from those for a pure Hubble flow because of this local perturbation called the "shear field" by Kristian and Sachs (1966). One method of procedure is to map the velocity field for nearby galaxies in order to separate the cosmological and shear components. Kristian and Sandage are attempting to do this using angular sizes of HII regions for galaxies and redshifts less than 2000 km sec^{-1}, but the data are not yet sufficient for a solution.

II. A NEW METHOD

A second procedure has recently become available from Racine's (1968) measurement of the luminosity function of 2000 globular clusters in the E galaxy M87 in the Virgo Cluster. On the assumption that the brightest globular clusters have the same absolute luminosity in galaxies which have sufficently large cluster population, the distance to M87, and hence to the E cloud of the Virgo Cluster, can be found. This distance can then be used to calibrate the absolute luminosity of the first-ranked E galaxy in this cluster.

It has previously been shown (Hubble 1936; Humason 1936; Humason, Mayall, and Sandage 1956, hereinafter referred to as "HMS"; Sandage 1968) that the brightest E galaxy in clusters have a remarkably small dispersion in absolute luminosity. Ths calibration of M_B(brightest) then permits calculation of H by entering the observed red-shift-apparent magnitude relation for the first-ranked cluster galaxy at redshifts larger than $\sim$4000 km sec^{-1}, which is the outer boundary of the local anistropy. It is important to note that the velocity of the Virgo Cluster is not required, and the effects of the local shear field are thereby circumvented.

III. M_B FOR THE BRIGHTEST VIRGO CLUSTER E GALAXY

a) Distance to M87

Racine's photometry of the globular clusters in M87 shows that the luminosity function at the bright end is steep and begins at P = 21.2 mag. Converting this to the B system of Johnson and Morgan by B $\simeq$ P + 0.1, we find that B = 21.3 for the brightest globular cluster in M87.

The absolute luminosity of the brightest globular cluster, M_B(first), can be estimated from the data from M31 and, less reliably, from the galactic system. Photoelectric photometry of the clusters in M31 by Kron and Mayall (1960), Hiltner (1960), and Kinman (1963) have been summarized by Kinman (1963). There is a well-defined upper luminosity limit at V = 14.1 if the cluster M II is excluded. This cluster is difficult to measure because of the close proximity of two moderately bright stars (see Pl. II of Mayall and Eggen 1953). Until the question of contamination can be re-examined, M II is excluded from further discussion here, and we adopt cluster B282 in M31 as the brightest, at V = 14.13 and B = 15.01. Clusters H12 and H42 are near this upper limit, at V = 14.23,

B = 15.04, and V = 14.22, B = 15.25, respectively. We choose to consider here only those clusters in M31 where absorption within this galaxy may be negligible. We have not considered clusters where M_B appears to be brighter than that for B282 after uncertain absorption corrections are applied (Kinman 1963, group C).

The apparent blue modulus of M31 is taken to be $(m - M)_{AB} = 24.84$, obtained from the photometry of Cepheids by Baade and Swope (1963) in their outlying field IV and using the calibration of the (P,L)-relation given elsewhere (Sandage and Tammann 1968). The calibration is based on Cepheids in open clusters in the galactic system, the absolute luminosities of which rest ultimately on photometric parallaxes via the distance to the Hyades (Wayman, Symms, and Blackwell 1965). The effect of a systematic error in this distance, following Hodge and Wallerstein (1966), is discussed later. The adopted data then give $M_B = -9.83$ for B282 in M31.

How constant M_B(first) may be from galaxy to galaxy is not known, either empirically or theoretically. On statistical grounds alone the assumption is likely to be invalid for galaxies with few globular clusters. But clusters in galaxies such as M31 or the galactic system may form an adequate sample. Mindful that future work may clarify the problem, we can proceed quantitatively by assuming that M_B(first) is a stable statistic independent of the total population N as long as N is large enough. It should be noted, as discussed later, that this assumption provides an upper limit to H. If some form of the Scott effect (1957) applies to globular clusters, it can only decrease the value of H in the present context, since M_B(first) in M87 with 2000 clusters would be brighter than M_B(first) in M31 with ~250 clusters.

Partial justification for the assumption of constant M_B may be found in the nearly vertical asymptote of Racine's luminosity function for clusters in M87. Another check comes from comparing M_B(first) in the galactic system with B282 in M31. Gascoigne and Burr's (1956) photometry of ω Cen gives the total apparent luminosity as $P_t = 4.25$, or $B_t = 4.35$. The horizontal branch occurs at V = 14.5 (Roy. Obs. Ann., No. 2, 1966). The middle of the RR Lyrae domain is at B = 14.85 (Saunders 1963), which gives an apparent blue modulus $(m-M)_{AB} = 14.05$ if $M_B(RR) = +0.8$ (i.e., $M_V = +0.5$, $\langle B - V \rangle = +0.3$). Hence $M_B = -9.7$ for ω Cen, which is the brightest known globular cluster in the galactic system.[1] M_B for ω Cen compares well with $M_B = -9.83$ for B282 in M31, but since the value is based on the somewhat less reliable route via M_B for the RR Lyrae stars rather than through the Cepheid (P,L)-relation, we adopt $M_B = -9.83$ in the following discussion.

Combining $M_B = -9.83$ with Racine's value B = 21.3 gives $(m-M)_{AB} = 31.1$ for M87. The possible range of uncertainty is

[1]According to Gascoigne and Burr's photometry, combined with a modulus of $(M - M)_{AB} = 13.50$ from Tifft's photometry (1963), 47 Tuc has $M_V = -8.70$, which does not rival ω Cen.

discussed later. The new modulus may be compared with an older value of $(m - M)_{AB} \simeq 30.8$, which follows from Baum's (1955) photoelectric data that gave a 6.0-mag difference between the first-ranked globular clusters in M31 and M87. Sandage (1958) obtained $(m - M)_{AB} \simeq 30.8$ from other considerations.

b) The First-ranked Cluster Galaxy

The brightest E galaxy in the Virgo Cluster is NGC 4472 (Holmberg 1958; de Vaucouleurs 1961b). These authors give a mean integrated visual magnitude of $V_t = 8.45$. The color index $B - V = +0.97$ (de Vaucouleurs 1961a) gives $B_t(4472) = 9.42$. If $(m - M)_{AB} = 31.1$, then $M_B(4472) = -21.68$, which is independent of the absorption between us and the Virgo Cluster.

IV. THE HUBBLE CONSTANT

Figure 1 shows the most recent data for the redshift-apparent magnitude relation for first-ranked cluster galaxies in forty-one clusters. All clusters in Humason's Table III of the Humason-Mayall catalogue (HMS 1956) are plotted. The redshift of the most distant cluster was taken from Minkowski (1960); the next two most distant, from Baum (1962). Also included are new clusters observed in a current program on radio sources which will be discussed elsewhere (Sandage 1969).

Dots represent galaxies that are radio-quiet to 9 flux units at 178 MHz, crosses are radio sources in the 3CR (Bennett 1962) that are brightest cluster members, and open triangles are Baum's data (1962) transformed to the zero point of the present magnitude system. The abscissa is the photoelectrically measured V magnitude corrected for (1) aperture effect to an isophote of about 25 mag per square second by a method which will be given elsewhere, (2) K-dimming taken from tables given by Oke and Sandage (1968), and (3) galactic absorption using $A_V = 0.18$ cosec (b - 1).

The Virgo Cluster is represented by four points (two dots and two triangles) corresponding to Baum and Holmberg's photometry and using $cz = 1136$ km sec^{-1} for the mean of all galaxies in the neighborhood of the cluster (HMS 1956, Table II), or $cz = 950$ km sec^{-1} for the E cloud alone (de Vaucouleurs 1961b). The Fornax Cluster is represented by two points at $cz = 1526$ km sec^{-1} and $cz = 1437$ km sec^{-1}, depending on whether NGC 1375 is included or excluded from the mean (HMS 1956, Table I; Mayall and de Vaucouleurs 1962). The box in the lower left-hand corner is the range over which Hubble (1929) first formulated the velocity-distance relation.

A listing of the data and a discussion of Figure 1 are in preparation. For the present purpose it is sufficient to note that the line has been drawn with the theoretical slope $dV_c/d\log z = 5$ for $z \to 0$ and has been fitted to the data only in zero point. The equation of the line is $V_c = 5 \log cz - 6.78$.

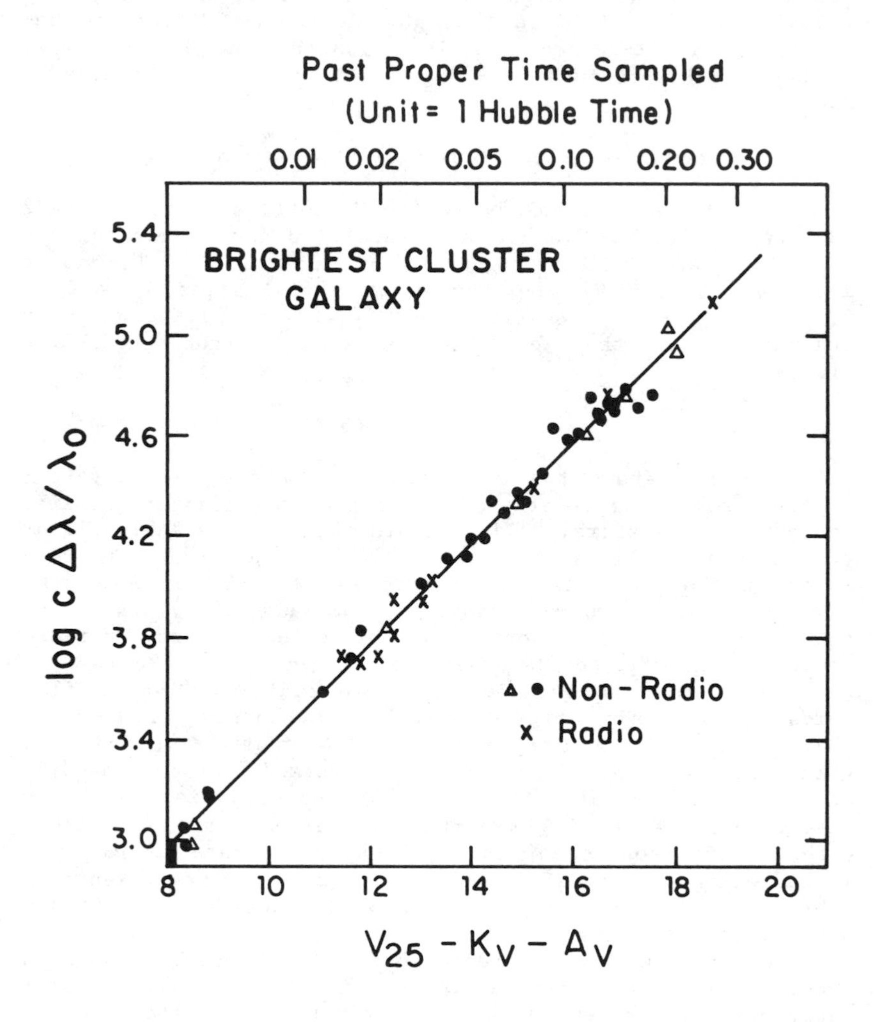

Figure 1. - Hubble diagram for first-ranked cluster galaxies in forty-one clusters. Abscissa is the photoelectric V magnitude corrected for aperture effect, K-dimming, and galactic absorption. Dots represent galaxies that are radio-quiet to 9 flux units at 178 MHz; crosses, radio sources from the 3CR catalogue that are first-ranked cluster members; triangles, Baum's data transformed to the V-magnitude system of the dots and crosses. Ordinate is redshift, corrected for galactic rotation.

The most striking feature of Figure 1 is the small scatter of the first-ranked cluster galaxies about this line. The formal dispersion, read as a magnitude difference, is $\sigma \simeq +0.3$ mag. The size of the dispersion shows that the absolute luminosity of the brightest galaxy in clusters with $N \gtrsim 30$ members is a remarkably stable statistic.

We can now read V_c from Figure 1 at a redshift of, say, 10^4 km sec^{-1}, which is well beyond the local anisotropy. The equation of the line gives $V_c = 13.22$ at $\log cz = 4.0$. Because $\langle B - V \rangle = +0.97$ for E galaxies corrected for K-effect (Oke and Sandage 1968), then $B_c = 14.19$. If we assume that $\langle M_B \rangle = -21.68$ from our calibration, then the apparent blue modulus for a cluster at $\log cz = 4.0$ at the galactic pole is $(m - M)_{AB} = 35.87$.

The true modulus must now be obtained. This requires correction for the galactic half-thickness layer.[2] Assuming $A_B(b = 90°) = 0.25$ mag gives $(m - M)_T = 35.62$ for a distance of 1.33×10^2 Mpc. No correction for different cosmological models is necessary at this redshift.

The Hubble constant is then $H = cz/D = 75.3$ km sec^{-1} Mpc^{-1}, or $H^{-1} = 12.9 \times 10^9$ years.

V. ESTIMATE OF ERRORS

The value is uncertain because of three random and four systematic effects. The random errors with their estimated uncertainties are (1) an uncertainty of perhaps ±0.3 mag in Racine's value of $B = 21.3$ for the brightest globular cluster in M87, due to a combination of transfer errors to SA 57 and the difficulty of photometry on the M87 background; (2) a possible error of ±0.2 mag in V_t for NGC 4472; (3) $\sigma = \pm 0.3$ mag for the brightest cluster galaxy about the mean line of Figure 1. If these combine in the usual way, the resulting uncertainty in the distance modulus would have a formal value of about ±0.5 mag. Neglecting the systematic errors for the moment, these random effects give $H = 75^{+19}_{-15}$ km sec^{-1} Mpc^{-1}, or $H^{-1} = 13^{+3.7}_{-2.7} \times 10^9$ years.

The systematic errors concern (1) the adopted value of M_B for the brightest globular cluster, (2) the value of the galactic half-thickness absorption, and (3) a possible difference in the distance to M87 and NGC 4472 because of the finite angular diameter of the E cloud of the Virgo Cluster. NGC 4472 lies near the edge of the projected distribution on the plane of the sky (de Vaucouleurs 1961b, Fig. 2) and therefore is at the distance of the core. The uncertainty arises because M87, seen projected on the core, could be located anywhere along the line of sight through the core. The maximum difference in distance, if M87 is at the periphery of the three-dimensional distribution, whose angular radius is about

[2] The absorption correction A_V in Fig. 1 was only to the pole and not to outside the galactic system, since A_V was adopted to be proportional to cosec b − 1.

5°, is $\delta r/r = \pm 0.09$, which introduces an uncertainty of 9 percent in the value of H. However, due to the strong density gradient of core members, M87 may be near the cluster center in space, and the back-to-front effect has been neglected.

The systematic effect in M_B is composed of three parts.

A. M_B(first) may be a function of the total number of globular clusters in a given galaxy, becoming brighter for a larger sample. This would make M_B(first) brighter by ΔM for the first-ranked cluster in M87 (2000 clusters), compared with M31 ($\sim$250 clusters), and will reduce H by a factor of $1 + 0.46\ \Delta M$ for small ΔM.

B. If highly absorbed clusters in M31 have M_B brighter than it is for B282, H will again be decreased.

C. If the modulus of the Hyades must be increased by ΔM_H (Hodge and Wallerstein 1966), the distance to M31 via the Cepheids increases, and M_B for cluster B282 becomes brighter by ΔM_H. This again reduces H by the factor given above.

If the galactic half-thickness absorption is, say, 0.50 mag rather than 0.25 mag in B, all distances obtained by the use of apparent-modulus methods are decreased, and H becomes larger by a factor of 1.12.

The value of H would be 55^{+14}_{-11} ($H^{-1} = 17.7 \times 10^9$ years) if M_B(brightest globular cluster) were, say, $M_B = -10.5$, and we still adopt $A_B(1/2) = 0.25$ mag. H would be 12 per cent higher if $A_B = 0.50$ mag.

VI. THE LOCAL ANISTROPY AT THE VIRGO CLUSTER

Adopting a consistent set of calibrations, one can obtain the deviation of the Virgo Cluster from the pure cosmological expansion field by finding H_{VC} from the cluster data alone and comparing it with H_∞ derived above. If $(m - M)_{AB} = 31.1$ and $(m - M)_T = 30.85$, then D = 14.8 Mpc for the Virgo Cluster. De Vaucouleurs (1958) gives $cz = 950 \pm 70$ km sec^{-1} for the redshift of the Virgo E cloud. Hence, $H_{VC} = 64 \pm 5$ km sec^{-1} Mpc^{-1}, where the quoted error is due to the uncertainty in cz alone. Therefore, the ratio of the asymptotic value of H to the local value is

$$H_\infty / H_{VC} = 75/64 = 1.17 \pm 0.09 \ .$$

The actual error is, of course, larger because of a possible difference of M_B(NGC 4472) from $\langle M_B(\text{first}) \rangle$.

The present determination can be compared with de Vaucouleurs' (1958) value of 1.58 based on an explicit model for the local anisotropy. The smallness of our value can be seen graphically from the smallness of the deviation of the Virgo Cluster from the line in Figure 1. It should be emphasized again that part of the observed deviation could be due to a difference of absolute magnitude of NGC 4472 from the mean of the other first-ranked cluster galaxies, rather than to a difference, Δz, due to the local gravitational perturbation over a scale of 10 Mpc. With this in mind, our present data provide no clear evidence that a signficant anisotropy occurs at the Virgo Cluster, but this does not change the importance of the de Vaucouleurs discovery (1958, 1966) for other directions in the sky.

It is a pleasure to thank Jerome Kristian, René Racone, Maarten Schmidt, and Sidney van den Bergh for useful discussions.

REFERENCES

Baade, W., and Swope, H. H. 1963, A.J., **68**, 435.
Baum, W. A. 1955, Pub. A.S.P., **67**, 328.
Baum, W. A. 1962, in Problems of Extragalactic Research, ed. G. C. McVittie (New York: Macmillan), p. 390.
Bennett, A. S. 1962, Mem. R.A.S., **68**, 163.
Bergh, S. van den. 1960, Zs. f. Ap., **49**, 198.
Gascoigne, S. C. B., and Burr, E. J. 1956, M.N.R.A.S., **116**, 570.
Hiltner, W. A. 1960, Ap.J., **131**, 163.
Hodge, P. W., and Wallerstein, G. 1966, Pub. A.S.P., **78**, 411.
Holmberg, E. 1958, Lund. Medd., Ser. II, No. 136.
Hubble, E. P. 1929, Proc. Nat. Acad. Sci., **15**, 168.
Hubble, E. P. 1936, Ap.J., **84**, 270.
Humason, M. L. 1936, Ap.J., **83**, 10.
Humason, M. L., Mayall, N. U., and Sandage, A. R. 1956, A.J., **61**, 97.
Kinman, T. D. 1963, Ap.J., **137**, 213.
Kristian, J. 1967 (private communication).
Kristian, J. and Sachs, R. K. 1966, Ap.J., **143**, 379.
Kron, G. E., and Mayall, N. U. 1960, A.J., **65**, 581.
Mayall, N. U., and Eggen, O. J. 1953, Pub. A.S.P., **65**, 24.
Mayall, N. U., and Vaucouleurs, A. de. 1962, A.J., **67**, 363.
Minkowski, R. 1960, Ap.J., **132**, 908.
Oke, J.B., and Sandage, A. 1968, Ap.J. (in press).
Racine, R. 1968, J.R.A.S. Canada (in press).
Sandage, A. 1958, Ap.J., **127**, 513.
Sandage, A. 1962, in Problems of Extragalactic Research, ed. G. C. McVittie (New York: Macmillan), p. 359.
Sandage, A. 1968, Observatory (in press).
Sandage, A. 1969 (in preparation).
Sandage, A., and Tammann, G. A. 1968, Ap.J., **151**, 531.
Saunders, J. 1963, Sky and Telescope, **26**, 133.
Scott, E. L. 1957, A.J., **62**, 248.
Sersic, J. L. 1960, Zs. f. Ap., **50**, 168.
Tifft. W. G. 1963, M.N.R.A.S., **126**, 209.
Vaucouleurs, G. de. 1958, A.J., **63**, 253.
Vaucouleurs, G. de. 1961a, Ap.J. Suppl., No. 48, **5**, 233.
Vaucouleurs, G. de. 1961b, Ap.J. Suppl., No. 56, **6**, 213.
Vaucouleurs, G. de. 1966, Proc. Meeting on Cosmology for the Galileo Conference, Vol. 2, Tomo 3, ed. L. Rosino (Florence: G. Barbèra), p. 37.
Wayman, P. A., Symms, L. S. T., and Blackwell, K. C. 1965, R.O.B., No. 98.

A HOMOGENEOUS UNIVERSE OF CONSTANT MASS AND INCREASING RADIUS ACCOUNTING FOR THE RADIAL VELOCITY OF EXTRA-GALACTIC NEBULAE[1]

Abbé G. Lemaître

I. INTRODUCTION

According to the theory of relativity, a homogeneous universe may exist such that all positions in space are completely equivalent; there is no center of gravity. The radius of space R is constant; space is elliptic, i.e. of uniform positive curvature $1/R^2$; straight lines starting from a point come back to their origin after having travelled a path of length πR; the volume of space has a finite value $\pi^2 R^3$; straight lines are closed lines going through the whole space without encountering any boundary.

Two solutions have been proposed. That of de Sitter ignores the existence of matter and supposes its density equal to zero. It leads to special difficulties of interpretation which will be referred to later, but it is of extreme interest as explaining quite naturally the observed receding velocities of extra-galactic nebulae, as a simple consequence of the properties of the gravitational field without having to suppose that we are at a point of the universe distinguished by special properties.

The other solution is that of Einstein. It pays attention to the evident fact that the density of matter is not zero, and it leads to a relation between this density and the radius of the universe. This relation forecasted the existence of masses enormously greater than any known at the time. These have since been discovered, the distances and dimensions of extra-galactic nebulae having become known. From Einstein's formulae and recent observational data, the radius of the universe is found to be some hundred times greater than the most distant objects which can be photographed by our telescopes.

Each theory has its own advantages. One is in agreement with the observed radial velocities of nebulae, the other with the existence of matter, giving a satisfactory relation between the radius and the mass of the universe. It seems desirable to find an intermediate solution which could combine the advantages of both.

At first sight, such an intermediate solution does not appear to exist. A static gravitational field for a uniform distribution of matter without internal stress has only two solutions, that of Einstein and that of de Sitter. De Sitter's universe is empty, that of Einstein has been de-

[1]Translated by permission from "Annales de la Société scientifique de Bruxelles," Tome XLVII, série A., premiere partie.)

scribed as "containing as much matter as it can contain." It is remarkable that the theory can provide no mean between these two extremes.

The solution of the paradox is that de Sitter's solution does not really meet all the requirements of the problem. Space is homogeneous with constant positive curvature; space-time is also homogeneous, for all events are perfectly equivalent. But the partition of space-time into space and time disturbs the homogeneity. The co-ordinates used introduce a center. A particle at rest at the center of space describes a geodesic of the universe; a particle at rest otherwhere than at the center does not describe a geodesic. The co-ordinates chosen destroy the homogeneity and produce the paradoxical results which appear at the so-called "horizon" of the center. When we use co-ordinates and a corresponding partition of space and time of such a kind as to preserve the homogeneity of the universe, the field is found to be no longer static; the universe becomes of the same form as that of Einstein, with a radius no longer constant but varying with the time according to a particular law.

In order to find a solution combining the advantages of those of Einstein and de Sitter, we are led to consider an Einstein universe where the radius of space or of the universe is allowed to vary in an arbitrary way.

II. EINSTEIN UNIVERSE OF VARIABLE RADIUS. FIELD EQUATIONS. CONSERVATION OF ENERGY

As in Einstein's solution, we liken the universe to a rarefied gas whose molecules are the extra-galactic nebulae. We suppose them so numerous that a volume small in comparison with the universe as a whole contains enough nebulae to allow us to speak of the density of matter. We ignore the possible influence of local condensations. Furthermore, we suppose that the nebulae are uniformly distributed so that the density does not depend on position. When the radius of the universe varies in an arbitrary way, the density, uniform in space, varies with time. Furthermore, there are generally interior stresses, which, in order to preserve the homogeneity, must reduce to a simple pressure uniform in space and variable with time. The pressure, being two-thirds of the kinetic energy of the "molecules," is negligible with respect to the energy associated with matter; the same can be said of interior stresses in nebulae or in stars belonging to them. We are thus let to put $p = 0$.

Nevertheless it might be necessary to take into account the radiation-pressure of electromagnetic energy travelling through space; this energy is weak but it is evenly distributed through the whole of space and might afford a notable contribution to the mean energy. We shall thus keep the pressure p in the general equations as the mean radiation-pressure of light, but we shall write $p = 0$ when we discuss the application to astronomy.

We denote the density of total energy by ρ, the density of radiation energy by $3p$, and the density of the energy condensed in matter by $\delta = \rho - 3p$. We identify ρ and $-p$ with the components $T_4^{\ 4}$ and $T_1^{\ 1} = T_2^{\ 2} = T_3^{\ 3}$ of the material energy tensor, and δ with T. Working out the contracted Riemann tensor for a universe with a line-element given by

$$ds^2 = - R^2 d\sigma^2 + dt^2 , \qquad (1)$$

where $d\sigma$ is the elementary distance in a space of radius unity, and R is a function of the time t, we find that the field equations can be written

$$3\frac{R'^2}{R^2} + \frac{3}{R^2} = \lambda + \kappa\rho \qquad (2)$$

and

$$2\frac{R''}{R} + \frac{R'^2}{R^2} + \frac{1}{R^2} = \lambda - \kappa\rho \qquad (3)$$

Accents denote derivatives with respect to t. λ is the unknown cosmological constant, and κ is the Einstein constant whose value is 1.87×10^{-27} in C.G.S. units (8π in natural units).

The four identities giving the expression of the conservation of momentum and of energy reduce to

$$\frac{d\rho}{dt} + \frac{3R'}{R}(\rho + p) = 0 \qquad (4)$$

which is the energy equation. This equation can replace (3). As $V = \pi^2 R^3$ it can be written

$$d(V\rho) + pdV = 0 , \qquad (5)$$

showing that the variation of total energy plus the work done by radiation-pressure in the dilation of universe is equal to zero.

III. UNIVERSE OF CONSTANT MASS

If $M = V\delta$ remains constant, we write, α being a constant,

$$\kappa\delta = \frac{\alpha}{R^3} \qquad (6)$$

As

$$\rho = \delta + 3p$$

we have

$$3d(pR^3) + 3pR^2 dR = 0 \tag{7}$$

and, β being a constant of integration,

$$\kappa p = \frac{\beta}{R^4} \tag{8}$$

and therefore

$$\kappa\rho = \frac{\alpha}{R^3} + \frac{3\beta}{R^4} \tag{9}$$

By substitution in (2) we have

$$\frac{R'^2}{R^2} = \frac{\lambda}{3} - \frac{1}{R^2} + \frac{\kappa\rho}{3} = \frac{\lambda}{3} - \frac{1}{R^2} + \frac{\alpha}{3R^3} + \frac{\beta}{R^4}\,. \tag{10}$$

and

$$t = \int \frac{dR}{\sqrt{\lambda R^2/3 - 1 + \alpha/3R + \beta/R^2}} \tag{11}$$

When a and β vanish, we obtain the de Sitter solution in Lanczos' form-

$$R = \sqrt{3/\lambda}\, \cosh \sqrt{\lambda/3}\, (t - t_0) \tag{12}$$

The Einstein solution is found by making $\beta = 0$ and R constant. Writing $R' = R'' = 0$ in (2) and (3) we find

$$\frac{1}{R^2} = \lambda \qquad \frac{3}{R^2} = \lambda + \kappa\rho \qquad \rho = \delta$$

or

$$R = \frac{1}{\sqrt{\lambda}} \qquad \kappa\delta = \frac{2}{R^2} \tag{13}$$

and from (6)

$$\alpha = \kappa\delta R^3 = \frac{2}{\sqrt{\lambda}} \tag{14}$$

The Einstein solution does not result from (14) alone; it also supposes that the initial value of R' is zero. If we write

$$\lambda = \frac{1}{R_0^{\ 2}} \tag{15}$$

we have for $\beta = 0$ and $\alpha = 2R_0$

$$t = R_0 \sqrt{3} \int \frac{dR}{R-R_0} \sqrt{R/(R+2R_0)} \tag{16}$$

For this solution the two equations (13) are of course no longer valid.

Writing

$$\kappa\delta = \frac{2}{R_E^{\ 2}} \tag{17}$$

we have from (14) and (15)

$$R^3 = R_E^{\ 2} R_0 \tag{18}$$

The value of R_E, the radius of the universe computed from the mean density by Einstein's equation (17), has been found by Hubble to be

$$R_E = 8.5\times 10^{28} \text{ cm} = 2.7\times 10^{10} \text{ parsec} \tag{19}$$

We shall see later that the value of R_0 can be computed from the radial velocities of the nebulae; R can then be found from (18).

Finally, we shall show that a serious departure from (14) would lead to consequences not easily acceptable.

IV. DOPPLER EFFECT DUE TO THE VARIATION OF THE RADIUS OF THE UNIVERSE

From (1) we have for a ray of light

$$\sigma_2 - \sigma_1 = \int_{t_1}^{t_2} \frac{dt}{R} \tag{20}$$

where σ_1 and σ_2 relate to spatial co-ordinates. We suppose that the light is emitted at the point σ_1 and observed at σ_2. A ray of light emitted slightly later starts from σ_1 at time $t_1 + \delta t_1$ and reaches σ_2 at time $t_2 + \delta t_2$. We have therefore

$$\frac{\delta t_2}{R_2} - \frac{\delta t_1}{R_1} = 0, \qquad \frac{\delta t_2}{\delta t_1} - 1 = \frac{R_2}{R_1} - 1 \tag{21}$$

where R_1 and R_2 are the values of the radius R at the time of emission t_1 and at the time of observation t_2. If δt_1 is the period of the emitted light, δt_2 is the period of the observed light. Now δt_1 is also the period of light emitted under the same conditions in the neighborhood of the observer, because the period of light emitted under the same physical conditions has the same value everywhere when reckoned in proper time. Therefore

$$\frac{v}{c} = \frac{\delta t_2}{\delta t_1} - 1 = \frac{R_2}{R_1} - 1 \tag{22}$$

is the apparent Doppler effect due to the variation of the radius of the universe. It equals the ratio of the radii of the universe at the instants of observation and emission, diminished by unity.

v is that velocity of the observer which would produce the same effect. When the light source is near enough, we have the approximate formulae

$$\frac{v}{c} = \frac{R_2 - R_1}{R_1} = \frac{dR}{R} = \frac{R'}{R}\, dt = \frac{R'}{R}\, r$$

where r is the distance of the source. We have therefore

$$\frac{R'}{R} = \frac{v}{cr} \tag{23}$$

From a discussion of available data, we adopt

$$\frac{R'}{R} = 0.68 \times 10^{-27} \text{ cm}^{-1} \tag{24}$$

and find from (16)

$$\frac{R'}{R} = \frac{1}{R_0\sqrt{3}}\sqrt{1 - 3y^2 + 2y^3} \tag{25}$$

where

$$y = \frac{R_0}{R} \tag{26}$$

Now from (18) and (26)

$$R_0{}^2 = R_E{}^2 y^3 \tag{27}$$

and therefore

$$3\left(\frac{R'}{R}\right)^2 R_E{}^2 = \frac{1 - 3y^2 + 2y^3}{y^3} \tag{28}$$

With the adopted numerical data (24) and (19), we have

$$y = 0.0465$$

giving

$$R = R_E\sqrt{y} = 0.215R_E = 1.83\times10^{28} \text{ cm} = 6\times10^9 \text{ parsecs}$$
$$R_0 = Ry = R_E y^{3/2} = 8.5\times10^{26} \text{ cm} = 2.7\times10^8 \text{ parsecs}$$
$$= 9\times10^8 \text{ light-years} .$$

Integral (16) can easily be computed. Writing

$$x^2 = \frac{R}{R + 2R_0} \tag{29}$$

it can be written

$$t = R_0\sqrt{3} \int \frac{4x^2 dx}{(1-x^2)(3x^2-1)}$$

$$= R_0 \sqrt{3} \log \frac{1+x}{1-x} + R_0 \log \frac{\sqrt{3}x-1}{\sqrt{3}x+1} + C \qquad (30)$$

If σ is the fraction of the radius of the universe travelled by light during time t, we have also

$$\sigma = \int \frac{dt}{R} = \sqrt{3} \int \frac{2dx}{3x^2 - 1} = \log \frac{\sqrt{3}x - 1}{\sqrt{3}x + 1} + C' \qquad (31)$$

The following table gives values of σ and t for different values of R/R_0: --

Table 1
Values of σ and t

$\frac{R}{R_0}$	$\frac{t}{R_0}$	σ Radians	σ Degrees	$\frac{v}{c}$
1	$-\infty$	$-\infty$	$-\infty$	19
2	-4.31	-0.889	-51°	9
3	-3.42	-0.521	-30	5 2/3
4	-2.86	-0.359	-21	4
5	-2.45	-0.266	-15	3
10	-1.21	-0.087	-5	1
15	-0.50	-0.029	-1.7	1/3
20	0.00	0.000	0.0	0
25	0.39	0.017	1	..
∞	∞	0.087	5	..

The constants of integration are adjusted to make σ and t vanish for R/R_0 = 20 in place of 21.5. The last column gives the Doppler effect computed from (22). The approximate formula (23) would make v/c proportional to r and to σ. The error is only 0.005 for v/c = 1. The approximate formula may therefore be used within the limits of the visible spectrum.

V. THE MEANING OF EQUATION (14)

The relation (14) between the two constants λ and a has been adopted following Einstein's solution. It is the necessary condition that the quartic under the radical in (II) may have a double root R_0 giving on integration a logarithmic term. For simple roots, integration would give a square root, corresponding to a minimum of R as in de Sitter's solution (12). This minimum would generally occur at time of the order of R_0, say 10^9 years - i.e. quite recently for stellar evolution. If the positive roots were to become imaginary, the radius would vary from zero upwards, the variation slowing down in the neighborhood of the modulus of the imaginary roots. In both cases the time of variation of R in the same sense would be of the order of R_0 if the relation between λ and α were seriously different from (14).

VI. CONCLUSION

We have found a solution such that

(1°) The mass of the universe is a constant related to the cosmological constant by Einstein's relation

$$\sqrt{\lambda} = \frac{2\pi^2}{\kappa M} = \frac{1}{R_0} .$$

(2°) The radius of the universe increases without limit from an asymptotic value R_0 for $t = -\infty$.

(3°) The receding velocities of extragalactic nebulae are a cosmical effect of the expansion of the universe. The initial radius R_0 can be computed by formulae (24) and (25) or by the approximate formula

$$R_0 = \frac{rc}{v\sqrt{3}} .$$

This solution combines the advantages of the Einstein and de Sitter solutions.

Note that the largest part of the universe is for ever out of our reach. The range of the 100-inch Mount Wilson telescope is estimated by Hubble to be 5×10^7 parsecs, or about R/200. The corresponding Doppler effect is 3000 km/sec. For a distance of 0.087 R it is equal to unity, and the whole visible spectrum is displaced into the infra-red. It is impossible to see ghost-images of nebulae or suns, as even if there were no absorption these images would be displaced by several octaves into the infra-red and would not be observed.

It remains to find the cause of the expansion of the universe. We have seen that the pressure of radiation does work during the expansion. This seems to suggest that the expansion has been set up by the radiation itself. In a static universe light emitted by matter travels round space, comes back to its starting-point, and accumulates indefinitely. It seems that this may be the origin of the velocity of expansion R'/R which Einstein assumed to be zero and which in our interpretation is observed as the radial velocity of extra-galactic nebulae.

REFERENCES

de Sitter, W. 1930, Proc. Nat. Acad. Sci., **16**, 474.
de Sitter, W. 1930, B.A.N., **5**, No. 185, 193.
de Sitter, W. 1930, B.A.N., **5**, No. 185, 200.
de Sitter, W. 1931, Scientia (January)
Du Val, P. 1924, Phil. Mag., 6, **47**, 930.
Eddington, A. S. 1930, M.N.R.A.S., **90**, 668.
Einstein, A. 1922, Z. f. Phys., **11**, 326.
Einstein, A. 1923, Z. f. Phys., **16**, 228.
Friedmann, A. 1922, Z. f. Phys., **10**, 377.
Lanczos, K. 1922, Phys. Zeits., **23**, 539.
Lemaître, G. 1925, Jour. of Math. and Phys., **4**, No. 3,
Lemaître, G. 1929, Revue des questions scientifiques, March.
Lemaître, G. 1930, B.A.N., **5**, No. 200.
Tolman, R. C. 1930, P.N.A.S., **16**, 320.
Weyl, H. 1923, Phys. Zeits., **24**, 230.

AN UPPER LIMIT ON THE NEUTRINO REST MASS[1]

R. Cowsik[2] and J. McClelland
Department of Physics, University of California
Berkeley, California

Received: July 17, 1982

In order that the effect of gravitation of the thermal background neutrinos on the expansion of the universe not be too severe, their mass should be less than 8 eV/c^2.

Recently there has been a resurgence of interest in the possibility that neutrinos may have a finite rest mass. These discussions have been in the context of weak-interaction theories (Tennakone and Pakvasa 1971; Tennakone and Pakvasa 1972), possible decay of solar neutrinos (Bahcall, Cabibbo and Yahil 1972), and enumerating the possible decay modes of the K_L^0 meson (Barshay 1972). Elsewhere, we have pointed out that the gravitational interactions of neutrinos of finite rest mass may become very important in the discussion of the dynamics of clusters of galaxies and of the universe (Cowsik and McClelland, no date). Considerations involving massive neutrinos are not new (Markov 1964; Bahcall and Curtis 1961); an excellent review of the early developments in the field is given by Kuchowicz (1967; 1969). Here we wish to point out that the recent measurement (Sandage 1972) of the deceleration parameter, q_0, implies an upper limit of a few tens of electron volts on the sum of the masses of all the possible light, stable particles that interact only weakly.

In discussing this problem we take the customary point of view that the universe is expanding from an initially hot and condensed state as envisaged in the "big-bang" theories (Peebles 1971). In the early phase of such a universe when the temperature was greater than ~ 1 MeV, processes of neutrino production, which have also been considered in the context of high-temperature stellar cores (Ruderman 1969), would lead to the generation of the various kinds of neutrinos. In fact, similar processes would generate populations of other fermions and boson as well, and conditions of thermal equilibrium allow us to estimate their number density (Landau and Lifshitz 1969):

$$n_{Fi} = \frac{2S_i+1}{2\pi^2\hbar^3} \int_0^\infty \frac{p^2 dp}{\exp[E/kT(z_{eq})] + 1}, \tag{1a}$$

[1]Work supported by the National Aeronautics and Space Administration under Grant Number NGR05-003-376.

[2]On leave from the Tata Institute of Fundamental Research, Bombay, India.

and

$$n_{Bi} = \frac{2S_i+1}{2\pi^2\hbar^3}\int_0^\infty \frac{p^2dp}{\exp[E/kT(z_{eq})] - 1} . \quad (1b)$$

Here n_{Fi} is the number density of fermions of the ith kind, n_{Bi} is the number density of bosons of the ith kind, S_i is the spin of the particle (notice that in writing the multiplicity of states of the particles we have not discriminated against the neutrinos; since we are discussing neutrinos of nonzero rest mass, we have assumed that both the helicity states are allowed), $E = c(p^2+m^2c^2)^{1/2}$, k is Boltzmann's constant, and $T(z_{eq}) = T_r(z_{eq}) = T_F(z_{eq}) = T_B(z_{eq}) = T_m(z_{eq}) = \ldots$ is the common temperature of radiation, fermions, bosons, matter, etc. at the latest epoch, characterized by the red shift Z_{eq}, when they may be considered to be in thermal equilibrium; $kT(z_{eq}) \approx 1$ MeV.

Since our discussion pertains to neutrinos and any hypothetical stable weak bosons (Bahcall, Cabibbo and Yahil 1972), we may assume that $kT(z_{eq}) \approx 1 \text{ MeV} >> mc^2$. In this limit Eqs. (1a) and (1b) reduce to

$$n_F(z_{eq}) \approx 0.0913(2S_i + 1)[T(z_{eq})/\hbar c]^3 , \quad (2a)$$

$$n_B(z_{eq}) \approx 0.122(2S_i + 1)[T(z_{eq})/\hbar c]^3 . \quad (2b)$$

As the universe expands and cools down, the neutrinos and such other weakly interacting particles survive without annihilation because of the extremely low cross sections (de Graff and Tolhoek 1966) for these processes. Consequently, the number density decreases simply as $\sim V(z_{eq})/V(z) = (1+z)^3/(1+z_{eq})^3$. Noticing that $1 + z = T_r(z)/T_r(0)$, the number densities of the various particles expected at the present epoch (z = 0) are given by

$$n_{Fi}(0) = n_{Fi}(z_{eq})\left[\frac{1}{1 + z_{eq}}\right]^3 \approx 0.0913(2S_i + 1)[T_r(0)/\hbar c]^3 \quad (3a)$$

and

$$n_{Bi}(0) \approx 0.122(2S_i + 1)[T_r(0)/\hbar c]^3 . \quad (3b)$$

Taking $T_r(0) \approx 2.7°$ K, we have

$$n_{Fi}(0) \approx 150(2S_i + 1) \text{ cm}^{-3} , \quad (4a)$$

and

$$n_{Bi}(0) \approx 200(2S_i + 1)\ \mathrm{cm}^{-3} \ . \qquad (4b)$$

These numbers are huge in comparison with the mean number density of hydrogen atoms in the universe; all the visible matter in the universe adds up to an average density of hydrogen atoms (Peebles 1971) of only $\sim 2\times 10^{-8}\ \mathrm{cm}^{-3}$. Notice that the expected density of the neutrinos and other weakly interacting particles is essentially independent of the temperature $T(z_{eq})$, of decoupling, and such other details; the measured temperature of the universal blackbody photons fixes the density of weak particles quite well.

Now consider Sandage's (1972) measurement of the Hubble constant H_0 and the deceleration parameter q_0 which together place a limit on ρ_{tot}, the density of all possible sources of gravitational potential in the universe (Peebles 1971). His results, $H_0 = 50\ \mathrm{km\ sec}^{-1}\ \mathrm{Mpc}^{-1} = 1.7\times 10^{-18}\ \mathrm{sec}^{-1}$ and $q_0 = +0.94\pm 0.4$, imply

$$\rho_{tot} = 3H_0^2 q_0/4\pi G = (10\pm 4)\times 10^{-30}\ \mathrm{g\ cm}^{-3}$$
$$\approx (6\pm 2)\times 10^3\ (\mathrm{eV}/c^2)\ \mathrm{cm}^{-3} < 10^4\ (\mathrm{eV}/c^2)\ \mathrm{cm}^{-3} \ . \qquad (5)$$

Here $G = 6.68\times 10^{-8}\ \mathrm{dyn\ cm}^2\ \mathrm{g}^{-2}$ is the gravitational constant. If m_i were to represent the mass spectrum of the various neutrinos and other stable weakly interacting particles, we can combine Eqs. (4a), (4b), and (5) to obtain the limit

$$\rho_{weak} \approx \Sigma n_{Bi} m_i + n_{Fj} m_j \gtrsim 150(2S_i + 1)\ m_i < \rho_{tot} \qquad (6)$$

or

$$\Sigma (2S_i + 1)\ m_i \lesssim 66\ \mathrm{eV}/c^2 \ .$$

Here the summation is to be carried out over all the particle and antiparticle states of both fermions and bosons. Considering only the neutrinos and antineutrinos of the muon and electron kind each having a mass of m_ν, Eq. (6) leads to the result $m_\nu < 8\ \mathrm{eV}/c^2$.

This limit is obtained assuming big-bang cosmology to be correct; however, it depends only very weakly on the value of the deceleration parameter and other details of the cosmology. Thus, even when one allows for a large uncertainty in the cosmological parameters, the limits on the masses of neutrinos and other stable weakly interacting particles derived in this paper are still much lower than the direct experimental limits (Bergkvist 1972; Shrum and Ziock 1971) of $m_{\nu\mu} < 1.5$ MeV/c^2 and $m_{\nu e} < 60$ eV/c^2.

Our thanks are due to Professor Eugene D. Commins, Professor J. N. Bahcall, Professor G. B. Field, and Professor P. Buford Price for many discussions.

REFERENCES

Bahcall, J. N., Cabibbo, N., and Yahil, A. 1972, Phys. Rev. Lett., **28**, 316.
Bahcall, J. N., and Curtis, R. B. 1961, Nuovo Cimento, **21**, 422.
Barshay, S. 1972, Phys. Rev. Lett., **28**, 1008.
Bergkvist, K. 1972, Nucl. Phys., **B39**, 317.
Cowsik, R., and McClelland, J. no date, Gravity of Neutrinos of Non-Zero Mass in Astrophysics (to be published).
de Graff, T., and Tolhoeck, H. A. 1966, Nucl. Phys., **81**, 596.
Kuchowicz, B. 1969, The Bibliography of the Neutrino (New York: Gordon and Breach).
Kuchowicz, B. 1969, Fortschr. Phys., **17**, 517.
Landau, L. D., and Lifshitz, E. M. 1969, Statistical Physics (Reading: Addison-Wesley), 2nd ed., p. 324.
Markov, M. A. 1964, The Neutrino (Moscow: Nauka).
Peebles, P. J. E. 1971, Physical Cosmology (Princeton: Princeton University Press).
Ruderman, M. A. 1969, in Topical Conference on Weak Interactions, CERN, Geneva, Switzerland, 1969 (Geneva: CERN Scientific Information Service), p. 111.
Sandage, A. 1972, Ap.J., **173**, 485.
Shrum, E. V., and Ziock, K. O. H. 1971, Phys. Lett., **37B**, 115.
Tennakone, K., and Pakvasa, S. 1971, Phys. Rev. Lett., **27**, 757.
Tennakone, K., and Pakvasa, S. 1972, Phys. Rev. Lett., **28**, 1415.

SECTION II

THREE DEGREES ABOVE ZERO

SECTION II

THREE DEGREES ABOVE ZERO

In his autobiography, My World Line (Gamow 1970), the late George Gamow, who was at the University of Petrograd (now the University of Leningrad) during Friedmann's tenure, writes that in 1924 Friedmann announced a series of lectures entitled "Mathematical Foundations of the Theory of Relativity." Gamow adds, "Naturally, I landed on the bench of the classroom for the first of his lectures." Indeed, Gamow had planned to do his Ph.D. thesis with Friedmann but before that could come about, Friedmann died. Historians of science can argue over who should be called the "father" of the Big Bang theory. Some will say it was Friedmann who did, indeed, write down the first equations that describe the expanding universe of general relativity, although, at least in his two papers, he did not speculate on the origin or consequences of the expansion. Some will argue that it was the Abbé Georges Lemaître who, in 1927 as we have seen, rediscovered Friedmann's equations and then went on to argue that the universe began as the explosion of a "cosmic egg" - again an egg - which originally contained all the matter and that this explosion was what set off the expansion. But certainly a good case could be made for Gamow.

It was Gamow and his collaborators, especially Ralph Alpher and Robert Herman, who turned the study of the early universe into a proper part of science. By the time they began working actively in the field - the late 1940s and early 1950s - nuclear physics had long since become a quantitative science. By comparison, Lemaitre's important papers were published before the neutron was discovered, let alone the nuclear fusion mechanisms which made stars shine. Gamow and his collaborators set out to use nuclear physics, standard quantum mechanics and relativity to account for what we see - the present distribution of matter. The basic idea was very simple. The Big Bang produced, they argued, a primordial stuff which Gamow called the "ylem" - Greek for "wood" or "matter." In the original version the ylem consisted of neutrons and gamma rays - later neutrinos were added. Because the interactions are very rapid, the gamma rays come into thermal equilibrium with the matter, and so their number density can be predicted by the standard methods of statistical mechanics once the matter temperature is known. The neutrons first decay and then fuse. That is, a neutron can become a proton via a beta decay process while a neutron and a proton can fuse with each other to form the deuteron in a radiative capture process. Two deuterons can make helium. One would like to say "and so on" except that after helium the process essentially stops. There are no stable mass number five or eight nuclei, and despite heroic efforts of great ingenuity, the collaboration was never really able to bridge these "mass gaps." These matters will be dealt with more fully in section III. What concerns us here is the fate of the primordial gamma rays. About these Gamow and his collaborators made, as we shall see, an absolutely prescient prediction which, for reasons we shall discuss, had no effect on the ultimate discovery of this radiation.

The first step came in a brief note (paper 11) that Gamow published in 1948. This note is a beautiful piece of quantitative nuclear physics applied to the early universe. The process that is the key to light element formation in the early universe is the formation of deuterium by radiative capture. This cannot take place, Gamow reasoned, until the temperature of the universe has fallen below the binding energy of the deuteron. For there to be appreciable deuterium formation the capture rate must be comparable to the expansion rate of the universe. This is essentially the condition that the mean free path for capture be less than or equal to the size of the universe. But the Friedmann equations give the expansion rate in terms of the energy density. At this stage of universal history the relevant energy density is that of the essentially massless particles that occur in numbers similar to that of the photon. In his paper Gamow considers only the photons' contribution to the density. We should also add neutrinos, which would have a small effect on the answer. The radiative capture rate can be computed from elementary nuclear theory in terms of the density of nucleons available for capture. Equating these two rates, one can solve for the nucleon density at this temperature - about a billion degrees. Gamow discovered that this density was about 10^{18} particles per cubic centimeter - about a million times less dense than ordinate matter. On the other hand the density of the radiation was about that of water, which means that at these temperatures one is dealing with matter in a very peculiar state indeed. Gamow, in this paper, does not follow the radiation to the present. Rather he concludes his paper with some remarks which give the impression that he thinks he has solved the problem of how galaxies form in the early universe. Given the fact that this is still one of the most difficult unsolved problems in modern cosmology, Gamow's optimism was, to say the least, premature.

The decisive step in predicting the present temperature of the ambient radiation left over from the Big Bang was taken by Gamow's two young associates Alpher and Herman. They first announced their result (Alpher and Herman 1948) in a 1948 letter to _Nature_ and then gave the arguments in detail in a 1949 _Physical Review_ article (paper 12). In this study Alpher and Herman improve on Gamow's determination of the nucleon density by actually integrating the relevant Boltzmann equation describing the time evolution of the nucleon density as a result of various interaction processes, including the possibility of the deuteron, once formed, being disassociated by photo-disintegration. They find an equilibrium concentration which they give us about 10^{-6} g/cm^3 which, translated into a volume density, gives a number comparable to Gamow's.

To use this number to find the present radiation temperature Alpher and Herman note that the nucleon number density varies as the reciprocal of the cube of the scale factor R. In other words, the number of nucleons per unit volume shrinks as the volume of the universe expands. It had been known since the work of R. C. Tolman in the 1930s that in an adiabatic expansion of the universe the photon temperature T and R are inversely proportional to each other. This follows from the fact the entropy per unit volume of a photon gas is proportional to the cube of its temperature, so that if the total entropy is to be constant the temperature and the scale factor must be reciprocal to each other. This result implies that the photon number density also decreases as R^{-3}, so that the ratio of the nucleon number to the photon number is a constant during the adiabatic expansion. However, the temperature of the matter does not stay the same as the photon temperature once the two go out of thermal equilibrium, which occurs after the temperature drops below a few thousand degrees Kelvin. Alpher and Herman took the _present_ density of nucleons from the measurements of Hubble, who found that on the average it was about 10^{-30} cm^3. Taking the cube root of the ratios of the matter density at about a billion degrees to the matter density at present gives the present temperature. This turned out to be 5° K, which may be compared to the measured temperature of 2.7° K. This is remarkable agreement considering the uncertainties in such quantities as the present matter density, which depends on the Hubble constant. Nevertheless, there is nothing serendipitous about this comparison. It is on as firm grounds as any calculation done in cosmology, and what we want to try to unravel briefly is why this remarkable prediction had no effect on the actual discovery of the radiation.

Much has been written about this (see, e.g., Weinberg 1977 and Bernstein 1984). In the end, one can only conjecture. Part of the reason was very probably the personality of Gamow himself. He was a larger-than-life Russian eccentric with an irrepressible and irreverant sense of humor. It was hard to know how seriously he himself took some of his scientific work, as brilliant as it often was. It is perhaps typical, and unfortunate, that the last paper which any member of the group published, which contained the prediction of the present ambient temperature, was a paper Gamow published (Gamow 1953) in the proceedings of the Royal Danish Academy. In this paper Gamow abandons the beautiful nuclear physics argument developed by Alpher and Herman for a new argument which, when examined, appears circular. If one knew the baryon density at _any_ temperature one could compare it to the density now and compute the present temperature. The temperature Gamow attempts to use is the "critical" temperature where matter begins to dominate radiation. There is such a temperature because, in the absence of reheating, the radiation energy falls off with the fourth power of the temperature while the matter energy density falls as only the third power. Gamow uses a baroque graphical extrapolation method to attempt to find this temperature. If one does the calculation straightforwardly one will find that the answer depends on knowing the present radiation temperature, which renders the method circular. Somehow Gamow managed to get a present photon temperature of 7° K out of this argument. What is more, a few physicists actually took this number as the present temperature of the universe.[1] But, by and large, this prediction like its predecessors was ignored.

Part of the reason may have been that in none of the papers in which the temperature of the ambient radiation was discussed did the members of the group make a specific experimental suggestion as to how it might be looked for. But it must have been clear to any experimenter that a 7° K blackbody spectrum was predominately in the microwave regime, the regime that had received so much attention during the war because of radar. There is some difference of opinion as to whether the ambient radiation could have been detected before the invention of the maser in 1954. This matter cannot be settled because the attempt was never made. Probably the most important reason was the ambience in which cosmology found itself prior to the actual discovery of the cosmic radiation. There were basically two competitive theories; the Big Bang theory championed by Gamow and his collaborators and the Steady State theory of Hoyle, Bondi and Gold. The two theories made diametrically opposed statements about the origin and evolution of the universe; the Steady State theory was based on the assumption that there was no evolution (see, e.g., Bondi 1961). Both theories could claim to fit, about equally well, such facts as were available to account for. It is, by the way, an interesting irony that because of a commitment to *their* cosmology, in which there were never high universal temperatures, the Steady State people developed a noncosmological theory of element formation in stars which is still the basic theory of heavy element formation. On the other hand, the Big Bang people because of their commitment to *their* cosmology developed a cosmological theory of the formation of light elements which is still the basic theory of light element formation. Nonparticipants found it difficult to believe either theory. Instead, physicists by and large were occupied with the very exciting new developments in fields like elementary particle physics and simply ignored cosmology, not taking seriously the notion that in the Big Bang one had the best laboratory for doing high energy physics ever devised. Unframed art is difficult to appraise in science as well as painting.

In any event, the first clear statement about the importance of looking for the ambient radiation after the Gamow group stopped publishing on the subject came in 1963 from the Soviet Union. In that year two Russian scientists, A. G. Doroshkevich and I. I. Novikov, published a paper in the Russian journal *Doklady* - it was translated and published in the

[1]See, e.g., E. Finlay-Freundlich (1954). We are grateful to Professor L. Grodzins for a discussion of this matter.

United States in 1964 (Doroshkevich and Novikov 1964) - entitled "Mean Density of Radiation in the Metagalaxy and Certain Problems in Relativistic Cosmology." Of particular interest to us is the last paragraph which we quote in part: "Measurements in the region of frequencies 10^9–5×10^{11} cps are extremely important for experimental checking of the Gamow theory. [Here, curiously, a reference is given to a paper by Gamow in which the temperature of the ambient radiation is not discussed.] . . . According to the Gamow theory, at the present time it should be possible to observe equilibrium Planck radiation with a temperature of 1-10° K . . . Measurement reported in [here reference is made to a paper by E. A. Ohm of the Bell Telephone Laboratories (Ohm 1961); we shall return shortly to a discussion of this paper] at a frequency of $\gamma = 2.4 \times 10^9$ cps give a temperature 2.3 ± 0.2° K which coincides with theoretically computed atmospheric noise . . . Additional measurements in this region (preferably on an artificial earth satellite) will assist in final solution of the problem of the correctness of the Gamow theory." On the face of it, the Russian scientists, who had misread the relevant table in Ohm's paper, were saying that Ohm's result was entirely accounted for by atmospheric radio noise. In fact, there was some unaccounted for noise in Ohm's radio telescope, but he did not have the accuracy to pin down the effect. By 1964, as it turned out, this discussion was academic, since the ambient radiation had already been discovered.

This is not the place to recount in detail how American Telephone - AT&T - found itself in the business of radio astronomy, a discipline, by the way, that had actually been invented at the Bell Telephone Laboratories in the 1930s.[2] Suffice it to say that in 1960 a program was begun at Bell Labs to use microwaves to collect signals bounced from a plastic weather balloon called Echo. The collector was an antenna shaped like a large alpine horn which had also been invented at the Laboratories. This was the antenna used by Ohm, and it was taken over by Arno Penzias and Robert Wilson when they joined the Laboratories in 1961 and 1963, respectively. The radio astronomy they had planned to do was frustrated by the fact that they found an unexplained noise in their telescope, a noise they spent an entire year trying to get rid of. Neither man had read the papers of Gamow and his collaborators. Neither had a group at Princeton University led by Robert Dicke, now Albert Einstein University Professor of Science emeritus. Dicke was intrigued by the variant of the Friedmann universe in which there are alternate contractions and expansions. He thought there might be radiation left over from the last contraction. He reasoned that the present energy density of this radiation had to be less than the visible matter density and thus put an upper limit to the temperature of the radiation of 40° K, putting the radiation in the microwave regime. He forgot that he had, in 1946,

[2]For a fuller account see, for example Bernstein (1984).

actually measured such an upper bound, which turned out to be 20° K. He recruited P. J. E. Peebles, a theorist, experimenters P. G. Roll and D. T. Wilkinson to begin an experiment to look for the very radiation that Penzias and Wilson had already, it turned out, been observing for almost a year at Bell Labs, not many miles away. The eventual meeting of the two groups resulted in the publication of the two fundamental papers we have reproduced as 13 and 14.

It must be emphasized that without a theoretical view of the kind that animated the Princeton group, it was not clear what to make of a single observation of excess antenna noise at one wavelength -7.35 cm. It took nearly a decade before a large part of black-body curve, covering a wave length ratio of nearly 2,000, was filled in experimentally. In paper 15, we give an example of a relatively recent discussion of these experiments. One would have to be sceptical, indeed, to try to interpret these results as other than a black-body spectrum. This result, one would have to say, gave a death knell to the Steady State cosmology. In that cosmology there is no special era during which the radiation could have thermalized. In order to account for the microwave background radiation in the Steady State theory, one would have to find some mechanism that produces radiation at 2.7° K, or one would have to say that the universe always existed, for some peculiar reason, with an ambient radiation background of 2.7° K. The latter notion reminds one of the creationist's way of accounting for fossils: i.e., they were created at the same time God created man. Both accounts take the discussion out of the domain of science.

A very important question is whether the background radiation is isotropic. The isotropy of the 3° radiation presumably reflects the isotropy of the last surface of charged particles - mainly electrons - from which the photons scattered. Until about 10^5 years after the Big Bang, photons were in equilibrium with charged particles. At that time, the electrons and protons "recombined" into neutral hydrogen atoms and the photons no longer scatter. From that time to the present the photons freely expand. This expansion preserves the shape of the black-body spectrum as the temperature falls. There is one anisotropy that is to be expected and, indeed, has been observed. One may imagine a frame of reference - not the Earth - in which the distribution is an isotropic black-body distribution. If there is a second frame moving uniformly with respect to the first, then in this frame the distribution of photons will not be isotropic, but will have an effective temperature depending linearly on the cosine of the angle between the direction of observation and the relative velocity vector of the two frames. This effect has been observed (Smurf, Gurenstein and Muller 1977), and seems to indicate that our galaxy is moving with a speed of about 640 km/s in a direction whose significance is obscure. What has <u>not</u> been observed is any other deviation from isotropy. Different parts of the sky have distributions that are the same to about one part in ten thousand. As we will see in the final section of the book, this is a puzzle, since it implies correlations of events in the distant past which seem to be outside each other's light cone. How this is explained is one of the subjects of the papers in the last section of the book.

REFERENCES

Alpher, R. A., and Herman, R. 1948, Nature, **162**, 774.
Bernstein, J. 1984, Three Degrees Above Zero (New York: Scribner's).
Bondi, H. 1961, Cosmology (Cambridge: Cambridge University Press).
Doroshkevich, A. G., and Novikov, I. I. 1964, Soviet Physics-Doklady, **9**, 111.
Finley-Freundlich, E. 1954, Phil. Mag., **45**, 303.
Gamow, G. 1953, Det. Kong. Danske Viden.; Mat.-Fys., **27** 10.
Gamow, G. 1970, My World Line (New York: Viking Press).
Ohm, E. A. 1961, Bell Syst. Techn. J., **40**, 1065.
Smurf, G. F., Gurenstein, M. V., and Muller, R. A. 1977, Phys. Rev. Lett., **39**, 898.
Weinberg, S. 1977, The First Three Minutes (New York: Basic Books).

THE ORIGIN OF ELEMENTS AND THE SEPARATION OF GALAXIES

G. Gamow
George Washington University, Washington, D.C.

Received June 21, 1948

The successful explanation of the main features of the abundance curve of chemical elements by the hypothesis of the "unfinished building-up process," (Gamow 1946; Alpher, Bethe, and Gamow 1948) permits us to get certain information concerning the densities and temperature which must have existed in the universe during the early stages of its expansion. We want to discuss here some interesting cosmological conclusions which can be based on these informations.

Since the building-up process must have started with the formation of deuterons from the primordial neutrons and the protons into which some of these neutrons have decayed, we conclude that the temperature at that time must have been of the order $T_0 \cong 10^{9\circ}$ K (which corresponds to the dissociation energy of deuterium nuclei), so that the density of radiation $\sigma T^4/c^2$ was of the order of magnitude of water density. If, as we shall show later, this radiation density exceeded the density of matter, the relativistic expression for the expansion of the universe must be written in the form:

$$\frac{d}{dt} \ln \ell = \left(\frac{8\pi G}{3} \frac{\sigma T^4}{c^2}\right)^{1/2} \quad (1)$$

where ℓ is an arbitrary distance in the expanding space, and the term containing the curvature is neglected because of the high density value. Since for the adiabatic expansion T is inversely proportional to ℓ, we can rewrite (1) in the form:

$$-\frac{d}{dt} \ln(T) = \frac{T^2}{c} \left(\frac{8\pi G\sigma}{3}\right)^{1/2} \quad (2)$$

or, integrating:

$$T^2 = \left(\frac{3}{32\pi G\sigma}\right)^{1/2} \times \frac{c}{t} . \quad (3)$$

For the radiation density we have:

$$\rho_{rad} = \frac{3}{32\pi G} \times \frac{1}{t^2} . \quad (4)$$

These formulas show that the time t_0, when the temperature dropped low enough to permit the formation of deuterium, was several minutes. Let us assume that at that time the density of matter (protons plus neutrons) was $\rho_{mat}°$. Since, in contrast to radiation, the matter is conserved in the process of expansion, ρ_{mat} was decreasing as $\ell^{-3} \sim T^3 \sim t^{-3/2}$. The value of $\rho_{mat}°$ can be estimated from the fact that during the time period Δt of about 10^2 sec (which is set by the rate of expansion), about one-half of original particles were combined into deuterons and heavier nuclei. Thus we write:

$$v \Delta t n \sigma \cong 1 \tag{5}$$

where $v = 5 \times 10^8$ cm/sec is the thermal velocity of neutrons at $10^{9}°$ K, n is the particle density, and $\sigma \cong 10^{-29}$ cm^2 the capture cross section of fast neutrons in hydrogen. This gives us $n \cong 10^{18}$ cm^{-3} and $\rho_{mat}° \cong 10^{-6}$ g/cm^2 substantiating our previous assumption that matter density was negligibly small compared with the radiation density. (Thus we have $\rho_{mat}° \times \Delta t \cong 10^{-4}$ g $\times$ cm^{-3} $\times$ sec and not 10^{+4} g $\times$ cm^{-3} sec as was given incorrectly in the previous paper (Alpher, Bethe and Gamow 1948) because of a numerical error in the calculations.)

Since $\rho_{rad} \sim t^{-2}$ whereas $\rho_{mat} \sim t^{-2/3}$ the difference by a factor of 10^6 which existed at the time 10^9 sec must have vanished when the age of the universe was $10^2 \times (10^6)^2 = 10^{14}$ sec $\cong 10^7$ years. At that time the density of matter and the density of radiation were both equal to $[(10^6)^2]^{-2} = 10^{-24}$ g/cm^3. The temperature at that epoch must have been of the order $10^9/10^6 \cong 10^{3}°$ K.

The epoch when the radiation density fell below the density of matter has an important cosmological significance since it is only at that time that the Jeans principle of "gravitational instability" (Jeans 1928) could begin to work. In fact, we would expect that as soon as the matter took over the principal role, the previously homogeneous gaseous substance began to show the tendency of breaking up into separate clouds which were later pulled apart by the progressive expansion of the space. The density of these individual gas clouds must have been approximately the same as the density of the universe at the moment of separation, i.e., 10^{-24} g/cm^3. The size of the clouds was determined by the condition that the gravitational potential on their surface was equal to the kinetic energy of the gas particles. Thus we have:

$$\frac{3}{2} kT = \frac{4}{3} \pi R^3 \rho \frac{G m_H}{R} = \frac{4\pi G m_H \rho}{3} R^2 . \tag{6}$$

With $T \cong 10^3$ and $\rho \cong 10^{-24}$ this gives $R \cong 10^{21}$ cm $\cong 10^3$ light years.

The fact that the above calculated density and radii correspond closely to the observed values for the stellar galaxies strongly suggests that we have here a correct picture of galactic formation. According to this picture the galaxies were formed when the universe was 10^7 years old, and were originally entirely gaseous. This may explain their regular shapes, resembling those of the rotating gaseous bodies, which must have been retained even after all their diffused material was used up in the process of star formation (as, for example, in the elliptic galaxies which consist entirely of stars belonging to the population II, Baade [1944]).

It may also be remarked that the calculated temperature corresponding to the formation of individual galaxies from the previously uniform mixture of matter and radiation, is close to the condensation points of many chemical elements. Thus we must conclude that some time before or soon after the formation of gaseous galaxies their material separated into the gaseous and the condensed (dusty) phase. The dust particles, being originally uniformly distributed through the entire cloud, were later collected into smaller condensations by the radiation pressure in the sense of the Spitzer-Whipple theory of star formation (Spitzer 1942; Whipple 1946). In fact, although there were no stars yet, there was still plenty of high intensity radiation which remained from the original stage of expanding universe when the radiation, and not the matter, ruled the things.

In conclusion I must express my gratitude to my astronomical friends, Dr. W. Baade, Dr. E. Hubble, Dr. R. Minkowski, and Dr. M. Schwartzschild for the stimulating discussion of the above topics.

REFERENCES

Alpher, R., Bethe, H., and Gamow, G. 1948, Phys. Rev., **73**, 803.
Baade, W. 1944, Ap.J., **100**, 137.
Gamow, G. 1946, Phys. Rev., **70**, 572.
Jeans, J. H. 1928, Astronomy and Cosmogony (Teddington: Cambridge University Press).
Spitzer, L. Jr. 1942, Ap.J., **95**, 329.
Whipple, F. L. 1946, Ap.J., **104**, 1.

REMARKS ON THE EVOLUTION OF THE EXPANDING UNIVERSE[1,2]

Ralph A. Alpher and Robert C. Herman
Applied Physics Laboratory, The Johns Hopkins University
Silver Spring, Maryland

Received: December 27, 1948

The relativistic energy equation for an expanding universe of non-interconverting matter and radiation is integrated. The above result, together with a knowledge of the physical conditions that prevailed during the element forming process in the early stages of the expansion, is used to determine the time dependences of proper distance as well as of the densities of matter and radiation. These relationships are employed to determine the mean galactic diameter and mass when formed as 2.1×10^3 light years and 3.8×10^7 sun masses, respectively. Galactic separations are computed to be of the order of 10^6 light years at the present time.

INTRODUCTION

With the experimental and theoretical information now available it is possible to give a tentative description of the structure and evolution of the universe. Investigations of cosmological models of various types have been carried out which explain many of the features of the observed universe (Tolman 1934). It does not appear to have been possible to complete these speculations principally for lack of sufficient physical data. Recent studies of the origin and relative abundances of the elements have yielded new information concerning the physical state of the universe at the very early time during which the elements were apparently formed (Gamow 1946; Alpher, Bethe, and Gamow 1948; Gamow 1948a; Alpher, Herman, and Gamow 1948; Alpher 1948; Alpher and Herman 1948a; Gamow 1948b; Alpher and Herman 1948b). According to this theory the ylem (the primordial substance from which the elements were formed) consisted of neutrons at a high density and temperature. Protons were formed by neutron decay, and the successive capture of neutrons led to the formation of the elements. In order to predict the observed relative abundances of the elements, it is necessary to stipulate the magnitude and the time dependence of the temperature, and density of matter during the period of element formation.

[1]The work described in this paper was supported by the Bureau of Ordinance, U.S. Navy, under Contract NOrd-7386.

[2]A preliminary account of this work was given at the New York meeting of the American Physical Society, January, 1949.

On the basis of a simplified version of the neutron capture theory, namely, one which involves the building up of deuterons only, Gamow (1948b) has examined the state of the universe at early times and traced the evolution of the universe through the formation of galaxies. For reasons which will be discussed later, Gamow's formulation gives rise to certain difficulties.

We have reformulated this problem from a somewhat different point of view, following some of Gamow's basic ideas (Gamow 1948a; Gamow 1948b). This reformulation, which is the main purpose of this paper, involves the use of the general non-static relativistic cosmological model together with knowledge of the physical conditions of matter and radiation which prevail now and also those which are required to predict the observed relative abundances of the nuclear species formed during the very early stages of the universe. As a consequence, it is possible to obtain the functional dependence of both the density of matter and radiation on time. On the basis of the foregoing, the formation of galaxies and other cosmological consequences are considered.

II. FORMULATION OF THE PROBLEM

The model of the expanding universe that we shall discuss is one in which there is a homogeneous and isotropic mixture of radiation and matter, assumed to be non-interconverting. This mixture is treated as a perfect fluid. If the pressure due to matter is neglected, one may write the relativistic energy equation for the non-static model in the following form (Tolman 1934):

$$d\left[\exp\left(\tfrac{1}{2}\, g(t)\right)\right]/dt = \pm\left[(8\pi/3)\rho\, \exp\left(g(t)\right) - R_0^{-2}\right]^{1/2}, \tag{1}$$

which is in relativistic units. The cosmological constant Λ is taken equal to zero. In Eq. (1), ρ is the density of mass and the radius of curvature, R, is given by $R = R_0 \exp\left(g(t)\right)$, where $\exp\left(g(t)\right)$ is the time-dependent factor in the spatial portion of the line element. Now,

$$\exp\left(\tfrac{1}{2}\, g(t)\right) = \ell/\ell_0 = R/R_0\,, \tag{2}$$

where ℓ is any proper distance, and ℓ_0, the unit of length, together with R_0, must be determined from the boundary conditions for Eq. (1). It should be pointed out that solutions of Eq. (1) involve ℓ/ℓ_0 and not ℓ alone. The density of mass ρ, which determines the geometry of the space, is the sum of the density of matter, ρ_m, and the density of radiation, ρ_r. If matter is to be conserved we must have

$$\rho_m \ell^3 = A = \text{constant} . \tag{3a}$$

Furthermore, if the universal expansion is adiabatic, the temperature, T, must vary (Tolman 1934) as ℓ^{-1}. If one assumes that the universe contains blackbody radiation, then

$$\rho_r \ell^4 = B = \text{constant} . \tag{3b}$$

It is to be noted that energy is not conserved in models of this type. Equations (3a) and (3b) obviously may be written as

$$\rho_r \rho_m^{-4/3} = \text{constant} . \tag{4}$$

It is clear that this relationship must hold throughout the universal expansion and that the density of mass at any time is

$$\rho = \rho_m + \rho_r = A\,\ell^{-3} + B\,\ell^{-4} , \tag{5}$$

providing, as stated earlier, there is no interconversion of matter and radiation. If we substitute Eqs. (2) and (5) into Eq. (1), and convert to c.g.s. units, we obtain

$$d\ell/dt = +\left[(8\pi G/3)(A\,\ell^{-3} + B\,\ell^{-4})\ell^2 - c^2 \ell_0^2/R_0^2\right]^{1/2} , \tag{6}$$

where the positive sign is taken to indicate expansion and c and G are the velocity of light and the gravitational constant, respectively. Equation (6) can be integrated and the result given in the form

$$t = K_1 + K_2^{-1}\left[\gamma\rho_{r''} + \gamma\rho_{m''}L + K_2 L^2\right]^{1/2}$$
$$-\left(\gamma\rho_{m''}/2K_2^{3/2}\right) \ln \left\{\left[\gamma\rho_{r''} + \gamma\rho_{m''}L + K_2 L^2\right]^{1/2} \tag{7}$$
$$+ K_2^{1/2} L + \left(\gamma\rho_{m''}/2K_2^{1/2}\right)\right\} ,$$

where

$$K_1 = (\gamma\rho_{m''}/2K_2{}^{3/2}) \ln [(\gamma\rho_{r''})^{1/2} + (\gamma\rho_{m''}/2K_2{}^{1/2})] - (\gamma\rho_{r''}/K_2{}^2)^{1/2} . \quad (8)$$

In Eqs. (7) and (8), $L = \ell/\ell_0$, $\gamma = (8\pi G/3)$, $K_2 = (c^2/|R_0^2|)$, and $\rho_{m''}$ and $\rho_{r''}$ are the densities of matter and radiation when $L = 1$. In order to integrate Eq. (6) and evaluate the integration constant, it is necessary to specify the parameter R_0 and consequently ℓ_0, which gives the units in which R_0 is measured. Examination of Eq. (6) indicates that R_0 can be determined only if it is possible to specify $[(d\ell/dt)/\ell]_{\ell=\ell_0}$, ρ_m, and ρ_r at any given time. Since $[(d\ell/dt)/\ell]_{\ell=\ell_0}$ is the expansion rate of space as determined by Hubble (1937) and known, therefore, only at the present time, since ρ_m is also known now, and if we assume that $\rho_m \gg \rho_r$ now, one may evaluate R_0 and K_2. Introducing the value of the present expansion rate of the universe $[(d\ell/dt)/\ell]_{\ell=\ell_0} = 1.8\times10^{-17}$ sec^{-1}, taking $\rho_{m''} = 10^{-30}$ g/cm^3 and $\ell=\ell_0=10^{10}$ cm, i.e., ℓ_0 is the side of a cube containing one gram of matter now, one obtains $R_0 = 1.7\times10^{27}(-1)^{1/2}$ cm and $K_2 = 3.2\times10^{-34}$ sec^{-2}. The constants appearing in Eqs. (7) and (8) involve the present densities of matter and radiation. Clearly, in utilizing Eqs. (7) or (8) one may introduce the density values at any other time providing one specifies a value of L at that time which leads to the present value of the density of matter. For convenience we have chosen ℓ_0 to be the side of a cube containing one gram of matter at the present time, so that L = 1 now. Furthermore, we have again for convenience assumed that L = 0 at t = 0. While Eq. (6) has a singularity at t = 0 which is physically unreasonable, we have employed the solutions in such a manner that the singularity is of no consequence.

For purposes of computation it is convenient to employ an approximate form for Eq. (7) which is valid for early t, i.e., when

$$L[(\rho_{m''}/\rho_{r''}) + (K_2/\gamma\rho_{r''})L] < 1 . \quad (9)$$

The expansion of Eq. (7) which satisfies the above inequality is

$$t = (4\gamma\rho_{r''})^{-1/2} L^2 + (\rho_{m''}/6\gamma^{1/2}\rho_{r''}{}^{3/2}) L^3 + (8\gamma^{3/2}\rho_{r''}{}^{3/2})^{-1} \times[(3\gamma\rho_{m''}{}^2/4\rho_{r''}) - K_2] L^4 + \ldots . \quad (10)$$

The validity of Eqs. (7) or (10) is questionable for very early times, i.e., in the vicinity of the singularity at t = 0, when the energy of light quanta was comparable to the rest mass of elementary particles. In fact, Einstein (1945) has pointed out that there is a difficulty at very early times because of the separate treatment of the metric field (gravitation) and electromagnetic fields and matter in the theory of relativity. For large densities of field and of matter, the field equations and even the field variables which enter into them will have no real significance. However, since we do not concern ourselves with the "beginning" this difficulty is obviated. In addition to the fact that the relativistic energy equation is not valid for very early times, there are the problems of angular momentum of matter in the universe, as well as certain physical factors involved in the formation of the elements, which we cannot handle satisfactorily at present.

In order to utilize the above equations, it is necessary to specify $\rho_{m''}$, $\rho_{r''}$, and K_2. While it may appear that one need specify the matter and radiation densities at the present time only, because of Eq. (4), specifying $\rho_{m''}$ and $\rho_{r''}$ is equivalent to specifying $\rho_{m'}$ and $\rho_{r'}$, these being the densities at a time during the period of element formation. This time is to be specified later. (The primed quantities should not be confused with the running variables.) It must be remembered that the value of R_0 employed is that calculated from the present value of dL/dt.

III. PHYSICAL CONDITIONS DURING THE EXPANSION

Some information is available regarding the values of the matter and radiation densities at the present time and, recently, studies of the relative abundances of the elements have indicated values for these densities prevailing very early in the universe during the period of element formation. Because of Eq. (4) a knowledge of $\rho_{m'}$ and $\rho_{r'}$ during the element forming period together with $\rho_{m''}$ fixes a value for $\rho_{r''}$, the present radiation density, which is perhaps the least well-known quantity.

In a recent paper Gamow (1948b), by considerations which are different than those we have employed, found a set of physical conditions which prevailed during the early stages of the universe. He studied the formation of deuterons only, by the capture of neutrons by protons, taking into account the universal expansion. Equations for the formation of deuterons were integrated from t=0, subject to the condition that there were neutrons at the start (unit concentration by weight) and that the final concentration by weight of protons and deuterons was 0.5. This solution determined a parameter α which in turn defined the magnitude of the matter density,[3] $\rho_m = \rho_0 t^{-3/2}$.

[3]The expression for the parameter α, as given by Gamow (1948b), has been found to be incorrect (Alpher and Herman 1948b). We find that $\alpha(= \rho_m v \sigma t/m$, where $\rho_m = \rho_0 t^{-3/2}$ is the density of matter, v is the mean velocity of particles of mass

We believe that a determination of the matter density on the basis of only the first few light elements is likely to be in error. Our experience with integrations required to determine the relative abundances of all elements (Alpher 1948; Alpher and Herman 1948a) indicated that these computed abundances are critically dependent upon the choice of matter density. Furthermore, all formulations of the neutron capture process which have been made thus far neglect the thermal dissociation of nuclei, which is one of the important competing processes during the element forming period if elements are formed from a very early time.

In order to clarify the difficulties associated with the singularity at t=0, we digress here for an examination of the equations employed to describe the formation of the elements. These equations, recently given by the authors (Alpher and Herman 1948a), include neutron decay and universal expansion but do not take into account the effects of nuclear evaporation or any processes other than radiative capture of neutrons. In terms of concentrations by weight, $x_j = m_j n_j/\rho_m$, rather than particle concentrations, n_j, Eqs. (6) - (8) of reference 7 may be written as

$$dx_0/dt = -\lambda x_0 - \sum_{j=1}^{J} (p_j\rho_m/m_j)x_j x_0 , \qquad (11a)$$

$$dx_1/dt = \lambda x_0 - (p_1\rho_m/m_1)x_1 x_0 , \qquad (11b)$$

and

$$dx_j/dt = j(p_{j-1}\rho_m/m_{j-1})x_{j-1}x_0 - j(p_j\rho_m/m_j)x_j x_0,$$

$$j = 2, 3, \ldots, J, \qquad (11c)$$

m, and σ is the capture cross section protons for neutrons) is correctly given by

$$\alpha = \frac{2^{9/4}\pi^{5/4}G^{1/4}a^{1/4}e^2\hbar}{3^{1/4}m^{9/2}c^{11/2}k} (|\mu_P| + |\mu_N|)^2(\varepsilon^{1/2} + \varepsilon_0{}^{1/2})\varepsilon^{3/2}\rho_0 .$$

In this expression all the quantities have been defined by Gamow (1948b) except μ_P and μ_N, the magnetic moments in nuclear magnetons of proton and neutron, respectively, ε_0, the binding energy of the virtual triplet state of the deuteron, and the radiation density constant $a = 7.65\times10^{-15}$ ergs cm^{-3} deg^{-4}. Our expression differs from that originally given by Gamow because of algebraic errors contained in his results and because he neglected the magnetic moment factor.

where x_0, x_1, and x_j are the concentrations by weight of neutrons, protons, and nuclei of atomic weight $2 \leq j \leq J$, respectively, m_j the nuclear mass, ρ_m the density of matter, λ the neutron decay constant, and p_j the effective neutron capture volume swept out per second by nuclei of species j. Gamow (1948b) has solved Eqs. (11a) and (11b) numerically, taking j=1, and thereby describing the building up of deuterons only. In general, Eqs. (11) have a singularity at the origin because when $t \to 0$, $\rho_m \to \infty$ as $t^{-3/2}$. In the approximation used by Gamow this singularity is reduced because a relation for the capture cross section of protons for neutrons is employed which makes $p_1\rho_m(= \sigma_1 v \rho_m)$ vary as t^{-1}.

It may be seen readily that Eq. (11c) can be written in the form

$$dx_j/dz = (p_{j-1}/\lambda m_{j-1})x_{j-1} - (p_j/\lambda m_j)x_j,$$

$$j = 2, 3, \ldots, J, \tag{11d}$$

where

$$z = \int_{\tau_0}^{\tau} j\rho_m(\tau)x_0(\tau)d\tau \ , \tag{11e}$$

and

$$\tau = \lambda t \ .$$

In general, the integrand in Eq. (11e) is singular at $\tau = 0$, so that one must take $\tau_0 > 0$. This implies the choice of an initial time at which the element forming process started. Physically, one may not speak of an intial time because there were competing processes which became unimportant as the neutron capture process became important. Competing processes such as photo-disintegration and nuclear evaporation fall off approximately exponentially with time so that neutron capture would become significant rather rapidly, say in a time of the order of 10^2 seconds. The inclusion of this type of competing process in principle could be handled and would yield a better estimate of the relative abundances of the elements. However, without a better knowledge of cosmology at very early t it does not appear to be possible to avoid the above mentioned difficulty. Finally, if Eqs. (11a), (11b), and (11c) are solved simultaneously for $J = 4$, the remaining equations for $j > 4$ are given by Eq. (11d) which is a simple first-order linear differential equation with constant coefficients. Nevertheless, Eqs. (11a) and (11b), which are the controlling equations for the process, are not reduced to a simple form and must still be solved in their present form. Because of the above difficulties we find it necessary to introduce the concept of a starting time for the element forming process. Equations (11) have not yet been solved but are given to illustrate the singularity. So far as we know, any formulation of a theory of element building which includes the type of cosmology discussed will reflect these same difficulties.

In what follows we continue the discussion of the physical conditions employed in the solutions of the relativistic energy equation. The mean density of matter in the universe at the present time has been determined by Hubble (1937) to be

$$\rho_{m''} \cong 10^{-30} \text{ g/cm}^3 . \qquad (12a)$$

An estimate of the density of matter, $\rho_{m'}$, prevailing at the start of the period of element formation is obtained by integration of the equations for the neutron capture theory of the formation of the elements. Integrations in which neutron decay is explicitly included, but in which the expansion of the universe is not included, yield a matter density of 5×10^{-9} g/cm^3. Preliminary investigations of the equations, including the universal expansion, indicate that this density should be increased by a factor roughly of the order of 100 in order that one may correctly determine the relative abundance of the elements with the universal expansion taken into account. In fact, we have numerically integrated for the light elements the complete equations [see Eqs. (11)] with an "initial" density about 100 times the density used in obtaining solutions without the universal expansion (Alpher and Herman 1948a). We find that the above factor of ~100 is roughly what might be required. Accordingly, we have taken

$$\rho_{m'} \cong 10^{-6} \text{ g/cm}^3 . \qquad (12b)$$

As discussed elsewhere (Alpher 1948; Alpher and Herman 1948a), the temperature during the element process must have been of the order of $10^8 - 10^{10}$° K. This temperature is limited, on the one hand, by photo-disintegration and thermal dissociation of nuceli and, on the other hand, by the lack of evidence in the relative abundance data for resonance capture of neutrons. For purposes of simplicity we have chosen

$$\rho_{r'} \cong 1 \text{ g/cm}^3 , \qquad (12c)$$

which corresponds to $T \cong 0.6\times10^{9}$° K at the time when the neutron capture process became important.

In accordance with Eq. (4), the specification of $\rho_{m''}$, $\rho_{m'}$, and $\rho_{r'}$ fixes the present density of radiation, $\rho_{r''}$. In fact, we find that the value of $\rho_{r''}$ consistent with Eq. (4) is

$$\rho_{r''} \cong 10^{-32} \text{ g/cm}^3 , \qquad (12d)$$

which corresponds to a temperature now of the order of 5° K. This mean temperature for the universe is to be interpreted as the background temperature which would result from the universal expansion alone. However, the thermal energy resulting from the nuclear energy production in stars would increase this value.

Since we have $\rho_{r'} \gg \rho_{m'}$ at early time the energy relation given in Eq. (6) may be integrated in a simpler form, with the result

$$T = \left[(32\pi Ga)/(3c^2) \right]^{-1/4} t^{-1/2} \text{ °K}$$

$$= 1.52\times 10^{10} t^{-1/2} \text{ °K} . \qquad (13a)$$

The density of radiation, ρ_r, may be found from $\rho_r = (a/c^2)T^4$, or

$$\rho_r = 4.48\times 10^5 t^{-2} \text{ g/cm}^3 . \qquad (13b)$$

These expressions for T and ρ_r at early time are the consequence of the assumption of an adiabatic universe filled with blackbody radiation. It can also be shown that with the densities chosen in Eq. (12) we have for early time

$$\rho_m = 1.70\times 10^{-2} t^{-3/2} \text{ g/cm}^3 . \qquad (13c)$$

Using ℓ and ℓ_0 as already defined, we may determine the constants A and B in Eq. (3). With the densities discussed above we find A = 1 g and B = 10^8 g cm. These values of A and B fix the dependence of ρ_m and ρ_r on time through $L(=\ell/\ell_0)$. Using these values of A and B, we have computed L, ρ_m, ρ_r, and T. These quantities are plotted on a logarithmic scale in Fig. 1. It should be noted in Fig. 1 that all the quantities plotted bear simple relationships with the time to within several orders of magnitude of the time when the universal expansion changes from one controlled by gravitation to one of free escape. This transition occurs in the region of about 10^{13} – 10^{14} sec. Following this transition the quantities L, ρ_m, ρ_r, and T again are simple functions of the time. The relations for large t are as follows:

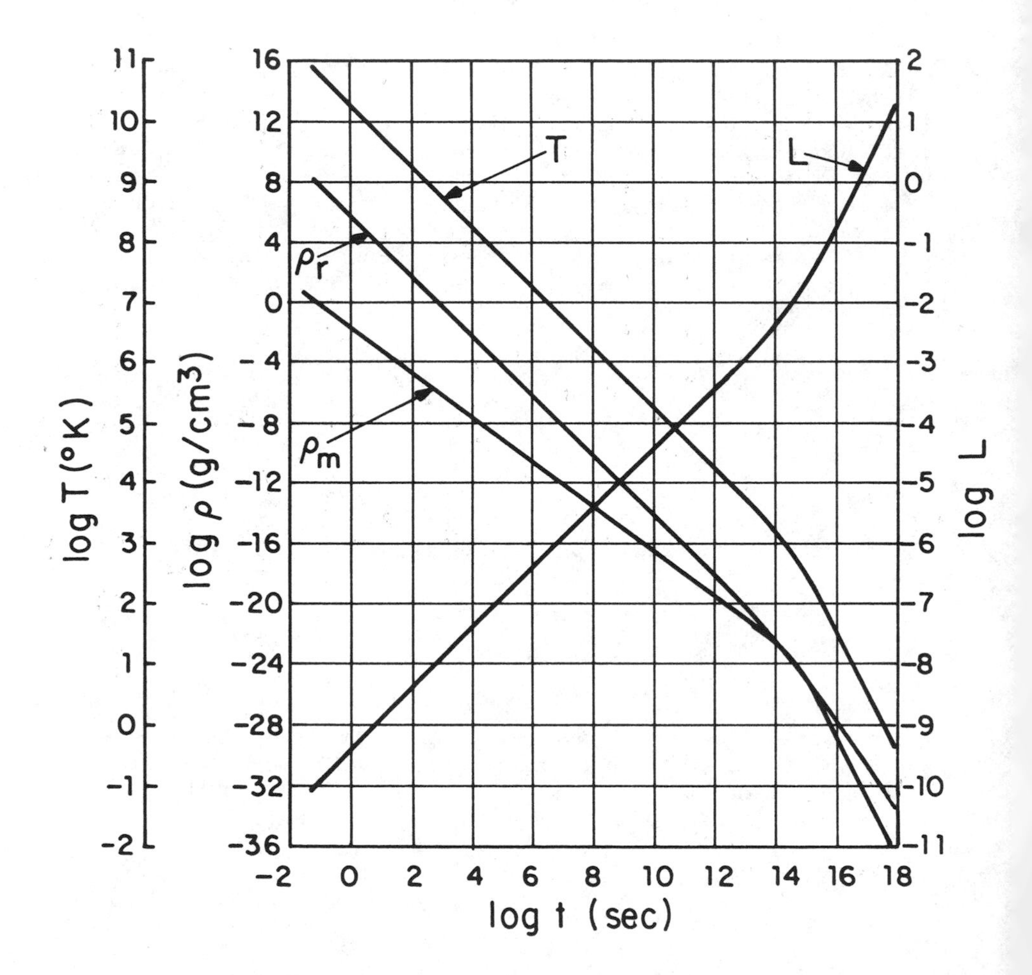

Figure 1.

The time dependence of the proper distance L, the densities of matter and radiation, ρ_m, and ρ_r, as well as the temperature, T, are shown for the case where $\rho_{m''} \cong 10^{-30}$ g/cm^3, $\rho_{r''} \cong 10^{-32}$ g/cm^3, $\rho_{m'} \cong 10^{-6}$ g/cm^3, and $\rho_{r'} \cong 1$ g/cm^3. [See Eq. (12).]

$$L = K_2^{1/2}\, t,$$

$$\rho_m = \left(\rho_{m''}/K_2^{3/2}\right) t^{-3}, \tag{14}$$

$$\rho_r = \left(\rho_{r''}/K_2^{2}\right) t^{-4},$$

and,

$$T = \left(c^2 \rho_{r''}/aK_2^{2}\right) t^{-1}.$$

It is to be noted that in the region of transition to free escape the densities of matter and radiation become equal so that, in fact, prior to the transition the expansion is controlled chiefly by radiation and subsequent to the transition by matter. The universe is now in the freely expanding state, and, since the radius of curvature is imaginary, is of the open, hyperbolic type.

In order to study how sensitive this model is to the choice of densities, we have considered the following additional set of density values which satisfy Eq. (4):

$$\rho_{m'} \cong 1.78\times 10^{-4}\ \mathrm{g/cm^3},$$

$$\rho_{r'} \cong 1\ \mathrm{g/cm^3}, \tag{15}$$

$$\rho_{m''} \cong 10^{-30}\ \mathrm{g/cm^3},$$

and

$$\rho_{r''} \cong 10^{-35}\ \mathrm{g/cm^3}.$$

The value obtained for $\rho_{r''}$ in this case corresponds to a present mean temperature of about 1° K. The constants A and B are found to be 1 g and 10^5 g cm, respectively. In Fig. 2 we have plotted the time dependence of the quantities of interest. One finds that the transition occurs at an earlier time than in the previous case, namely, at $\sim 10^{10}$ sec, which implies that this universe would have been in a state of free expansion for a considerably longer time. Apparently the behavior of the model is extremely sensitive to the choice of density conditions. However, the simple type of relations for L, ρ_m, ρ_r, and T that were given previously still apply, but with different constants and different regions of validity.

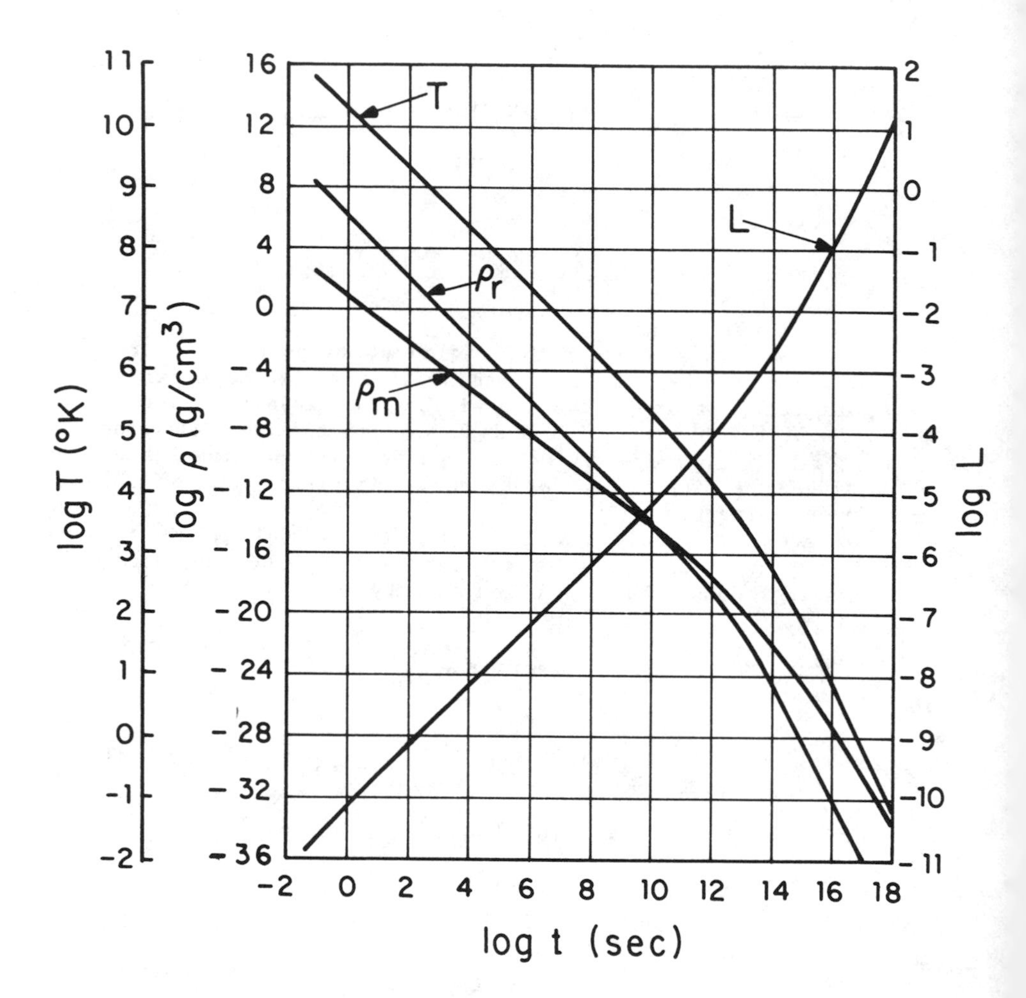

Figure 2.

The time dependence of the proper distance L, the densities of matter and radiation, ρ_m, and ρ_r, as well as the temperature, T, are shown for the case where $\rho_{m''} \cong 10^{-30}$ g/cm^3, $\rho_{r''} \cong 10^{-35}$ g/cm^3, $\rho_{m'} \cong 1.8\times10^{-4}$ g/cm^3, and $\rho_{r'} \cong 1$ g/cm^3. [See Eq. (15).]

The time at which $\rho_m = \rho_{m'}$ and $\rho_r = \rho_{r'}$ for both sets of densities given in Eqs. (12) and (15) are found from Eq. (13b) to be 6.7×10^2 seconds, with a corresponding temperature of 0.59×10^9 °K. We have chosen $\rho_r \cong 1$ g/cm^3 in both cases because the corresponding temperature is seen by independent considerations to be that required for the element forming process. As will be seen later, the densities given in Eq. (15) with $\rho_{m'} \cong 1.78\times 10^{-4}$ g/cm^3 do not yield a satisfactory description of the size and mass of galaxies. On the other hand, as already stated a density $\rho_{m'} \cong 100$ (5×10^{-9} g/cm^3) is apparently enough to overcome the effect of the universal expansion and give the correct relative abundances of the elements. Thus, on the basis of these considerations one is led to the conclusion that when $t \cong 6.70\times 10^2$ sec, and $\rho_{r'} \cong 1$ g/cm^3 we have

$$5.0\times 10^{-7}\ \mathrm{g/cm^3} \lesssim \rho_{m'} \lesssim 1.8\times 10^{-4}\ \mathrm{g/cm^3} .$$

While it is not particularly germane to the study reported in this paper, it is interesting to note that one may find the dependence of the universal expansion rate on the time in this type of model. This rate is the percentage change in proper distance per unit time determined by Hubble (1937) from the red-shift in spectra of nebulae, and is given in $V = Hd$, where V is the velocity of recession of a nebula at a distance d. In our notation, we have, in general,

$$H = \left(dL/dt\right)/L = L^{-1}\left(\gamma\rho L^2 + K_2\right)^{1/2} . \tag{16}$$

For early time this reduces to

$$H = (2t)^{-1} , \tag{16a}$$

and, for late time, to

$$H = t^{-1} . \tag{16b}$$

For early and late t, the value of H does not depend upon the choice of densities. However, in the transition region where the functional form of H changes, the manner of change does depend on the existing density conditions. The universal expansion rate is the reciprocal of the age of the universe if measured during the period of free expansion.

IV. THE FORMATION OF GALAXIES

In his discussion of the evolution of the universe, Gamow (1948b) suggested that galactic formation occurred at the time when the densities of matter and radiation were equal. He assumed that the Jeans' criterion of gravitation instability may be applied at this time and as a consequence derives expressions for the galactic diameter and mass.[4] We have carried out calculations (Alpher and Herman 1948) based on Gamow's formulation using the corrected expressions for D and M given in footnote 4. We find that $\rho_m = \rho_r$ when $t_c \cong 0.86\times 10^{18}$ sec, which is greater than the age of the universe. This arises out of the fact that, in addition to the difficulties with density determinations mentioned earlier, there is involved an extrapolation of relations valid only for early t past their region of validity. It is evident from our choice of densities that the densities of matter and radiation must be equal at a time t_c which is earlier than now. We have retained Gamow's basic idea of galactic condensation at t_c and have applied the Jeans' criterion,[5]

$$D^2 = (5\pi kT_c)/(3G m\rho_{m,c}) , \tag{17}$$

where T_c and $\rho_{m,c}$ are taken at $t = t_c$. We may write

[4]Using the corrected form of α described in footnote 3, we find for the galactic diameter, D, and mass, M, the following corrected expressions according to Gamow's formulation:

$$D^2 = \frac{10\pi^2 e^2 \hbar}{3Gm^{11/2}c^5\alpha} (|\mu_P| + |\mu_N|)^2(\varepsilon^{1/2} + \varepsilon_0{}^{1/2})\,\varepsilon^{3/2} t_c ,$$

and

$$M \cong \rho_m D^3 =$$

$$\frac{5^{3/2}\pi^{7/4} e\hbar^{1/2} k}{2^{3/4}3^{5/4}G^{7/4}a^{1/4}m^{15/4}c^2\alpha^{1/2}} (|\mu_P|+|\mu_N|)(\varepsilon^{1/2}+\varepsilon_0{}^{1/2})\varepsilon^{3/4} ,$$

where t_c is the time at which the densities of matter and radiation were equal.

[5]Except for a numerical factor, Jeans' criterion (1929) may be obtained by equating the average thermal energy of a particle with the gravitational potential energy of this particle on the surface of a sphere of diameter D.

$$D = KB^{9/8} , \quad (18)$$

where

$$K = [(5\pi kc^{1/2})/(3a^{1/4}Gm)]^{1/2} , \quad (18a)$$

$$B = \rho_r \rho_m^{-4/3} , \quad (18b)$$

and

$$M \cong \rho_m D^3 = K^3 B^{3/8} . \quad (19)$$

For the set of density conditions as given in Eq. (12), we obtain for D and M the values 2.1×10^3 light years and 3.8×10^7 sun masses, respectively. When the densities of matter and radiation were equal, $t_c \cong 3.5\times 10^{14}$ sec $\cong 10^7$ years, $\rho_{m,c} \cong 10^{-24}$ g cm^3 and $T_c \cong 5.9\times 10^2$ °K. For the set of densities given in Eq. (15) we obtain $D \cong 1$ light year, $M \cong 2.8\times 10^5$ sun masses, $t_c \cong 1.8\times 10^{10}$ sec $\cong 6\times 10^2$ years, $\rho_{m,c} \cong 10^{-15}$ g/cm^3, and $T_c \cong 10^5$° K. In the former case we find values for the galactic mass, diameter and density which are roughly of the order of magnitude observed for the average nebula. In the latter case the values differ by many orders of magnitude. Thus, the values one obtains for the galactic mass and diameter appear to be extremely critical to the choice of densities. One might interpret the large discrepancy in the latter case as arising from the fact that the density conditions chosen appear to be incompatible with the neutron-capture theory of the formation of the elements.

The Jeans' criterion of gravitational instability was derived by the consideration of an acoustic wave propagating in a static medium. If the Jeans' criterion is satisfied, regions of condensation whose size is of the order of D, D being the acoustic wavelength, would have separated and would have been gravitationally stable. The separation between condensations would then also have been of the order of D. The separation distance would increase with time, however, because of the universal expansion, whereas the condensations, being gravitationally stable units, would not expand. Subsequently, stars would evolve in these condensations and nebular configurations would be established (Gamow and Teller 1939).[6]

[6]In this paper it is shown that if galaxies were formed during a period of free expansion then

$$G\rho_m[(4\pi/3)(D/2)^3]/(D/2) > (H^2/2)(D/2)^2 ,$$

where H is Hubble's expansion rate and D is the diameter of the condensation. This condition sets a lower limit to ρ_m, namely $\rho_m > (3H^2/8\pi G) = 0.6\times 10^{-27}$ g/cm^3 and is satisfied by the density value we obtain for galaxies at the time of formation.

From the time variation of proper distance the separation between galaxies is computed to be about 10^6 light years at the present time, in general agreement with observed separations.

The applicability of Jeans' criterion of gravitational instability to this situation must be seriously questioned since it does not contain the possible effects of universal expansion, radiation, relativity, and low matter density. However, it seems reasonable to attach some significance to the time at which radiation and matter densities are equal, because beyond this time the expansion is free and it would become increasingly difficult to form condensations (see footnote 6). It should be mentioned that Lifshitz (1946) has considered the problem of gravitational instability associated with infinitesimal perturbations of an arbitrary nature in a general relativistic expanding universe and has found that the system is stable and the perturbations do not grow. Until such time as a physically satisfactory criterion for the formation of galaxies is found, it does not appear to be profitable to delve further into such questions as the variation in galactic mass and size with time of formation.

V. ACKNOWLEDGMENTS

Our thanks are due to Dr. G. Gamow for his interest in this work and for many stimulating discussions. We wish to express our appreciation to Dr. R. B. Kershner for his interest and invaluable aid concerning some mathematical questions. We wish also to thank Drs. J. W. Follin, Jr., and F. T. McClure for valuable discussions, and Miss S. Thomas for her aid during the preparation of the manuscript.

REFERENCES

Alpher, R. A., 1948, Phys. Rev., **74**, 1577.
Alpher, R. A., and Herman, R. C. 1948a, Phys. Rev., **74**, 1737.
Alpher, R. A., and Herman, R. C. 1948b, Nature, **162**, 774.
Alpher, R. A., Herman, R. C., and Gamow, G. 1948, Phys. Rev., **74**, 1198.
Alpher, R. A., Bethe, H. A., and Gamow, G. 1948, Phys. Rev., **73**, 803.
Einstein, A. 1945, The Meaning of Relativity (Princeton: Princeton University Press).
Gamow, G. 1946, Phys. Rev., **70**, 572.
Gamow, G. 1948a, Phys. Rev. **74**, 505.
Gamow, G., 1948b, Nature, **162**, 680.
Gamow, G., and Teller, E. 1939, Phys. Rev. **55**, 654.
Hubble, E. P. 1937, The Observational Approach to Cosmology (Oxford: Clarendon Press).
Jeans, J. H., 1929, Astronomy and Cosmogony (London: Cambridge University Press).
Lifshitz, E. 1946, J. Phys. U.S.S.R., **10**, 116.
Tolman, R. C. 1934, Relativity, Thermodynamics and Cosmology (Oxford: Clarendon Press).

COSMIC BLACK-BODY RADIATION[1]

R. H. Dicke, P. J. E. Peebles, P. G. Roll, D. T. Wilkinson
Palmer Physical Laboratory
Princeton, New Jersey

Received May 7, 1965

One of the basic problems of cosmology is the singularity characteristic of the familiar cosmological solutions of Einstein's field equations. Also puzzling is the presence of matter in excess over antimatter in the universe, for baryons and leptons are thought to be conserved. Thus, in the framework of conventional theory we cannot understand the origin of matter or of the universe. We can distinguish three main attempts to deal with these problems.

1. The assumption of continuous creation (Bondi and Gold 1948; Hoyle 1948), which avoids the singularity by postulating a universe expanding for all time and a continuous but slow creation of new matter in the universe.

2. The assumption (Wheeler 1964) that the creation of new matter is intimately related to the existence of the singularity, and that the resolution of both paradoxes may be found in a proper quantum mechanical treatment of Einstein's field equations.

3. The assumption that the singularity results from a mathematical over-idealization, the requirement of strict isotropy or uniformity, and that it would not occur in the real world (Wheeler 1958; Lifshitz and Khalatnikov 1963).

If this third premise is accepted tentatively as a working hypothesis, it carries with it a possible resolution of the second paradox, for the matter we see about us now may represent the same baryon content of the previous expansion of a closed universe, oscillating for all time. This relieves us of the necessity of understanding the origin of matter at any finite time in the past. In this picture it is essential to suppose that at the time of maximum collapse the temperature of the universe would exceed 10^{10} °K, in order that the ashes of the previous cycle would have been reprocessed back to the hydrogen required for the stars in the next cycle.

Even without this hypothesis it is of interest to inquire about the temperature of the universe in these earlier times. From this broader viewpoint we need not limit the discussion to closed oscillating models. Even if the universe had a singular origin it might have been extremely hot in the early stages.

Could the universe have been filled with black-body radiation from this possible high-temperature state? If so, it is important to notice that as the universe expands the cosmolog-

[1]This research was supported in part by the National Science Foundation and the Office of Naval Research of the U.S. Navy.

ical redshift would serve to adiabatically cool the radiation, while preserving the thermal character. The radiation temperature would vary inversely as the expansion parameter (radius) of the universe.

The presence of thermal radiation remaining from the fireball is to be expected if we can trace the expansion of the universe back to a time when the temperature was of the order of 10^{10} °K ($\sim m_e c^2$). In this state, we would expect to find that the electron abundance had increased very substantially, due to thermal electron-pair production, to a density characteristic of the temperature only. One readily verifies that, whatever the previous history of the universe, the photon absorption length would have been short with this high electron density, and the radiation content of the universe would have promptly adjusted to a thermal equilibrium distribution due to pair-creation and annihilation processes. This adjustment requires a time interval short compared with the characteristic expansion time of the universe, whether the cosmology is general relativity or the more rapidly evolving Brans-Dicke theory (Brans and Dicke 1961).

The above equilibrium argument may be applied also to the neutrino abundance. In the epoch where $T > 10^{10}$ °K, the very high thermal electron and photon abundance would be sufficient to assure an equilibrium thermal abundance of electron-type neutrinos, assuming the presence of neutrino-antineutrino pair-production processes. This means that a strictly thermal neutrino and antineutrino distribution, in thermal equilibrium with the radiation, would have issued from the highly contracted phase. Conceivably, even gravitational radiation could be in thermal equilibrium.

Without some knowledge of the density of matter in the primordial fireball we cannot predict the present radiation temperature. However, a rough upper limit is provided by the observation that black-body radiation at a temperature of 40 °K provides an energy density of 2×10^{-29} gm/cm^3, very roughly the maximum total energy distribution compatible with the observed Hubble constant and acceleration parameter. Evidently, it would be of considerable interest to attempt to detect this primevel thermal radiation directly.

Two of us (P. G. R. and D. T. W.) have constructed a radiometer and receiving horn capable of an absolute measure of thermal radiation at a wavelength of 3 cm. The choice of wavelength was dictated by two considerations, that at much shorter wavelengths atmospheric absorption would be troublesome, while at longer wavelengths galactic and extragalactic emission would be appreciable. Extrapolating from the observed background radiation at longer wavelengths (~100 cm) according to the power-law spectra characteristic of synchrotron radiation or bremsstrahlung, we can conclude that the total background at 3 cm due to the Galaxy and the extragalactic sources should not exceed 5×10^{-3} °K when averaged over all directions. Radiation from stars at 3 cm is $< 10^{-9}$ °K. The contribution to the background due to the atmosphere is expected to be approximately 3.5 °K, and this can be accurately measured by tipping the antenna (Dicke, Beringer, Kyhl, and Vane 1946).

While we have not yet obtained results with our instrument, we recently learned that Penzias and Wilson (1965) of the Bell Telephone Laboratories have observed background radiation at 7.3-cm wavelength. In attempting to eliminate (or account for) every contribution to the noise seen at the output of their receiver, they ended with a residual of 3.5° ± 1 °K. Apparently this could only be due to radiation of unknown origin entering the antenna.

It is evident that more measurements are needed to determine a spectrum, and we expect to continue our work at 3 cm. We also expect to go to a wavelength of 1 cm. We understand that measurements at wavelengths greater than 7 cm may be filled in by Penzias and Wilson.

A temperature in excess of 10^{10} °K during the highly contracted phase of the universe is strongly implied by a present temperature of 3.5 °K for black-body radiation. There are two reasonable cases to consider. Assuming a singularity-free oscillating cosmology, we believe that the temperature must have been high enough to decompose the heavy elements from the previous cycle, for there is no observational evidence for significant amounts of heavy elements in outer parts of the oldest stars in our Galaxy. If the cosmological solution has a singularity, the temperature would rise much higher than 10^{10} °K in approaching the singularity (see, e.g., Fig. 1).

It has been pointed out by one of us (P. J. E. P.) that the observation of a temperature as low as 3.5 °K, together with the estimated abundance of helium in the protogalaxy, provides some important evidence on possible cosmologies (Peebles 1965). This comes about in the following way. Considering again the epoch $T \gg 10^{10}$ °K, we see that the presence of the thermal electrons and neutrinos would have assured nearly equal abundances of neutrons and protons. Once the temperature has fallen so low that photodissociation of deuterium is not too great, the neutrons and protons can combine to form deuterium, which in turn readily burns to helium. This was the type of process envisioned by Gamow, Alpher, Herman, and others (Alpher, Bethe, and Gamow 1948; Alpher, Follin, and Herman 1953; Hoyle and Tayler 1964). Evidently the amount of helium produced depends on the density of matter at the time helium formation became possible. If at this time the nucleon density were great enough, an appreciable amount of helium would have been produced before the density fell too low for reactions to occur. Thus, from an upper limit on the possible helium abundance in the protogalaxy we can place an upper limit on the matter density at the time of helium formation (which occurs at a fairly definite temperature, almost independent of density) and hence, given the density of matter in the present universe, we have a lower limit on the present radiation temperature. This limit varies as the cube root of the assumed present mean density of matter.

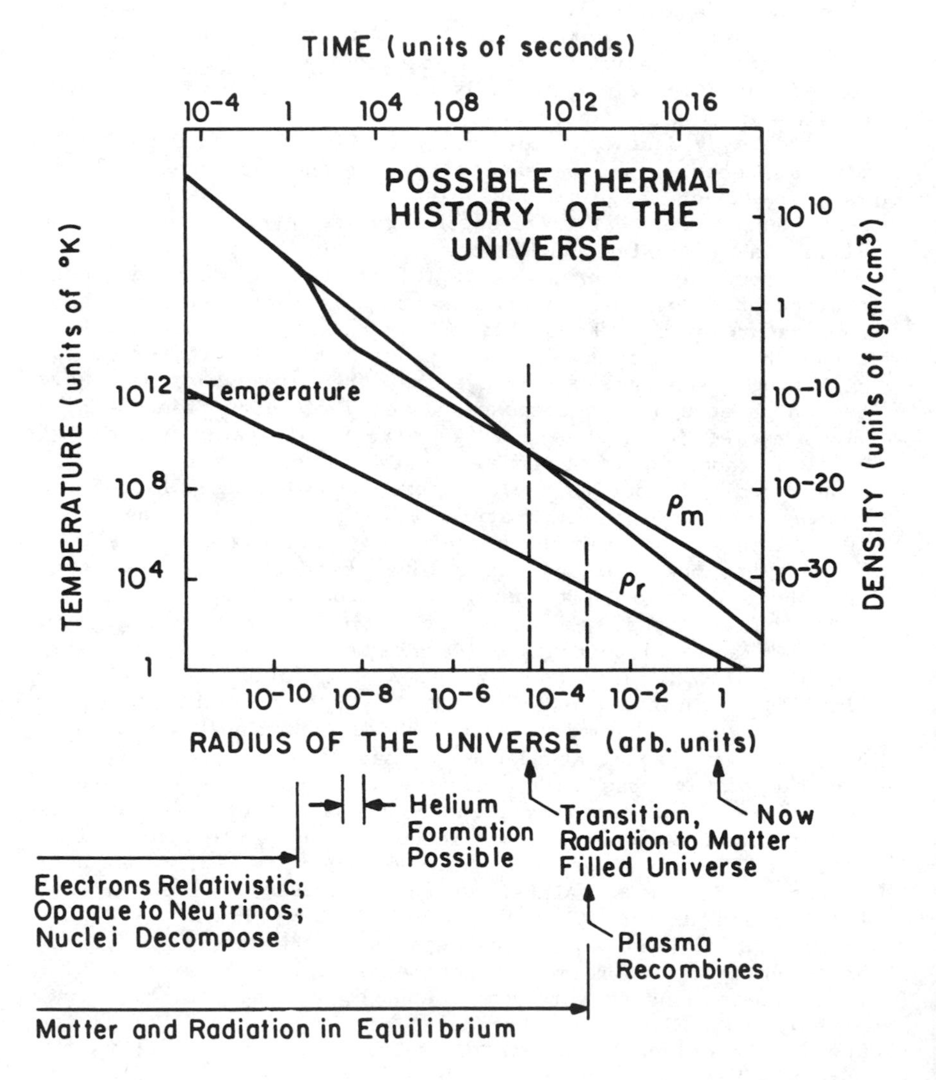
TIME (units of seconds)
10^{-4} 1 10^{4} 10^{8} 10^{12} 10^{16}
POSSIBLE THERMAL HISTORY OF THE UNIVERSE
TEMPERATURE (units of °K)
10^{12}
10^{8}
10^{4}
1
Temperature
DENSITY (units of gm/cm^3)
10^{10}
1
10^{-10}
10^{-20}
10^{-30}
ρ_m
ρ_r
10^{-10} 10^{-8} 10^{-6} 10^{-4} 10^{-2} 1
RADIUS OF THE UNIVERSE (arb. units)
Helium Formation Possible
Transition, Radiation to Matter Filled Universe
Now
Electrons Relativistic; Opaque to Neutrinos; Nuclei Decompose
Plasma Recombines
Matter and Radiation in Equilibrium

Fig. 1. - Possible thermal history of the Universe. The figure shows the previous thermal history of the Universe assuming a homogeneous isotropic general-relativity cosmological model (no scalar field) with present matter density 2×10^{-29} gm/cm^3 and present thermal radiation temperature 3.5° K. The bottom horizontal scale may be considered simply the proper distance between two chosen fiducial co-moving galaxies (points). The top horizontal scale is the proper world time. The line marked "temperature" refers to the temperature of the thermal radiation. Matter remains in thermal equilibrium with the radiation until the plasma recombines, at the time indicated. Therafter further expansion cools matter not gravitationally bound faster than the radiation. The mass density in radiation is ρ_r. At present ρ_r is substantially below the mass density in matter, ρ_m, but, in the early Universe ρ_r exceeded ρ_m. We have indicated the time when the Universe exhibited a transition from the chracteristics of a radiation-filled model to those of a matter-filled model.

Looking back in time, as the temperature approaches 10^{10}° K the electrons become relativistic, and thermal electron-pair creation sharply increases the matter density. At temperatures somewhat greater than 10^{10}° K these electrons should be so abundant as to assure a thermal neutrino abundance and a thermal neutron-proton abundance ratio. A temperature of this order would be required also to decompose the nuclei from the previous cycle in an oscillating Universe. Notice that the nucleons are non-relativistic here.

The thermal neutrons decay at the right-hand limit of the indicated region of helium formation. There is a left-hand limit on this region because at higher temperatures photodissociation removes the deuterium necessary to form helium. The difficulty with this model is that most of the matter would end up in helium.

While little is reliably known about the possible helium content of the protogalaxy, a reasonable upper bound consistent with present abundance observations is 25 percent helium by mass. With this limit, and assuming that general relativity is valid, then if the present radiation temperature were 3.5 °K, we conclude that the matter density in the universe could not exceed 3×10^{-32} gm/cm^3. (See Peebles 1965 for a detailed development of the factors determining this value.) This is a factor of 20 below the estimated average density from matter in galaxies (Oort 1958), but the estimate probably is not reliable enough to rule out this low density.

CONCLUSIONS

While all the data are not yet in hand we propose to present here the possible conclusions to be drawn if we tentatively assume that the measurements of Penzias and Wilson (1965) do indicate black-body radiation at 3.5 °K. We also assume that the universe can be considered to be isotropic and uniform, and that the present energy density in gravitational radiation is a small part of the whole. Wheeler (1958) has remarked that gravitational radiation could be important.

For the purpose of obtaining definite numerical results we take the present Hubble redshift age to be 10^{10} years.

Assuming the validity of Einstein's field equations, the above discussion and numerical values impose severe restrictions on the cosmological problem. The possible conclusions are conveniently discussed under two headings, the assumption of a universe with either an open or a closed space.

Open universe. - From the present observations we cannot exclude the possiblity that the total density of matter in the universe is substantially below the minimum value 2×10^{-29} gm/cm^3 required for a closed universe. Assuming general relativity is valid, we have concluded from the discussion of the connection between helium production and the present radiation temperature that the present density of material in the universe must be $\lesssim 3 \times 10^{-32}$ gm/cm^3, a factor of 600 smaller than the limit for a closed universe. The thermal-radiation energy density is even smaller, and from the above arguments we expect the same to be true of neutrinos.

Apparently, with the assumption of general relativity and primordial temperature consistent with the present 3.5 °K, we are forced to adopt an open space, with very low density. This rules out the possibility of an oscillating universe. Furthermore, as Einstein (1950) remarked, this result is distinctly non-Machian, in the sense that, with such a low mass density we cannot reasonably assume that the local inertial properties of space are determined by the presence of matter, rather than by some absolute property of space.

Closed universe. - This could be the type of oscillating universe visualized in the introductory remarks, or it could be a universe expanding from a singular state. In the framework of the present discussion the required mass density in excess of 2×10^{-29} gm/cm^3 could not be due to thermal radiation, or to neutrinos, and it must be presumed that it is due to ordinary matter, perhaps intergalactic gas uniformity distributed or else in large clouds (small protogalaxies) that have not yet generated stars (see Fig. 1).

With this large matter content, the limit placed on the radiation temperature by the low helium content of the solar system is very severe. The present black-body temperature would be expected to exceed 30 °K (Peebles 1965). One way that we have found reasonably capable of decreasing this lower bound to 3.5 °K is to introduce a zero-mass scalar field into the cosmology. It is convenient to do this without invalidating the Einstein field equation, and the form of the theory for which the scalar interaction appears as an ordinary matter interaction (Dicke 1962) has been employed. The cosmological equation (Brans and Dicke 1961) was originally integrated for a cold universe only, but a recent investigation of the solutions for a hot universe indicates that with the scalar field the universe would have expanded through the temperature range $T \sim 10^9$ °K so fast that essentially no helium would have been formed. The reason for this is that the static part of the scalar field contributes a pressure just equal to the scalar-field energy density. By contrast, the pressure due to coherent electromagnetic radiation or to relativistic particles is one third of the energy density. Thus, if we traced back to a highly contracted universe, we would find that the scalar-field energy density exceeded all other contributions, and that this fast increasing scalar-field energy caused the universe to expand through the highly contracted phase much more rapidly than would be the case if the scalar field vanished. The essential element is that the pressure approaches the energy density, rather than one third of the energy density. Any other interaction which would cause this, such as the model given by Zel'dovich (1962), would also prevent appreciable helium production in the highly contracted universe.

Returning to the problem stated in the first paragraph, we conclude that it is possible to save baryon conservation in a reasonable way if the universe is closed and oscillating. To avoid a catastrophic helium production, either the present matter density should be $< 3 \times 10^{-32}$ gm/cm^3, or there should exist some form of energy content with very high pressure, such as the zero-mass scalar, capable of speeding the universe through the period of helium formation. To have a closed space, an energy density of 2×10^{-29} gm/cm^3 is needed. Without a zero-mass scalar, or some other "hard" interaction, the energy could not be in the form of ordinary matter and may be presumed to be gravitational radiation (Wheeler 1958).

One other possibility for closing the universe, with matter providing the energy content of the universe, is the assumption that the universe contains a net electron-type neutrino abundance (in excess of antineutrinos) greatly larger than the nucleon abundance. In this case, if the neutrino abundance were so great that these neutrinos are degenerate, the degeneracy would have forced a negligible equilibrium neutron abundance in the early, highly contracted universe, thus removing the possibility of nuclear reactions leading to helium formation. However, the required ratio of lepton to baryon number must be $> 10^9$.

We deeply appreciate the helpfulness of Drs. Penzias and Wilson of the Bell Telephone Laboratories, Crawford Hill, Holmdel, New Jersey, in discussing with us the result of their measurements and in showing us their receiving system. We are also grateful for several helpful suggestions of Professor J. A. Wheeler.

REFERENCES

Alpher, R. A., Bethe, H. A., and Gamow, G. 1948, Phys. Rev., **73**, 803.
Alpher, R. A., Follin, J. W., and Herman, R. C. 1953, Phys. Rev., **92**, 1347.
Bondi, H., and Gold, T. 1948, M.N.R.A.S., **108**, 252.
Brans, C., and Dicke, R. H. 1961, Phys. Rev., **124**, 925.
Dicke, R. H. 1962, Phys. Rev., **125**, 2163.
Dicke, R. H., Beringer, R., Kyhl, R. L., and Vane, A. B. 1946, Phys. Rev., **70**, 340.
Einstein, A., 1950, The Meaning of Relativity (3d ed.; Princeton, N.J.: Princeton University Press), p. 107.
Hoyle, F. 1948, M.N.R.A.S., **108**, 372.
Hoyle, F., and Tayler, R. J. 1964, Nature, **203**, 1108.
Liftshitz, E. M., and Khalatnikov, I.M. 1963, Adv. in Phys., **12**, 185.
Oort, J. H., 1958, La Structure et l'évolution de l'universe (11th Solvay Conf. [Brussels: Editions Stoops]), p. 163.
Peebles, P. J. E. 1965, Phys. Rev. (in press).
Penzias, A. A., and Wilson, R. W. 1965, private communication.
Wheeler, J. A., 1958, La Structure et l'évolution de l'universe (11th Solvay Conf. [Brussels: Editions Stoops]), p. 112.
Wheeler, J. A., 1964, in Relativity, Groups, and Topology, ed. C. DeWitt and B. DeWitt (New York: Gordon & Breach).
Zel'dovich, Ya. B. 1962, Soviet Phys. - J.E.T.P., **14**, 1143.

A MEASUREMENT OF EXCESS ANTENNA TEMPERATURE AT 4080 Mc/s

A. A. Penzias and R. W. Wilson
Bell Telephone Laboratories, Inc.
Holmdel, New Jersey

Received May 13, 1965

Measurements of the effective zenith noise temperature of the 20-foot horn-reflector antenna (Crawford, Hogg, and Hunt 1961) at the Crawford Hill Laboratory, Holmdel, New Jersey, at 4080 Mc/s have yielded a value about 3.5° K higher than expected. This excess temperature is, within the limits of our observations, isotropic, unpolarized, and free from seasonal variations (July, 1964 - April, 1965). A possible explanation for the observed excess noise temperature is the one given by Dicke, Peebles, Roll, and Wilkinson (1965) in a companion letter in this issue.

The total antenna temperature measured at the zenith is 6.7° K of which 2.3° K is due to atmospheric absorption. The calculated contribution due to ohmic losses in the antenna and back-lobe response is 0.9° K.

The radiometer used in this investigation has been described elsewhere (Penzias and Wilson 1965). It employs a traveling-wave maser, a low-loss (0.027-db) comparison switch, and a liquid helium-cooled reference termination (Penzias 1965). Measurements were made by switching manually between the antenna input and the reference termination. The antenna, reference termination, and radiometer were well matched so that a round-trip return loss of more than 55 db existed throughout the measurement; thus errors in the measurement of the effective temperature due to impedance mismatch can be neglected. The estimated error in the measured value of the total antenna temperature is 0.3° K and comes largely from uncertainty in the absolute calibration of the reference termination.

The contribution to the antenna temperature due to atmospheric absorption was obtained by recording the variation in antenna temperature with elevation angle and employing the secant law. The result, 2.3° ± 0.3° K, is in good agreement with published values (Hogg 1959; DeGrasse, Hogg, Ohm, and Scovil 1959; Ohm 1961).

The contribution to the antenna temperature from ohmic losses is computed to be 0.8° ± 0.4° K. In this calculation we have divided the antenna into three parts: (1) two non-uniform tapers approximately 1 m in total length which transform between the 2 1/8-inch round output waveguide and the 6-inch-square antenna throat opening; (2) a double-choke rotary joint located between these two tapers; (3) the antenna itself. Care was taken to clean and align joints between these parts so that they would not significantly increase the loss in the structure. Appropriate tests were made for leakage and loss in the rotary joint with negative results.

The possibility of losses in the antenna horn due to imperfections in its seams was eliminated by means of a taping test. Taping all the seams in the section near the throat and most of the others with aluminum tape caused no observable change in antenna temperature.

The backlobe response to ground radiation is taken to be less than 0.1° K for two reasons: (1) Measurements of the response of the antenna to a small transmitter located on the ground in its vicinity indicate that the average back-lobe level is more than 30 db below isotropic response. The horn-reflector antenna was pointed to the zenith for these measurements, and complete rotations in azimuth were made with the transmitter in each of ten locations using horizontal and vertical transmitted polarization from each position. (2) Measurements on smaller horn-reflector antennas at these laboratories, using pulsed measuring sets on flat antenna ranges, have consistently shown a back-lobe level of 30 db below isotropic response. Our larger antenna would be expected to have an even lower back-lobe level.

From a combination of the above, we compute the remaining unaccounted-for antenna temperature to be 3.5° ± 1.0° K at 4080 Mc/s. In connection with this result it should be noted that DeGrasse *et al.* (1959) and Ohm (1961) give total system temperatures at 5650 Mc/s and 2390 Mc/s, respectively. From these it is possible to infer upper limits to the background temperatures at these frequencies. These limits are, in both cases, of the same general magnitude as our value.

We are grateful to R. H. Dicke and his associates for fruitful discussions of their results prior to publication. We also wish to acknowledge with thanks the useful comments and advice of A. B. Crawford, D. C. Hogg, and E. A. Ohm in connection with the problems associated with this measurement.

Note added in proof. - The highest frequency at which the background temperature of the sky had been measured previously was 404 Mc/s (Pauliny-Toth and Shakeshaft 1962), where a minimum temperature at 16° K was observed. Combining this value with our result, we find that the average spectrum of the background radiation over this frequency range can be no steeper than $\lambda^{0.7}$. This clearly eliminates the possibility that the radiation we observe is due to radio sources of types known to exist, since in this event, the spectrum would have to be much steeper.

REFERENCES

Crawford, A. B., Hogg, D. C., and Hunt, L. E. 1961, Bell System Tech. J., **40**, 1095.

DeGrasse, R.W., Hogg, D. C., Ohm, E. A., and Scovil, H. E. D. 1959, "Ultra-low Noise Receiving System for Satellite or Space Communication," Proceedings of the National Electronics Conference, **15**, 370.

Dicke, R. H., Peebles, P. J. E., Roll, P. G., and Wilkinson, D. T. 1965, Ap.J., **142**, 414.

Hogg, D. C. 1959, J. Appl. Phys., **30**, 1417.

Ohm, E. A. 1961, Bell System Tech. J., **40**, 1065.

Pauliny-Toth, I. I. K. and Shakeshaft, J. R. 1962, M.N.R.A.S., **124**, 61.

Penzias, A. A. 1965, Rev. Sci. Instr., **36**, 68.

Penzias, A. A., and Wilson, R. W. 1965, Ap.J. (in press).

MEASUREMENT OF THE SPECTRUM OF THE SUBMILLIMETER COSMIC BACKGROUND[1]

D. P. Woody, J. C. Mather[2], N. S. Nishioka, and P. L. Richards
Department of Physics, University of California
Berkeley, California
and
Inorganic Materials Research Division
Lawrence Berkeley Laboratory
Berkeley, California

Received: February 24, 1975

The spectrum of the night sky has been measured in the frequency range from 3 to 40 cm^{-1} using a fully calibrated liquid-helium-cooled balloon-borne spectrophotometer at an elevation of 39 km. A model based on the known molecular parameters was used to subtract the atmospheric emission. In the frequency range from 4 to 17 cm^{-1}, the spectrum of the background radiation is that of a blackbody with a temperature of $2.99^{+0.07}_{-0.14}$ K.

Direct microwave measurements (Peebles 1971) have established the spectrum of the cosmic background radiation at seven frequencies from 0.02 to 3 cm^{-1}, and indirect optical measurements (Peebles 1971; Hegyi, Traub, and Carleton 1974) provide points at 3.8 and 7.6 cm^{-1}. Existing direct measurements from balloon (Muehlner and Weiss 1973a; Muehlner and Weiss 1973b) and rocket (Williamson et al. 1973; Houck, Soifer, and Harwit 1972) platforms provide limits on the broad-band energy flux in various bands between 1 and 20 cm^{-1}. All of these measurements are consistent with a blackbody spectrum with a temperature of ≈ 3 K. These measurements do not, however, establish the spectral shape of the background radiation beyond the peak of the blackbody curve at ≈ 6 cm^{-1}. We report in this Letter a direct measurement of this spectrum over the frequency range from 4 to 17 cm^{-1}.

A description of the apparatus used for this measurement has been published elsewhere (Mather, Richards, and Woody 1974; Mather 1974). The radiation was collected by a liquid-helium-cooled conical antenna with an apodizing horn (Muehlner and Weiss 1973a; Muehlner and Weiss 1973b) at the input to minimize diffraction side lobes. The geometrical beam had a full width of 7.6° and the antenna pattern was measured out to

[1]This work was supported in part by the U.S. Energy Research and Development Administration, and in part by the Space Sciences Laboratory, University of California, Berkeley, under National Aeronautics and Space Administration Grant No. NGL05-003-497.

[2]Present address: Goddard Institute for Space Studies, 2880 Broadway, New York, New York 10025.

70° off axis. A liquid-helium-cooled polarizing interferometer (Martin and Puplett 1969) was used as a Fourier spectrometer to measure the spectrum of the collected radiation. The detector was a germanium bolometer illuminated with germanium "immersion optics" (Mather, Richards, and Woody 1974; Mather 1974). The cryostat was vented to the atmosphere and reached a temperature of 1.65 K at float elevation.

The spectral flux responsivity of the apparatus was calibrated both in the laboratory and during the flight. The flight calibration obtained from a movable, ambient-temperature blackbody which filled 17% of the beam is shown in Figure 1(a). This calibration agreed with laboratory calibrations to within a few percent.

The cryostat containing the antenna and the spectrometer was mounted in a gondola with the required telemetry and launched from Palestine, Texas, by the National Center for Atmospheric Research (NCAR) at 2008 CDT on 24 July 1974. The gondola was suspended 0.6 km below the 3.3×10^5-m^3 balloon and was free to rotate about the vertical axis. 4 h of the data were obtained at a float altitude of ~ 39 km.

Figure 1(b) shows the night-sky spectrum measured by observation at a zenith angle of 24° with no window over the optics. The flow of helium boil-off gas from the cryostat was vented through the antenna. This was sufficient to prevent condensation of atmospheric gases into the cooled optics. 23 interferograms with an (unapodized) resolution of 1.4 cm^{-1} were obtained during 69 min of observing, as well as 2 interferograms with a resolution of 0.28 cm^{-1} during 24 min of observing. These data were averaged together, apodized, and Fourier transformed with linear phase correction to obtain the spectrum shown in Figure 1(b). The spectral power can be obtained from it by dividing out the instrumental flux responsivity and adding a correction for the spectrometer temperature. The noise level shown is the rms detector noise for the high-resolution interferograms alone.

These data were analyzed by fitting them with a model which contained four adjustable parameters. The cosmic background radiation was modeled by a blackbody spectrum with adjustable temperature. The model used for the atmospheric emission was based on tabulated line parameters for water, ozone, and oxygen (McClatchey et al. 1973; Gebbie et al. 1969). The spectrum was computed by assuming an isothermal atmosphere; an exponential pressure profile; altitude-dependent, pressure-broadened, Lorentzian line shapes; and a constant mixing ratio for the three gases. The pressure at float altitude varied from 3.2 to 3.4 mbar. The ambient temperature measured during the flight was 215±10 K. The column densities of water, ozone, and oxygen were treated as adjustable parameters. The measured interferogram was fitted by the Fourier transform of the product of the model spectrum and the responsivity of the apparatus. In this way, the experimental resolution was included correctly and problems with unresolved lines were avoided. In addition, the more precise data were automatically weighted more strongly in the fit.

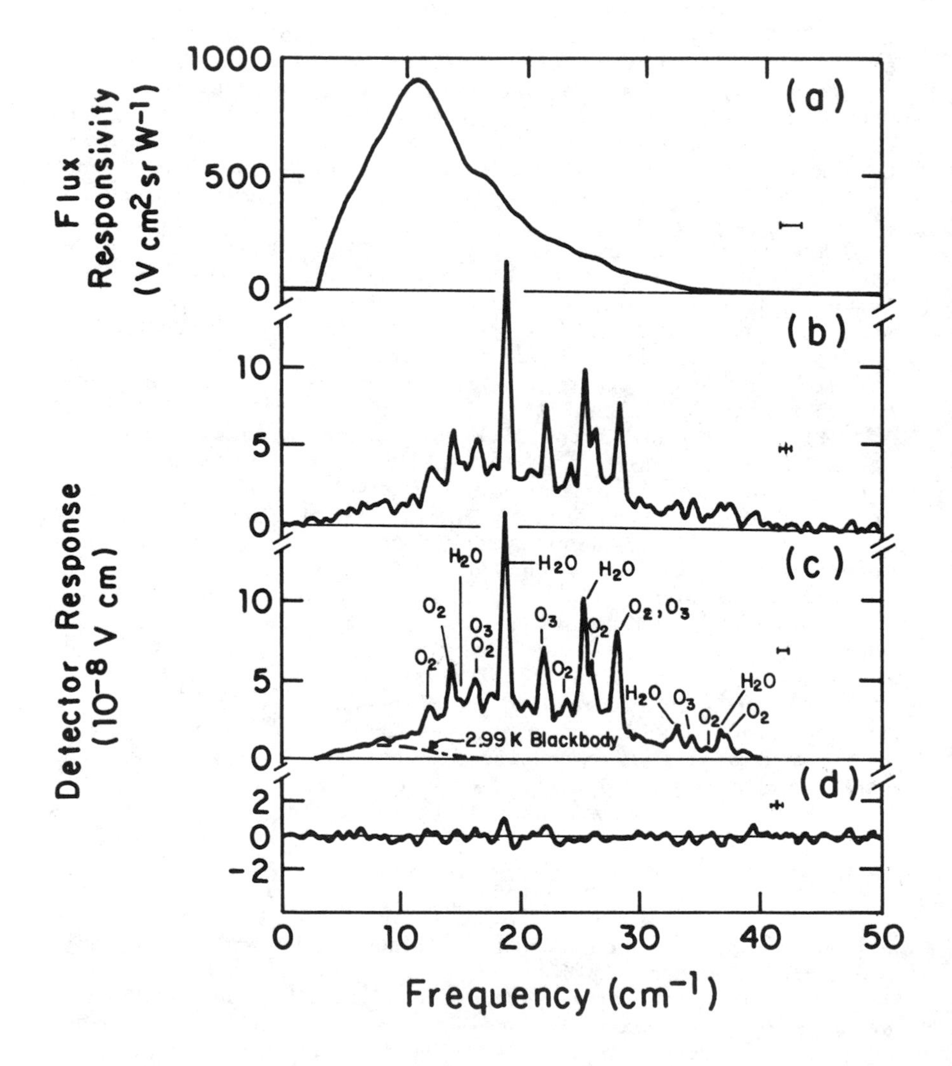

Figure 1. (a) Instrumental flux responsivity as a function of frequency. (b) Observed instrumental response to the night sky. (c) The fitted-model spectrum. The origins of some of the stronger atmospheric emission lines are shown. (d) The difference between the curves of (b) and (c).

The small, but finite, side-lobe response of the antenna meant that some earthshine could have contributed to the measured signal. An attempt was made to measure this earthshine directly by comparing spectra obtained at a zenith angle of 45° with those obtained at 24°. The column densities of atmospheric emitters computed from the 24° spectra were scaled as the secant of the zenith angle and used to subtract the atmospheric contribution from the data obtained at 45°. The residual was assumed to be earthshine and was scaled back to 24° using the measured angular dependence of the antenna pattern (Mather, Richards, and Woody 1974; Mather 1974). The spectrum of the residual thus obtained was less than the noise and no earthshine correction was included in the model.

Information about the radiation emitted by the warm portion of the optical system was obtained by varying its temperature during the flight. The junction between the cone and the horn was heated from 3.4 to 16 K. The emission of the antenna at the lower temperature was estimated from the increase in the observed signal to be substantially less than the detector noise, so no correction was made. No other warm parts of the appartus are expected to contribute significantly to the collected radiation.

Table 1
MODEL PARAMETERS AND ERRORS

	Value with 90% confidence limits	Error in blackbody temperature[a] (K)
Fixed parameters		
Atmospheric temperature	215^{+35}_{-10} K	+0.05,-0.02
Calibration factor[b]	$33.2^{+3.3}_{-3.3}$ K	+0.05,-0.06
Earthshine	$0^{+6}_{-0} \times 10^{-13} \nu^{1/2}$ W/cm^2 sr cm^{-1}	-0.13,+0.00
Fitted parameters		
H_2O vertical column density	$3.92^{+0.20}_{-0.20} \times 10^{17}$ molecules/cm^2	-0.001,+0.001
O_3 vertical column density	$3.50^{+0.18}_{-0.18} \times 10^{17}$ molecules/cm^2	-0.02,+0.02
O_2 vertical column density	$1.67^{+0.17}_{-0.17} \times 10^{22}$ molecules/cm^2	-0.01,+0.01
Blackbody temperature[c]	$2.99^{+0.07}_{-0.14}$ K	

[a] Error induced in fitted blackbody temperature by parameter errors quoted in column 2.
[b] Product of calibrator temperature and filling factor.
[c] Error determined by the rms sum of the detector noise plus the errors shown in column 3.

The calculated spectrum which gave the best fit to the observed data is shown in Fig. 1(c). The spectrum of a blackbody at the best-fit temperature of 2.99 K is shown as a dashed line. The values for the free parameters obtained from the fit are given in Table I. The fitted value for the column density of oxygen agrees with the value 1.54×10^{22} molecules/cm^2 computed from a mixing ratio of 21% and an altitude of 39 km. The values for all three gases are in good agreement with the results of other measurements at the same elevation (Muehlner and Weiss 1973a; Muehlner and Weiss 1973b). To estimate the errors in our determination of the cosmic blackbody temperature the derivatives of the fitted blackbody temperature with respect to the most sensitive fixed and free parameters were calculated. The uncertainty in these parameters and the implied errors in the blackbody temperature are shown in Table I.

Figure 1(d) shows the difference between the observed spectrum in Figure 1(b) and the calculated spectrum in Figure 1(c). The magnitude of the noise can be estimated from the residual above 40 cm^{-1} where there is no optical signal. Since this residual is comparable in regions with and without optical signal it is dominated by random detector noise. No significant deviations between the model and the observed night-sky spectrum are apparent.

The spectrum of the cosmic background radiation is obtained by subtracting the atmospheric contribution from the measured night-sky spectrum. Figure 2 shows the measured spectrum of the cosmic background radiation compared with that of 2.99-K blackbody. Both curves are plotted with a constant fractional resolution of 20%. The 2σ error limits were computed by assuming that the residual in Figure 1(d) was entirely random noise. The dramatic reduction of the noise compared with Figure 1 is due to the large amount of low-resolution data. The measurement establishes that the cosmic background radiation has a thermal spectrum from 4 to ~ 17 cm^{-1}, where the curve has fallen to ≈ 10% of its peak value.[3]

We have plotted our data for the thermodynamic temperature as a function of frequency in Figure 3 along with selected narrow-band results of other experiments. A standard statistical analysis (χ^2) test suggests that all the referenced measurements (plus ours) are consistent with the value

[3]No significant bias is introduced by assuming a blackbody background when fitting the atmospheric parameters. If we assume no background radiation in the fit, or if we assume that it has no sharp features and fit the atmospheric model to the data with large path difference, the estimate of the amount of O_2 is increased and the derived spectra fall below those in Figure 2 (between 12 and 17 cm^{-1}), but within the error limit. The procedure used for Figure 2 is preferable because the rms residual is smaller and the O_2 density agrees with the accurate value known from the pressure and the mixing ratio.

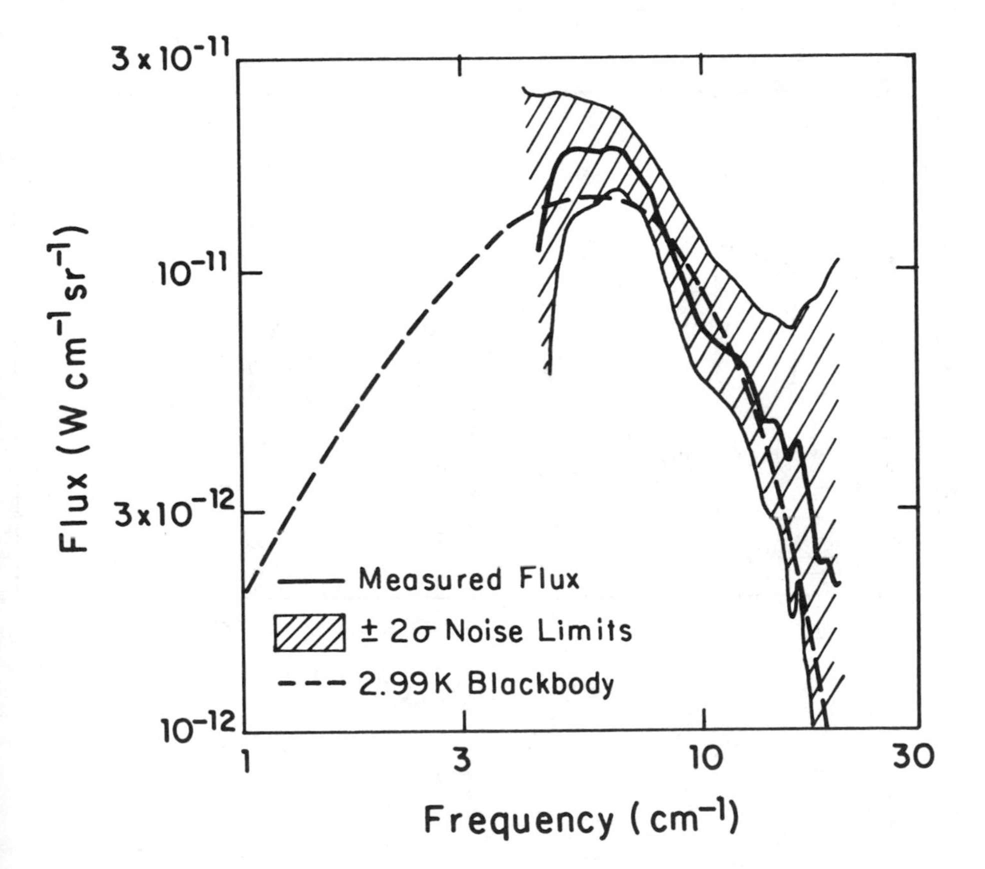

Figure 2. The present measurement of the cosmic background radiation compared with a 2.99-K blackbody curve. Both curves are plotted with a constant fractional resolution of 20%.

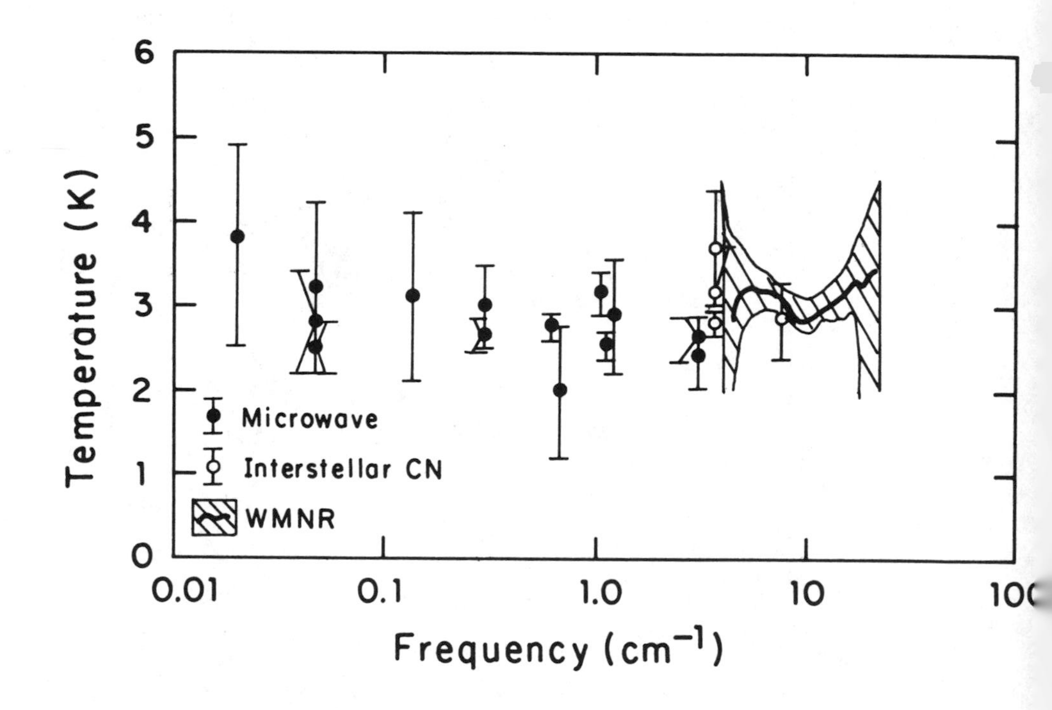

Figure 3. The present measurement (±2σ) of the thermodynamic temperature of the cosmic background radiation compared with the selected results of other experiments. The data for frequencies ≤ 3 cm^{-1} were obtained using ground-based microwave radiometers (Peebles 1971). The data at 3.8 and 7.6 cm^{-1} were obtained from optical measurements of cyanogen (Hegyi, Traub, and Carleton 1974; Thaddeus 1972).

2.90 ± 0.08 K (2σ). The results from our experiment alone are $2.99^{+0.07}_{-0.14}$ K (90% confidence).

The authors are greatly indebted to many persons for assistance with this experiment. Professor C.S. Bowyer suggested the project and provided a balloon for the first flight. Professor K. A. Anderson provided the gondola and a nearly ideal array of telemetry equipment. Mr. J. H. Primbsch gave invaluable assistance in all areas in the art of ballooning. Mr. B. W. Andrews helped with design calculations, and the NCAR staff at Palestine, Texas, provided us with two successful balloon flights.

REFERENCES

Gebbie, H. A. et al. 1969, Proc. Roy. Soc., Ser. A, **310**, 579.
Hegyi, W. A., Traub, W. A., and Carleton, N. P. 1974, Ap.J., **190**, 543.
Houck, J. R., Soifer, B. T., and Harwit, M. 1972, Ap.J. (Lett.), **178**, L29.
McClatchey, R. A. et al. 1973, Air Force Cambridge Research Laboratory Atmospheric Absorption Line Parameters Compilation, AFCRL-TR-73-0096, Environmental Research Papers No. 434 (unpublished).
Martin, D. H., and Puplett, F. 1969, Infrared Phys., **10**, 105.
Mather, J. C. 1974, Ph.D. Thesis (University of California, Berkeley).
Mather, J. C., Richards, P. L., and Woody, D. P. 1974, IEEE Trans. Microwave Theory Tech., **22**, 1046, 1974.
Muehlner, D. J., and Weiss, R. 1973a, Phys. Rev. D, **7**, 326.
Muehlner, D. J., and Weiss, R. 1973b, Phys. Rev. Lett., **30**, 757.
Peebles, P. J. E. 1971, Physical Cosmology (Princeton: Princeton University Press).
Thaddeus, P. 1972, Ann. Rev. Astron. Astrophys., **10**, 305.
Williamson, K. D. et al. 1973, Nature, **241**, 79.

SECTION III

FORMATION OF THE LIGHT ELEMENTS

SECTION III

FORMATION OF THE LIGHT ELEMENTS

Studies by astronomers and astrophysicists have shown that the major part of the baryonic matter in the present universe, both in the stars and in the interstellar medium, is in the form of hydrogen-1 and helium-4. Small amounts of the other stable isotopes of these elements also occur, as well as heavier elements, but the total mass of all atoms other than H^1 and He^4 is only about 1 percent of the mass of all the atoms in the universe as a whole (Allen 1973). The high abundances of oxygen, iron, silicon and other heavier elements that we observe on earth is the result of complex separation and concentration processes, rather than an indication of the actual abundances.

Throughout the period covered by this book, astrophysicists have been concerned with the question of how the observed elements arose, presumably from a simpler situation that once existed. It was recognized quite early that because of the coulomb barrier to nuclear reactions involving charged particles, high temperatures and densities were needed to produce the transmutations involved in converting one element into another. An obvious source of such high temperatures and densities exists in the deep interiors of stars, and that provides a natural environment for some element formation. In addition, we have seen that in the early stages of the expanding universe, the temperatures and densities everywhere in space were even higher than in the interior of stars now, providing another possible environment for element synthesis.

We mentioned in the introduction to section II that arguments raged in the 1940s and 1950s as to which of these environments was the actual birthplace of the elements. From the present perspective we can say that the dispute was resolved by compromise, in that both stellar interiors and the Big Bang have played important roles in producing the observed distribution of elements. Furthermore, the production of the chemical elements is now recognized as being just two stages of an ongoing evolutionary process, in which the matter content of the universe has undergone numerous changes in the past history of the universe, and may continue to do so into the distant future. Since in this book we are mainly concerned with the universe as a whole, we concentrate here on cosmological production of isotopes of the light elements, hydrogen and helium. The reader will find a detailed discussion of the production of heavier elements in stars in the paper of Burbidge *et al.* (1957).

Early work on the synthesis of light elements was done by Hayashi (1950). However, the most detailed analysis, essentially along lines that are those presently accepted, was carried out by Alpher, Follin and Herman in paper 16. In this paper, they give a qualitatively accurate picture of conditions in the universe when the temperature was between 100 MeV and a few eV. Although there are some numerical differences between their results and what is now believed, the main features of their work have survived. (They did not have the benefit of the V-A theory of weak interactions.) For a more up-to-date treatment of some of the same questions, see Peebles' work in paper 17. Alpher *et al.* begin with assumptions about the conditions that applied when the temperature of the universe was the equivalent of 100 MeV. This occurred at a time some 10^{-4} seconds after the Big Bang. At that time, most of the energy density of the universe was in the form of photons, neutrinos and electron-positron pairs. Nucleons were present in much smaller numbers than these particles, and in spite of their larger rest mass, contributed very little to the overall energy density. This condition remained true until the temperature dropped below a few thousand degrees Kelvin, which occurred some 10^5 years later, well after the period of concern to us here. This early period is referred to as being "radiation dominated," as opposed to the present "matter dominated" era. An integration of the Friedmann equation relating dR/dt to the energy density shows that the radiation dominated era R varies as $t^{1/2}$ power, whereas in the matter dominated era R goes as $t^{2/3}$. In either case, it is then possible to compute R and the temperature T for any time. Some values of this relation, and a few important events that took place are given in table 1.

The light particles that were present at the beginning of the period were all in thermal and chemical equilibrium with one another because the reactions between them took place much more quickly than the expansion rate of the universe. However, as time went on, and the temperature of the universe dropped, several events occurred that destroyed the condition of thermal equilibrium, first for the neutrinos, then for the electrons and positrons. Because of the great sensitivity of neutrino reaction rates to density and temperature, at a temperature of 10 MeV or so neutrino reaction rates became slower than the expansion rate of the universe, given by 1/R dR/dt. After this time, it is a good approximation to treat the neutrinos as if they were a freely expanding Fermi-Dirac gas of massless particles, whose temperature therefore scales as 1/R, just as that of the photons does. Indeed, the main difference in behavior between photons and neutrinos is that the temperature of the photons, which remain in equilibrium longer because of their much more rapid *electromagnetic* interactions, is influenced by what happens to the electron-positron pairs, while that of the neutrinos is not.

As the temperature continued to drop it eventually reached an equivalent of about 100 keV. At this temperature, there are not enough photons with energy greater than twice the electron rest mass to induce significant electron pair creation. Therefore, essentially all the positrons and most of the electrons will annihilate, and not be recreated. Their energy goes into heating up the photons, whose temperature then becomes about 1.4 times that of the neutrinos. So when the temperature of the universe dropped below about 10^9

Table 1

History of the Early Universe

Time (sec.)	Thermal Energy (eV)	R/R_o	Events
2×10^{-4}	10^8	2×10^{-12}	muons annihilate
10^{-2}	10^7	2×10^{-11}	neutrinos decouple
1.1	10^6	2×10^{-10}	
200	10^5	2.6×10^{-9}	$e^+ - e^-$ annihilate He synthesis
10^{13}	1	6×10^{-4}	matter domination begins

This table is adapted from Peebles, paper 17 and from Weinberg (1972). The precise time at which matter domination begins depends on the assumed values for the present nucleon density. Thermal energies can be converted to temperature by the factor 1 eV = 11, 605° K.

degrees, what remained of the abundance of particle types of earlier times were photons and neutrinos in similar large numbers, together with a leavening of electrons and nucleons, in concentrations about 10^9 times smaller than the photons. It is in this environment that the synthesis of light elements took place.

One important aspect of this synthesis, discussed by Hayashi (1950) and in papers 16 and 17 concerns the neutron to proton number ratio. During the time period when the density is high enough, it is possible for neutrons and protons to interconvert through neutrino emission and absorption reactions. At high temperatures of 100 MeV or so the conversion rates are fast enough that neutrons and protons are in equilibrium, and their number ratio just depends on the n-p mass difference. At these temperatures, the neutron-proton ratio is about 1. However, as the temperature drops below an

MeV, the rates drop rapidly, the two species are no longer in equilibrium, and the n-p ratio drops off considerably, as shown in figures 2 and 3 of paper 16 and in table 1 of paper 17. The important conclusion is that this ratio can be _predicted_ on the basis of fundamental physics rather than taken as an unknown initial condition, and then used as the basis of further calculations of synthesis of light nuclei.

It should be mentioned, however, that two tacit assumptions went into the calculations in these papers, which have sometimes been questioned. The result for the neutron-proton ratio depends both on the number of neutrinos present and on the time period over which the interconversion takes place. The former is only determined if one assumes that there is no chemical potential for electron neutrinos, or equivalently, that the numbers of neutrinos and antineutrinos are equal. There is no direct observational evidence to this effect. If the chemical potentials were not equal, the neutrino numbers would not be those used in the calculations described, and the n-p ratio predictions would change. For a discussion of this effect, see Dimopoulos and Feinberg (1979).

The other uncertainty in the calculation has to do with the expansion rate during the period of n-p conversion. In Peebles' paper, this rate is taken as an unknown quantity, which in his paper is called S. Neutron-proton conversion can take place only when the densities are sufficiently high. The density decreases rapidly with time owing to the universal expansion, and the time period during which the density is great enough for conversion is inversely proportional to the expansion rate. This rate in turn depends on how many different massless particles exist in numbers comparable to photons during this period. At present, the species of this type known to us include photons and three different neutrino types. However, it is conceivable that other neutrino types, or other massless particles yet undiscovered in the laboratory, exist and contribute to the energy density and increase the expansion rate. It is clear that this effect goes in the direction of decreasing the time available for n-p conversion and so increasing the number of neutrons available for nucleosynthesis. Since the number of neutrons in the present universe can be estimated from the amount of helium, it is possible to use this argument to limit the number of unknown massless particle species (see Yang _et al._ 1979). An often quoted result is that there cannot be more than one unknown massless species with the same interactions as neutrinos. However, this effect could be compensated for by a suitable neutrino chemical potential, and in the absence of independent knowledge of that quantity, these arguments must be considered inconclusive. The number of massless neutrino species will soon be determined by laboratory measurements of the decay width of the Z° particle, in which case it may be possible to invert the argument, and use the n-p ratio to constrain the neutrino chemical potential.

If no further events occurred, the neutrons would all decay in a few minutes, and none would survive into the pres-

cued from falling at the last minute by the appearance of a new rung to grasp, so were the neutrons saved by the formation of deuterium, and then of helium. As discussed in the introduction to section II, deuterium formation was taking place even when the temperatures were above a few MeV, but at such high temperatures the concentration would be very small. However, when the temperature drops to an energy equivalent of about 100 keV, there are no longer enough high energy photons present to disintegrate deuterium, and any deuterons that now form survive for a while, which ensures that some bound neutrons will exist.

The deuterium itself does not survive to the present in large amounts, because deuterons can combine with the free protons that still exist in large numbers at these temperatures to form tritium and various helium isotopes, which are more stable. Peebles, in paper 17, and Wagoner, Fowley and Hoyle (1967) have solved the rate equations for the formation of the various isotopes. In addition to hydrogen and helium isotopes, small amounts of lithium are also formed. Some results for H and He are given in tables 3 and 4 of Peebles paper. It can be seen there that only a small amount of deuterium survives the competition with the heavier nuclei, and that starting at a temperature of about 10^9 degrees almost all of the neutrons that exist end up bound into He^4. The precise amounts of He^4 and other isotopes that survive depend, among other factors, on the expansion rates, as discussed above, and on the actual density of nucleons. Some attempts have been made to use such calculations to determine the actual nucleon densities in this period, rather than extrapolating these densities backwards from the present universe (see Yang _et al._ 1979). The result is that the nucleon density required to produce the "observed" amounts of deuterium is more or less consistent with the amount that would be inferred from presently visible matter, and is much smaller than the critical density discussed in section I. However, in view of questions concerning the possible production and destruction of small amounts of deuterium in stars, and the uncertainties in estimates of the actual amounts in the present universe, these arguments must be taken as suggestive, rather than conclusive.

While a few open questions remain, it is fair to conclude that the analysis of the formation of isotopes of the light elements has been a major success of the hot Big Bang theory, and furnishes convincing evidence that our ideas about subatomic physics and cosmology can be applied as far back in the history of the universe as 10^{-4} seconds after the Big Bang. In the next section, we will see how this success has emboldened physicists to try to extend these ideas to even earlier times, when the phenomena that took place involved ideas in both disciplines that are much less well understood.

REFERENCES

Allen, C. W. 1973, Astrophysical Quantities (London: Athlone).
Burbidge, E. M., Burbidge, G. R., Fowler, W. A., and Hoyle, F. 1957, Rev. Mod. Phys., **29**, 547.
Dimopoulos, S., and Feinberg, G. 1979, Phys. Rev., **D20**, 1283.
Hayashi, C. 1950, Prog. Theor. Phys., **5**, 224.
Wagoner, R. V., Fowler, W., and Hoyle, F. 1967, Ap.J., **148**, 3.
Weinberg, S. 1972, Gravitation and Cosmology (New York: Wiley).
Yang, J., Schramm, D. N., Steigmann, G., and Rood, R. T. 1979, Ap.J., **227**, 697.

PHYSICAL CONDITIONS IN THE INITIAL STAGES OF THE EXPANDING UNIVERSE[1,2]

Ralph A. Alpher, James W. Follin, Jr., and Robert C. Herman
Applied Physics Laboratory, The Johns Hopkins University, Silver Spring, Maryland

Received: September 10, 1953

The detailed nature of the general nonstatic isotropic cosmological model as derived from general relativity is discussed for early epochs in the case of a medium consisting of elementary particles and radiation which can undergo interconversion. The question of the validity of the description afforded by this model for the very early super hot state is discussed. The present model with matter-radiation interconversion exhibits behavior different from non-interconverting models, principally because of the successive freezing-in or annihilation of various constituent particles as the temperature in the expanding universe decreased with time. The numerical results are unique in that they involve no disposable parameters which would affect the time dependence of pressure, temperature, and density.

The study of the elementary particle reactions leads to the time dependence of the proton-neutron concentration ratio, a quantity required in problems of nucleogenesis. This ratio is found to lie in the range of ~4.5:1 - ~6.0:1 at the onset of nucleogenesis. These results differ from those of Hayashi mainly as a consequence of the use of a cosmological model with matter-radiation interconversion and of relativistic quantum statistics, as well as a different value of the neutron half-life.

I. INTRODUCTION

The nonstatic homogeneous isotropic cosmological model which satisfies the equations of general relativity has received a great deal of attention. However, the detailed nature of the model does not appear to have been examined at the extremely high temperatures and densities characteristic

[1] This work was supported by the U.S. Bureau of Ordinance, Department of the Navy, under Contract NOrd-7386.

[2] Preliminary accounts of this work were presented at the Symposium on the Abundance of the Elements held at Yerkes Observatory, Williams Bay, Wisconsin, November 6-8, 1952, under the joint sponsorship of the National Science Foundation and the University of Chicago, and at the 1953 Washington Meeting of the American Physical Society (Alpher 1953).

of the very early stages of the expanding universe. This question has been examined in the present paper and the dependence of the temperature and density on time has been determined for the case where the radiation density (taken to include photons, neutrinos, electrons, positrons, and mesons) is much greater than the density of matter (nucleons). For initial conditions compatible with present astrophysical observations, one can demonstrate that the radiation density exceeded the density of matter for about the first hundred million years in the expansion.

We have carried our study of this problem back to a temperature of ~100 Mev ($\sim 1.2 \times 10^{12}$ °K), corresponding to an epoch of $\sim 10^{-4}$ sec. For temperatures below this value one can treat reactions among elementary particles with some confidence. Furthermore, below ~100 MeV the energy stored in the gravitational field is a negligible part of the total energy so that the question of using a correct unified field theory, including the quantization of the field equations, can be avoided. Finally, at ~100 MeV one has a state of thermodynamic equilibrium among all the known constituent particles and radiation so that a knowledge of the previous history of the universe is not required. As part of the detailed study of the cosmological model we have examined the reactions among the elementary particles present and followed their course in the universal expansion. As will be seen, all reaction rates, except those involving the neutrino, are sufficiently high to maintain thermodynamic equilibrium. An examination of the kinetics of the reactions between nucleons and neutrinos has yielded the relative concentrations of protons and neutrons as a function of time. The only parameters involved in the cosmological model are the nucleon density and radius of curvature. At the early times prior to element formation, neither of these parameters affects the course of events because of the very high total density and because the nucleon density is negligible compared with the density of radiation. The nucleon density becomes of importance at later times in considering element formation, while the radius of curvature becomes of interest only at times of the order of a hundred million years.

The foregoing detailed considerations of the early stages of the universal expansion bear significantly on the neutron-capture theory of element formation. This theory has been concerned principally with understanding the general trend in the distribution of the cosmic abundances of the chemical elements with atomic weight (Alpher and Herman 1950; 1951; 1953).[3]

[3]Alpher and Herman (1953). The simple-neutron capture theory has satisfactorily reproduced for all but the lightest elements the observed approximately exponential decrease in abundance with increasing atomic weight up to A ~100, as well as the approximate constancy of abundance for the heavier elements. Briefly, in the neutron-capture theory, as thus far

For the lightest elements the processes of neutron capture and β decay, while adequate to explain the formation of the heavier elements, must be supplemented by the thermonuclear reactions involving protons, deuterons, and other light nuclei. The very light element reactions were examined in some detail by Fermi and Turkevich (Alpher and Herman 1950; 1951), using the cosmological model previously employed for the neutron-capture theory, with a finite starting time and a primordial mixture of neutrons and radiation. This improved light-element calculation did not satisfactorily resolve what remains the principle difficulty of the theory, namely, the deduction of the specific nuclear reactions and physical conditions which might carry the formation chain of reactions through and beyond atomic weight 5. To resolve this and other difficulties in the theory, it will apparently be necessary to remove many of the simplifying restrictions. In particular, the assumption of a starting time must be replaced by detailed consideration of element-building reactions increasing in importance from very early times in the universal expansion as the rates of various dissociative processes diminish with decreasing temperature. Moreover, one should include all possible reactions among the elementary particles, since these reactions, which are important at very high temperatures may influence the physical conditions that control the element-building processes.

developed, the various nuclear species were supposed to have been formed from nucleons by the successive radiative capture of fast neutrons with adjustment of nuclear charge by intervening β decay during the early stages of the expansion of the universe. The primordial material or ylem was taken to be a mixture of neutrons and radiation. As the universal expansion proceeded the neutrons underwent free decay, so that by the time the universal temperature had decreased to a value where nuclei would be thermally stable, an appreciable number of protons had been generated. Then the capture of neutrons by protons provided the first step in the formation of the successively heavier elements. More specifically, the temperature for the beginning of building-up reactions was taken to be ~0.1 MeV (corresponding to a specific starting time for element building in the cosmological model used, in which $T = 1.52\times10^{10}\ t^{-1/2}$ °K). This choice was dictated by the magnitude of the binding energy of the deuteron on the one hand and by the lack of evidence in the abundance data for any resonance neutron capture on the other hand. At the starting time, neutron decay had led to a proton-neutron ratio of ~1:7.

One of the appoximations involved thus far in calculations with the neutron-capture theory has been the smoothing of available data on fast neutron radiative capture cross sections as a function of atomic weight. Moreover, reactions other than radiative neutron capture among the very lightest elements have been ignored.

The elementary-particle reactions determine the ratio of the relative concentrations of protons and neutrons, a quantity which plays a vital role in predicting the general trend of abundances according to the neutron-capture theory. As has already been mentioned, in previous calculations the proton-neutron abundance ratio has been taken to be that resulting from free decay of the primordial neutrons during the period from the start of the expansion up to the starting time selected for element-building reactions. A more detailed calculation was made by Hayashi (195), who determined the value of the proton-neutron ratio resulting from spontaneous and induced β processes among protons and neutrons in the presence of electron pairs and neutrinos in the early stages of the expansion. Whereas on the basis of the crude assumption of neutron decay only, one obtains a proton-neutron ratio of ~1:7, Hayashi's calculation gave ~4:1 by the starting time for element-building reactions. With this latter value of the ratio, it has not yet proven possible to represent the cosmic abundance distribution in atomic weight on the basis of the simple neutron-capture theory (Alpher and Herman 1953; see footnote 3), a theory which contains only one arbitrary parameter, viz., the density of matter at the start of the element-building epoch, and which involves only neutron-capture reactions. In part because of this difficulty and because it seemed worth while to investigate the effect of certain modifications on Hayashi's calculation of the final value of the proton-neutron ratio, the work described in the remainder of this paper was carried out. Among the changes involved in the present study are the use of relativistic quantum statistics instead of Boltzmann statistics, a modified cosmological model for early epochs as required by the interconversion of matter and radiation, which as we have already indicated is of considerable interest for its own sake, and the use of the value of the neutron half-life recently reported by Robson (1951) which differs materially from the older value employed by Hayashi.

It seems most likely that element synthesis is intimately connected with questions of cosmology. In the present work we consider the sequence of events up to the time when the rate of element formation became significant. As we shall see later in detail, all the constituents remained in thermodynamic equilibrium as the universe expanded and cooled to a temperature of ~10 MeV. At ~10 MeV the neutrinos were essentially frozen out of the equilibrium. By ~0.3 MeV the proton-neutron ratio was almost entirely determined by the free decay of the neutron. It remains for future study to re-examine the formation of the elements by thermo-nuclear reactions as a subsequent part of the picture developed here. A detailed chronology is given in a later portion of this paper [see Section V].

II. THE COSMOLOGICAL MODEL

The theory of element formation by non-equilibrium thermonuclear reactions has been developed as an integral part of the very early stages of the expanding universe. Detailed calculations of the necessary rate processes require a knowledge of the temporal behavior of temperature, density, and rate of expansion during these early epochs. The cosmological model that has been used previously for this purpose is the most general nonstatic model satisfying the requirements of general relativity, exhibiting homogeneity and isotropy, and which is composed of a perfect fluid with no interconversion of matter and radiation (Alpher and Herman 1950; 1951; 1953; also see footnote 3). The rate of expansion and, implicitly, the rate of change of temperature in the expansion for this model, with no restrictions on the composition of the perfect working fluid are given in relativistic units by the following differential equations (Tolman 1934):

$$- \frac{e^{-g}}{R_0^2} - \frac{d^2g}{dt^2} - \frac{3}{4}\left(\frac{dg}{dt}\right)^2 + \Lambda = 8\pi p_0, \tag{1a}$$

$$\frac{3e^{-g}}{R_0^2} + \frac{3}{4}\left(\frac{dg}{dt}\right)^2 - \Lambda = 8\pi\rho_{00}, \tag{1b}$$

and

$$e^{\frac{1}{2}g(t)} = \ell/\ell_0 = R/R_0, \tag{1c}$$

where ℓ and R are proper distance and radius of curvature, respectively, given in units of ℓ_0 and R_0, Λ is the cosmological constant, and p_0 and ρ_{00} are proper pressure and density. The quantities p_0 and ρ_{00} are functions of temperature and of ℓ, and hence implicitly of time. Equation (1b) may also be rewritten, by using Eq. (1c), in the following form:

$$\frac{d\ell}{dt} = + \left(\frac{8\pi}{3}\rho_{00}\ell^2 - \frac{\ell_0^2}{R_0^2} + \frac{\Lambda\ell^2}{3}\right)^{1/2}, \tag{2}$$

with the plus sign taken to indicate expansion. We have taken $\Lambda = 0$ in keeping with current practice[4] As can be easily

[4] A very small value of Λ may be used to adjust the present age of this model, although it is of no consequence during the early epochs of interest here (Gamow 1949). In this connection it may be of interest to note that while Eqs. (1) and (2) contain a density singularity at zero time, they

shown, the constant term $\ell_0{}^2/R_0{}^2$ in Eq. (2) may be neglected in the application of this model to early epochs. This is equivalent to neglecting $\ell^2 e^{-g}/R_0{}^2$ in Eq. (1). If $\rho_{00} \propto \ell^{-n}$ where $n > 2$, then for sufficiently early times ℓ will be so small that one has $8\pi\rho_{00}\ell^2/3 \propto 8\rho\ell^{2-n}/3 \gg \ell_0{}^2/R_0{}^2$. Hence, for early epochs in the expansion one may replace Eq. (2) by

$$\frac{d\ell}{dt} = \frac{\ell}{2}\frac{dg}{dt} = +\left(\frac{8\pi}{3}\rho_{00}\ell^2\right)^{1/2} . \tag{3}$$

As has already been mentioned, the cosmological model, which is discussed in this paper, taken together with the presently observed smoothed-out matter density in the universe as well as the estimated age, are consistent with the supposition that during the early epochs of interest the matter density was much smaller than the radiation density (i.e., $\sim 1{:}10^6$). The neutron-capture theory of element formation (Alpher and Herman 1950; 1951) requires that the radiation density greatly exceeded the density of matter during the early epochs of the universal expansion. In this previous work it was not necessary to consider the interconversion of matter and radiation since for the epochs con-

also implicitly contain the conclusion that the duration or age of the expansion from this singularity is finite. Taking $\Lambda = 0$ and neglecting terms containing $1/R_0{}^2$ for early epochs, one can show that this age is given in cgs units by the following integral:

$$a = \int_0^a dt = \left(\frac{3}{8\pi G}\right)^{1/2} \int_{\rho(a)}^{\infty} \frac{c^2 d\rho}{\rho^{1/2}(p+\rho c^2)} ,$$

where p and ρ are total pressure and density. Since the pressure is positive,

$$a \leqslant \left(\frac{3}{8\pi G}\right)^{1/2} \int_{\rho(a)}^{\infty} \frac{d\rho}{\rho^{1/2}} = \left(\frac{3}{2\pi G\rho}\right)^{1/2} .$$

A lower bound on the duration may be obtained by noting that for a relativistic fluid $0 \leqslant p \leqslant \rho c^2/3$. Hence

$$a \leqslant \left(\frac{3}{8\pi G}\right)^{1/2} \int_{\rho(a)}^{\infty} \frac{3d\rho}{4\rho^{1/2}} = \left(\frac{8}{3\pi G\rho}\right)^{1/2} ,$$

so that a is bounded.

sidered the temperature was already below that required to maintain a significant electron-pair density. Hence, the working fluid for the cosmological model was taken as black-body radiation, containing a trace of matter, and expanding adiabatically according to $T \propto 1/\ell$. It has been shown (Alpher and Herman 1950; 1951) that for early epochs Eq. (3) leads to the following expressions for the radiation density, ρ_γ, the matter density, ρ_m, the total density, ρ_{total}, the temperature, T, and proper distance, ℓ, with $\rho_\gamma \gg \rho_m$:

$$\rho_{total} \cong \rho_\gamma \cong [3/(32\pi G)]t^{-2} = 4.48\times 10^5 t^{-2} \text{ g/cm}^3 , \quad (4)$$

$$\rho_m = \rho_0 \, t^{-\frac{3}{2}} \text{ g/cm}^3 , \quad (5)$$

$$T = (c^2 \rho_\gamma / a_\gamma)^{1/4} = 1.52\times 10^{10} t^{-1/2} \text{ °K} \quad (6)$$

and

$$\ell = (32\pi G \rho_{\gamma''} \ell_0{}^4/3)^{1/4} t^{1/2} , \quad (7)$$

where G is the gravitational constant, c is the velocity of light, a_λ is the Stefan-Boltzmann constant, t is the time in seconds from the "start" of the expansion, $\rho_{\gamma''}$ is the density of radiation when $\ell = \ell_0$, and ρ_0 is a constant. As will be seen, the above equations are still valid in the case discussed in the present paper providing that $t \gtrsim 100$ sec, ρ_λ is eliminated from Eq. (4), the constant 1.52 in Eq. (6) becomes 1.45 for Majorana neutrinos and 1.38 for Dirac neutrinos, and $\rho_{\gamma''}$ in Eq. (7) is replaced with the value of ρ_{total} when $\ell = \ell_0$. The quantity ρ_0 is the one arbitrary parameter in the simple neutron-capture theory. It has been adjusted in previous calculations (Alpher and Herman 1953; see footnote 3) so that the density of matter at the start of the element-forming processes would lead to the observed cosmic abundance distribution. The cosmological model at early epochs described by Eqs. (4) and (6) was adopted by Hayashi as a basis for his calculation of the proton-neutron ratio.

While we have assumed a homogeneous and isotropic model of the universe in agreement with present astronomical evidence, it should be pointed out that this restriction is not necessary in the present considerations. Homogeneity is required only over a region of radius equal to ct since nothing further away can affect the cosmology or the elementary particle reactions to be discussed. At the universal age of ~600 seconds corresponding to the end of the period of this study, the nuclear mass enclosed in the sphere of influence is ~10^{34} g, that is, ~5 solar masses, and is much less at earlier epochs. Another way of looking at this result is that lengths greater than ct, in particular R_0 and any gradient of R_0, must be negligible because of the finite velocity of propagation of disturbances.

As already mentioned, the cosmological model outlined above was a sufficient approximation in previous calculations (Alpher and Herman 1950; 1951) because the temperature taken for the start of element formation was well below the electron rest mass equivalent and, therefore, reactions among elementary particles and photons could be ignored. One has only to consider that the nucleons and nuclei formed remained in thermal equilibrium with the expanding radiation field. These previous calculations, which were based on the time scale described by Eq. (6), continue to be valid provided ρ_0 is adjusted as required to fit the time scale to be described in this paper. The adjustment required is insignificant.

The study of the induced and inverse β processes involving neutrons and protons prior to any appreciable element formation concerns much earlier epochs and therefore much higher temperatures. In this case one must consider positrons, electrons, neutrinos, antineutrinos (if distinguishable from neutrinos), and radiation. The equation of state for radiation only, implicit in Eqs. (4)-(6), may no longer be an adequate approximation. We shall suppose that this mixture of elementary particles and photons is at sufficiently high temperature for equilibrium to be maintained, but we shall not require temperatures so high as to require nucleon pairs. Furthermore, the nucleons present are assumed to have a negligible effect on pressure, density, and temperature, since even for temperatures as low as ~0.1 MeV the nucleon density is many orders of magnitude less than the radiation density.

The density and pressure of the constituents of the medium may be obtained from the Fermi-Dirac and Bose-Einstein distribution laws for the number of particles in the energy range dE at E, viz.,

$$N(E)dE = \frac{4\pi}{h^3} \Sigma' |\mathbf{p}| E[\exp(E/kT) \pm 1]^{-1} dE \ , \qquad (8)$$

where $|\mathbf{p}|$ is the momentum, and the summation, Σ', is over charge and spin states. In the present calculation the number of particles and photons is not conserved so that a degeneracy parameter is not required. The density and pressure, according to Eq. (8), are given by

$$\rho(T) = \frac{4\pi}{c^2h^3} \Sigma' \int_{mc^2}^{\infty} |\mathbf{p}| E^2 [\exp(E/kT) \pm 1]^{-1} dE \ , \qquad (9a)$$

and

$$p(T) = \frac{4\pi}{3h^3} \Sigma' \int_{mc^2}^{\infty} |\mathbf{p}|^3 [\exp(E/kT) \pm 1]^{-1} \, dE \ , \qquad (9b)$$

where

$$|\mathbf{p}| = (1/c)\ (E^2 - m^2c^4)^{1/2}\ . \tag{9c}$$

In particular, for electrons and positrons one obtains, with the transformation $E = m_e c^2 \cosh\theta$, the following:

$$\rho_{e^-} = \rho_{e^+} = \frac{a_e}{c^2}\int_0^\infty \frac{\sinh^2\theta\ \cosh^2\theta\, d\theta}{1 + \exp(x\cosh\theta)}\ \mathrm{g/cm^3}\ , \tag{10a}$$

$$p_{e^-} = p_{e^+} = \frac{a_e}{3}\int_0^\infty \frac{\sinh^4\theta\, d\theta}{1 + \exp(x\cosh\theta)}\ \mathrm{dynes/cm^2}\ . \tag{10b}$$

In these equations

$$a_e = 8\pi m_e{}^4 c^5/h^3\ ; \tag{11a}$$

m_e and h are the electron rest mass and Planck's constant, respectively;

$$x = m_e c^2/(kT) \tag{11b}$$

defines temperature in units of the electron rest mass, and k is Boltzmann's constant. Spin states have been counted in the above expressions, and the total electron energy includes rest mass.

In order to carry out numerical calculations, it should be noted that the definite integrals in the expressions for $p_e(x)$ and $\rho_e(x)$ can be expanded in series of modified Bessel functions $K_i(nx)$. One can write

$$f_0 = \int_0^\infty [1 + \exp(x\cosh\theta)]^{-1} d\theta = \sum_{n=1}^{\infty} (-1)^{n+1}\ K_0(nx), \tag{12}$$

$$\begin{aligned} f_1 &= \int_0^\infty \sinh^2\theta\,[1 + \exp(x\cosh\theta)]^{-1} d\theta \\ &= x^{-1} \sum_{n=1}^{\infty} (-1)^{n+1} n^{-1} K_1(nx)\ , \end{aligned} \tag{13}$$

and

$$f_2 = \int_0^\infty \sinh^4\theta [1 + \exp(x \cosh\theta)]^{-1} d\theta$$

$$= 3x^{-2} \sum_{n=1}^{\infty} (-1)^{n+1} n^{-2} K_2(nx) \,, \tag{14}$$

so that

$$\rho_e(x) = \rho_{e^-} + \rho_{e^+} = (2a_e/c^2)(f_1 + f_2) \,, \tag{15}$$

and

$$p_e(x) = p_{e^-} + p_{e^+} = (2a_e/3)\, f_2 \,. \tag{16}$$

In the high temperature limit, $kT \gg mc^2$, which is equivalent to setting $m = 0$ in Eqs. (9), the density and pressure for all Bose-Einstein particles approach those for photons except for factors which depend on spin and charge states. Similarly, for Fermi-Dirac particles the density and pressure approach those for neutrinos, again except for a factor which accounts for differing charge and spin states.

For radiation, taking into account the two states of polarization, one obtains the following from the Bose-Einstein integral:

$$\rho_\gamma = \frac{a_\gamma}{c^2} T^4 = \left(\frac{\pi^4 a_e}{15c^2}\right) x^{-4} \text{ g/cm}^3 \,, \tag{17}$$

and

$$p_\gamma = \frac{1}{3} \rho_\gamma c^2 \text{ dynes/cm}^2 \,, \tag{18}$$

where

$$a_\gamma = 8\pi^5 k^4 / 15 c^3 h^3 \,. \tag{18a}$$

For neutrinos we consider two cases,[5] namely, neutrinos and antineutrinos indistinguishable ($\nu \equiv \nu^*$) and distinguishable

[5] Recently, theoretical arguments in factor of distinguishability, i.e., against the Majorana theory of neutral particles, have been given by E. R. Caianiello (1952). However, we consider both cases throughout this paper because it does not appear to be a settled question at this time. (See also Wu [1952]).

($\nu \not\equiv \nu^*$). For the temperature range in which the neutrinos are in thermal equilibrium with the other constituents of the medium, the Fermi-Dirac integral gives

for $\nu \equiv \nu^*$:

$$\rho_\nu = \frac{7}{8} \rho_\gamma \ , \tag{19}$$

and

$$p_\nu = \frac{7}{8} p_\gamma \ ; \tag{20}$$

for $\nu \not\equiv \nu^*$:

$$\rho_\nu = \rho_\nu{}^* = \frac{7}{8} \rho_\gamma \ , \tag{21}$$

and

$$p_\nu = p_\nu{}^* = \frac{7}{8} p_\gamma \ , \tag{22}$$

so that the neutrino pressure and density in the latter case are twice those in the former. It should be noted that the results stated in Eqs. (19)-(22) are predicated on the assumption that no type of particle is degenerate in the present problem. The simple expressions for the neutrino density and pressure given in Eqs. (19)-(22) hold for all Fermi-Dirac particles in the limit of sufficiently high temperature, i.e., there is a contribution to the density of $(7/16)\rho_\gamma$ for each degree of freedom. Similarly for Bose-Einstein particles there is a contribution to the density of $(1/2)\rho_\gamma$ for each degree of freedom.

It can be shown from Eq. (9) that for Fermi-Dirac particle of mass, m_i,

$$\rho_i(x) \propto \left(\frac{m_i}{m_e}\right)^4 \rho_e[(m_i/m_e)x] \ ,$$

with the proportionality factor depending on the previously mentioned spin and charge states. Thus all Fermi-Dirac particles exhibit the same behavior provided that an appropriate shift is made in the temperature scale. A similar result can be obtained for Bose-Einstein particles. The qualitative behavior of ρ_i versus T after suitable normalization of the temperature scales is essentially the same for fermions and bosons.

The neutrino contribution given by Eqs. (19)-(22) to the total pressure and density requires modification for the temperature range of interest in calculating the proton-neutron ratio as a function of the time. At very high temperatures the neutrino component maintains itself in equilibrium with the other constituents of the medium through interaction with mesons. When the medium has expanded and cooled somewhat below a temperature equivalent to the rest mass of the lightest meson, the neutrinos freeze in and continue to expand and cool adiabatically as would a pure radiation gas. After this freeze-in the neutrino temperature will differ from that of the other components of the medium. It will be seen that the freeze-in must have occurred at a temperature higher than is required for neutrons and protons to be nearly in thermodynamic equilibrium. For the temperature region of interest, then, we must deal with nucleons, electrons, positrons, and radiation at one temperature and neutrinos at another temperature. The calculation of the neutron-proton ratio does not require that a specific freeze-in temperature be given, but only that neutrinos be frozen in before an appreciable fraction of the electron pairs start to annihilate.

It is of some interest to examine in more detail the freezing in of neutrinos during the period from ~15 to ~5 MeV. Non-equilibrium reactions involving neutrinos become important only below ~5 MeV. When the temperature was well above the rest mass equivalent of mesons, the neutrinos maintained equilibrium through interaction with mesons. At such temperatures the contribution of mesons to the density was 3.25 ρ_γ, while the total contribution due to photons, electrons, positrons, and neutrinos was 3.625 ρ_γ or 4.50 ρ_γ, for $\nu \equiv \nu^*$ and $\nu \not\equiv \nu^*$, respectively.[6] Since the meson rest energy is distributed uniformly among the lighter particles when the mesons annihilate, it is clear that the number of neutrinos will about double when meson annihilation occurs. Now the bulk of mesons will annihilate when the temperature in the universal expansion has dropped significantly below that equivalent to $m_\mu c^2$ (~108 MeV) or $m_\pi c^2$ (~138 MeV), or down to 10 MeV. At 10 MeV the Boltzmann factors for μ and π mesons are $\sim 2\times 10^{-5}$ and $\sim 10^{-6}$, respectively. This temperature decrease, as will be seen later when the time scale for the cosmological model is calculated, requires a duration of $\sim 10^{-2}$ sec in the universal expansion.

[6]As has been shown, in the high temperature limit the Fermi-Dirac μ^+ and μ^- mesons each contribute $(7/8)\rho_\gamma$, while the Bose-Einstein π^+, π^-, and π^0 mesons each contribute $(1/2)\rho_\gamma$ for a total of $3.25\rho_\gamma$. Electrons and positrons each contribute $(7/8)\rho_\gamma$, neutrinos contribute $(7/8)\rho_\gamma$ or $2(7/8)\rho_\gamma$ according as $\nu \equiv \nu^*$ or $\nu \not\equiv \nu^*$, and photons contribute ρ_γ for a total of 3.625 ρ_γ or 4.50 ρ_γ. The numerical factors obtained here depend on the discussion following Eq. (17).

The meson reactions $\pi^{\pm} \rightleftarrows \mu^{\pm} + \nu$ and $\mu^{\pm} \rightleftarrows e^{\pm} + 2\nu$ are very fast even if one neglects induced decay, having lifetimes of $\sim 2\times 10^{-8}$ sec and $\sim 2\times 10^{-6}$ sec, respectively. Since the concentrations of neutrinos and mesons are comparable, the reaction rate $1/(2\times 10^{-8})$ per second per neutrino is $\sim 10^{6}$ times the equilibrium rate (due to annihilation) of $\sim 1/10^{-2}$ per second per neutrino. Hence between ~100- and ~10-MeV thermal equilibrium holds. By 5 MeV, however, the Boltzmann factor $\exp(-m_{\mu}c^{2}/kT) \cong \exp(-138/5)$ has reduced the reaction rate to insignificance even though there is a good deal more time available for reactions to take place due to the reduced rate of cooling in the universal expansion. Hence the residual mesons cannot transfer a significant amount of rest mass energy to the neutrino gas, although almost all the meson rest mass energy is uniformly distributed.

Having described the nature of the medium, we can now proceed to determine the universal expansion rate for the period of interest in this problem. The rate of expansion for early times, Eq. (3), can be written in cgs units as

$$\frac{1}{2}\frac{dg}{dt} = \frac{1}{\ell}\frac{d\ell}{dt} = \frac{d\ \ln\ell}{dt} = \left(\frac{8\pi G}{3}\right)^{1/2} \rho^{1/2} , \tag{23}$$

where ρ, the total density, may now be written

$$\rho(x,\ell) = \rho(x) + \rho(\ell) , \tag{24}$$

for the following reason. The quantity

$$\rho(x) = \rho_{e^+} + \rho_{e^-} + \rho_{\gamma} \tag{24a}$$

depends only on the temperature, while

$$\rho(\ell) = \rho_{\nu}\ (\text{or}\ \rho_{\nu} + \rho_{\nu}^{*}) \tag{24b}$$

depends only on the proper distance ℓ, since the neutrinos are expanding adiabatically, as a radiation gas after freeze-in. We can, in fact, write $\rho_{\nu} \propto \ell^{-4}$, so that Eq. (23) can be rewritten as follows, after differentiation with respect to time:

$$\frac{d^{2}g}{dt^{2}} = \left(\frac{8\pi G}{3\rho}\right)^{1/2} \left(\frac{\partial\rho}{\partial\ \ln x}\frac{d\ \ln x}{dt} + \frac{\partial\rho}{\partial\ \ln\ell}\frac{d\ \ln\ell}{dt}\right) . \tag{25}$$

Substituting for $\rho^{1/2}$ from Eq. (23) and using Eq. (24) yields

$$\frac{d^2g}{dt^2} = \frac{8\pi G}{3}\left[\frac{d\rho(x)}{d\ \ln x}\frac{d\ \ln x}{d\ \ln \ell} - 4\rho(\ell)\right] . \tag{26}$$

If now we add Eqs. (1a) and (1b), neglect terms containing $1/R_0{}^2$, and convert to cgs units, we obtain

$$\frac{d^2g}{dt^2} = -\left(\frac{8\pi G}{c^2}\right)(p+\rho c^2)$$

$$= -\left(\frac{8\pi G}{c^2}\right)[p(x)+p(\ell)+\rho(x)c^2+\rho(\ell)c^2] . \tag{27}$$

If we equate Eqs. (26) and (27) and note that $(4/3)\rho(\ell)c^2 = p(\ell) + \rho(\ell)c^2$, then we obtain

$$\frac{d\ \ln\ell}{d\ \ln x} = \frac{-c^2}{3[p(x)+\rho(x)c^2]}\frac{d\rho(x)}{d\ \ln x} , \tag{28}$$

where

$$\frac{d\rho(x)}{d\ \ln x} = \frac{d\rho_e(x)}{d\ \ln x} - \frac{3}{c^2}[p_\gamma(x)+\rho_\gamma(x)c^2] . \tag{28a}$$

This result is independent of the presence of the frozen-in neutrinos. Equation (28) can now be integrated in the following manner. From Eqs. (10) one obtains

$$\frac{dp_e(x)}{d\ \ln x} = -(p_e + \rho_e c^2) , \tag{29}$$

or, more conveniently,

$$c^2x^4\frac{d\rho_e}{d\ \ln x} = -3x^4(p_e + \rho_e c^2) + \frac{d}{d\ \ln x}[x^4(p_e + \rho_e c^2)] . \tag{29a}$$

Employing Eq. (29a) in Eqs. (28) yields the desired result, namely,

$$\frac{d\ \ln\ell}{d\ \ln x} = 1 - \frac{\frac{d}{d\ \ln x}\{x^4[p(x) + \rho(x)c^2]\}}{3x^4[p(x) + \rho(x)c^2]} . \tag{30}$$

Since the adiabatic expansion of a radiation universe leads (Alpher and Herman 1950; 1951) to $d\ \ln\ell/d\ \ln x = 1$, the second term in Eq. (30) represents a correction to the description of the cosmological model previously used with the neutron-capture theory, a correction which acounts for the interconversion of matter and radiation. Equation (30) may be integrated to yield

$$\ln\ell = \ln x - \frac{1}{3}\ln\{x^4[p(x) + \rho(x)c^2]\} + \text{constant} \,. \quad (31)$$

Finally, using Eqs. (10) - (14), one can write Eq. (25) as

$$\frac{d\ln\ell}{d\ln x} = 1 + \frac{2a_e(f_0 + 2f_1)}{3[p(x) + \rho(x)c^2]} \,. \quad (32)$$

As will become evident, Eq. (32) is required in order to obtain the explicit time dependence of the temperature.

The neutrino temperature T_ν (or $x_\nu = m_e c^2/kT_\nu$) may be determined from Eq. (30) by recalling that during the period of interest the neutrinos expand and cool adiabatically, so that $x_\nu = f(\ell)$ only, and in fact, $x_\nu \propto \ell$. Then it follows that Eq. (31) can be written as

$$\ln x_\nu = \ln x - \frac{1}{3}\ln\{x^4[p(x) + \rho(x)c^2]\} + \text{constant} \,. \quad (33)$$

The constant of integration can be evaluated by noting that for small x (high temperatures) the neutrino temperature approaches the temperature of the rest of the medium. In fact, as $x \to 0$, $x_\nu \to x$ so that the integration constant becomes

$$\text{constant} = \frac{1}{3}\ln\{x^4[p(x) + \rho(x)c^2]\}\Big|_{x=0} \,. \quad (34)$$

From the definition of p_γ and ρ_γ it is evident [see discussion following Eq. (22)] that, for any x, $\rho_\gamma c^2 x^4$ = constant = $\pi^4 a_e/15$, and $p_\gamma x^4 = \frac{1}{3}\rho_\gamma c^2 x^4$. Since

$$\lim_{x\to 0} p_e x^4 = 2(7/8)p_\gamma x^4, \quad \lim_{x\to 0} \rho_e c^2 x^4 = 2(7/8)\rho_\gamma c^2 x^4 \,,$$

it follows that Eq. (33) can be written as

$$\left(\frac{x_\nu}{x}\right)^3 = \left(\frac{T}{T_\nu}\right)^3 = \frac{2.75(p_\gamma + \rho_\gamma c^2)}{p(x) + \rho(x)c^2} \,, \quad (35)$$

from which the neutrino temperature can be determined for any value of x. For the sake of completeness it should be noted that

$$\lim_{x\to\infty} p_e x^4 = 0, \qquad \lim_{X\to\infty} \rho_e c^2 x^4 = 0,$$

while the quantities $p_\gamma x^4$ and $\rho_\gamma c^2 x^4$ are constants for all x, as just described.

One other relationship which we shall require is that between temperature and time. This is obtained by multiplying Eq. (23) by d lnx/d lnℓ, as evaluated from Eq. (32), with the following result:

$$\frac{d\ \ln x}{dt} = \frac{d\ \ln x}{d\ \ln \ell}\ \frac{d\ \ln \ell}{dt} = \left(\frac{8\pi G}{3}\rho\right)^{1/2} \frac{d\ \ln x}{d\ \ln \ell}\ . \tag{36}$$

TABLE 1a
TIME SCALES, RATE COEFFICIENTS, AND QUANTUM STATISTICAL INTEGRALS[a]

T(°K)	x	t(sec) $\nu \equiv \nu^*$	t(sec) $\nu \not\equiv \nu^*$	t(sec) rad. model[b]	$\frac{15x^4}{\pi^4} f_0$
∞	0	0	0	0	0
5.930×10^{10}	0.1	0.04	0.04	0.06	2.2105×10^{-5}
2.965×10^{10}	0.2	0.14	0.12	0.26	2.6866×10^{-4}
1.482×10^{10}	0.4	0.55	0.50	1.05	2.9583×10^{-3}
9.884×10^{9}	0.6	1.27	1.14	2.36	1.1131×10^{-2}
7.413×10^{9}	0.8	2.26	2.02	4.20	2.6998×10^{-2}
5.930×10^{9}	1.0	3.58	3.21	6.57	5.1303×10^{-2}
2.965×10^{9}	2.0	15.65	14.12	26.28	0.25584
1.977×10^{9}	3.0	39.72	36.05	59.11	0.41847
1.482×10^{9}	4.0	78.99	72.04	105.20	0.43426
1.186×10^{9}	5.0	134.15	122.64	164.25	0.35354
9.884×10^{8}	6.0	203.84	186.56	236.49	0.24783
8.472×10^{8}	7.0	285.68	261.76	321.90	0.15696
7.413×10^{8}	8.0	378.40	347.03	420.44	9.2363×10^{-2}
6.589×10^{8}	9.0	481.59	442.06	532.17	5.1402×10^{-2}
5.930×10^{8}	10.0	595.56	546.96	657.02	2.7379×10^{-2}
0	∞	∞	∞	∞	0

[a] The universal constants employed in these calculations are those given by Bearden and Watts (1951). Note that the limiting values at high temperatures do not include any contributions from mesons.

[b] This column gives the time scale for the pure radiation model described by Eqs. (4)-(7).

TABLE 1b
TIME SCALES, RATE COEFFICIENTS, AND QUANTUM STATISTICAL INTEGRALS

T(°K)	$\frac{15x^4}{x^4} f_1$	$\frac{15x^4}{\pi^4} f_2$	K_n	K_p
∞	0	0.87500	∞	∞
5.930×10^{10}	1.2535×10^{-3}	0.87311	5.07×10^{6}	3.94×10^{6}
2.965×10^{10}	4.9011×10^{-3}	0.86754	1.91×10^{5}	1.12×10^{5}
1.482×10^{10}	1.8300×10^{-2}	0.84626	7.58×10^{3}	2.51×10^{3}
9.884×10^{9}	3.7625×10^{-2}	0.81339	1.17×10^{3}	2.34×10^{2}
7.413×10^{9}	6.0185×10^{-2}	0.77190	3.09×10^{2}	37.8
5.930×10^{9}	8.3590×10^{-2}	0.72406	1.18×10^{2}	9.37
2.965×10^{9}	0.16512	0.46121	6.88	4.00×10^{-2}
1.977×10^{9}	0.16423	0.25402	2.38	9.49×10^{-4}
1.482×10^{9}	0.12227	0.12828	1.91	2.69×10^{-5}
1.186×10^{9}	7.7674×10^{-2}	6.1253×10^{-2}	1.74	...
9.884×10^{8}	4.4664×10^{-2}	2.8128×10^{-2}	...	...
8.472×10^{8}	2.3982×10^{-2}	1.2552×10^{-2}	...	...
7.413×10^{8}	1.2248×10^{-2}	5.4786×10^{-3}	...	...
6.589×10^{8}	6.0209×10^{-3}	2.3499×10^{-3}	...	...
5.930×10^{8}	2.8716×10^{-3}	9.9368×10^{-4}	...	...
0	0	0	1.63	0

The integration of Eq. (36) (performed to an accuracy better than 0.1 percent on a Maddida, a digital differential analyzer built by Northup Aircraft, Inc.) for the two cases $\nu \equiv \nu^*$ and $\nu \not\equiv \nu^*$ gives the time in the universal expansion as a function of x, and these quantities are given in Tables 1a,b. For comparison, Tables 1a,b also contain the time as a function of x for the expanding cosmological model containing radiation only [see Eq. (6)]. Other quantities given in Table 1 are the series of modified Bessel functions f_0, f_1, and f_2, as defined in Eqs. (12) - (14), which are used in computing pressure and density. In Tables 2a,b are given the total density ρ (i.e., the density of electrons, positrons, neutrinos, and radiation) and the density of electrons plus positrons ρ_e, in units of the radiation density, ρ_γ, the neutrino temperature expressed as $x_\nu/x = T/T_\nu$, the differential quotient $d\ln\ell/d\ln x$, and finally the expansion rate $d\ln\ell/dt$.

Several interesting features of this cosmological model are evident upon examination of Tables 1 and 2. First, the temperature drops much more rapidly in the nonstatic model with interconversion of matter and radiation than it does in the model of adiabatically expanding radiation only.[7] However the cases of distinguishable and indistinguishable neutrinos differ very little in this respect. The total density ρ does not drop off very greatly until the universe has cooled to about the electron rest mass equivalent. At this point the large density contribution of electron pairs begins to decrease sharply [see Tables 2a,b] as the pairs disappear by

Table 2a
NEUTRINO TEMPERATURE; TOTAL, RADIATION, AND ELECTRON-PAIR DENSITIES; AND UNIVERSAL EXPANSION RATES[a]

T(°K)	x	x_ν/x	ρ_γ (g/cm^3)	ρ/ρ_γ ($\nu \equiv \nu^*$)	ρ/ρ_γ ($\nu \not\equiv \nu^*$)
∞	0	1.0000	∞	3.625	4.500
5.930×10^{10}	0.1	1.0002	7.226×10^7	3.623	4.497
2.965×10^{10}	0.2	1.0009	4.516×10^6	3.617	4.488
1.482×10^{10}	0.4	1.0037	2.822×10^5	3.591	4.454
9.884×10^9	0.6	1.0082	5.575×10^4	3.549	4.396
7.413×10^9	0.8	1.0145	1.764×10^4	3.490	4.317
5.930×10^9	1.0	1.0224	7.226×10^3	3.416	4.217
2.965×10^9	2.0	1.0821	4.516×10^2	2.891	3.529
1.977×10^9	3.0	1.1616	89.20	2.317	2.798
1.482×10^9	4.0	1.2407	28.22	1.870	2.240
1.186×10^9	5.0	1.3044	11.56	1.580	1.882
9.884×10^8	6.0	1.3478	5.575	1.411	1.676
8.472×10^8	7.0	1.3736	3.009	1.319	1.564
7.413×10^8	8.0	1.3876	1.764	1.271	1.508
6.589×10^8	9.0	1.3947	1.101	1.248	1.479
5.930×10^8	10.0	1.3981	0.7226	1.237	1.466
0	∞	1.4010	0	1.227	1.454

[a] The universal constants employed in these calculations are those given by J. A. Bearden and H.M. Watts (1951). Note that the limiting values at high temperatures do not include any contributions from mesons.

[7]There are perhaps shifts in the expansion time scale, much too small to appear with the number of significant figures given in Tables 1 a,b and 2 a,b. These are caused by the presence of and annihilation of mesons, nucleons, gravitons, etc., between x = 0 and x = 0.1, should such particles exist during this epoch, and should the relativistic cosmology apply at the extreme conditions existing during this very brief early period. There is, however, serious doubt that the cosmology applies and, since we are interested only in the epochs of temperature lower than x = 0.1, we can for the present perhaps ignore this problem and accept the insignificant additive constant in the time scale. See (Alpher and Herman [1950; 1951] as well as Einstein [1945]).

Table 2b
NEUTRINO TEMPERATURE; TOTAL, RADIATION, AND ELECTRON-PAIR DENSITIES; AND UNIVERSAL EXPANSION RATES[a]

T(°K)	ρ_e/ρ_γ	$\frac{d\ \ln\ell}{d\ \ln x}$	$\frac{d\ \ln\ell}{dt}$ (sec^{-1}) ($\nu \equiv \nu^*$)	$\frac{d\ \ln\ell}{dt}$ (sec^{-1}) ($\nu \not\equiv \nu^*$)
∞	1.750	1.0000	∞	∞
2.930×10^{10}	1.749	1.0005	14.51	16.17
2.965×10^{10}	1.745	1.0018	3.625	4.039
1.482×10^{10}	1.729	1.0073	0.9031	1.006
9.884×10^{9}	1.702	1.0161	0.3990	0.4441
7.413×10^{9}	1.664	1.0280	0.2226	0.2475
5.930×10^{9}	1.615	1.0424	0.1409	0.1566
2.965×10^{9}	1.253	1.1350	3.241×10^{-2}	3.581×10^{-2}
1.977×10^{9}	0.836	1.2129	1.290×10^{-2}	1.417×10^{-2}
1.482×10^{9}	0.501	1.2357	6.517×10^{-3}	7.132×10^{-3}
1.186×10^{9}	0.278	1.2054	3.834×10^{-3}	4.184×10^{-3}
9.884×10^{8}	0.146	1.1501	2.516×10^{-3}	2.742×10^{-3}
8.472×10^{8}	0.073	1.0966	1.787×10^{-3}	1.946×10^{-3}
7.413×10^{8}	0.035	1.0568	1.343×10^{-3}	1.463×10^{-3}
6.589×10^{8}	0.017	1.0313	1.052×10^{-3}	1.145×10^{-3}
5.930×10^{8}	0.008	1.0165	8.480×10^{-4}	9.231×10^{-4}
0	0	1.0000	0	0

[a] The universal constants employed in these calculations are those given by J. A. Bearden and H.M. Watts (1951). Note that the limiting values at high temperatures do not include any contributions from mesons.

annihilation into the radiation field which has fewer degrees of freedom. This behavior is demonstrated in Figure 1. Next, if one recalls that the expanding model of radiation only is represented by $d\ln\ell/d\ \ln x = 1$, then one can seen in Tables 2a,b that the maximum deviation from this model due to the interconversion of matter and radiation is ~24 percent, and this occurs at 0.125 MeV ($\sim m_ec^2/4$). This deviation represents a more rapid expansion rate than in the pure radiation model and in fact, the expansion rate $d\ln\ell/dt$ is higher in the model with interconversion by just the factor $(\rho/\rho_\gamma)^{1/2}$. Finally, it should be noted that x_ν/x (a quantity which does not depend on whether $\nu \equiv \nu^*$ or $\nu \not\equiv \nu^*$) differs from unity by less than one percent until x had increased to about 0.7 or kT ≅ 0.73 MeV. At x = 0.1, where kT = 5 MeV, the deviation is 0.02 percent. It is then quite clear that selecting say 5-10 MeV as the freeze-in temperature for neutrinos is not only reasonable but quite an adequate approximation. It should be noted that the mathematical limits approached as $x \to \infty$ for all the quan-

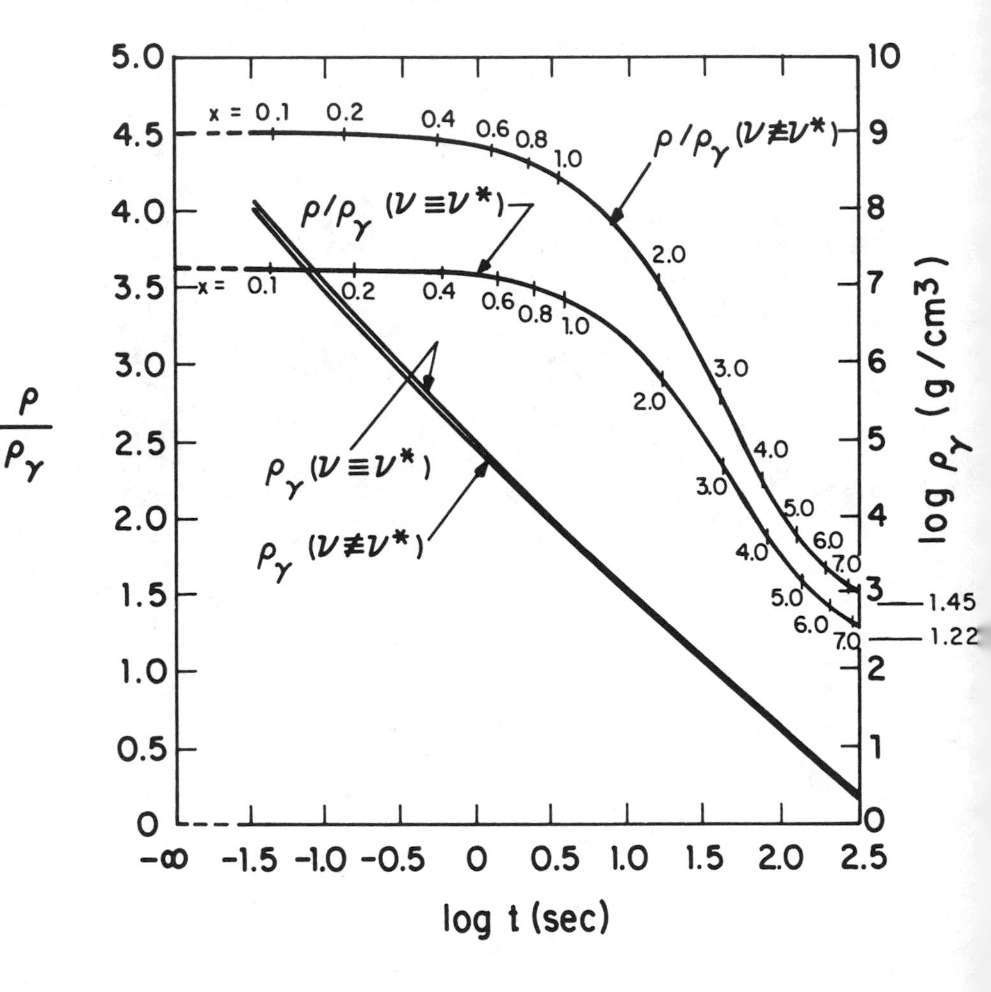

Figure 1

Total density ρ, in units of photon density ρ_γ, and ρ_γ versus time during the very early epochs of the expanding universe for Majorana and Dirac neutrinos. The corresponding temperatures are given in terms of $x = m_e c^2/kT$. The ρ/ρ_γ curves are extrapolated to $t = 0$ without regard to the presence of mesons and other elementary particles.

tities given in Tables 1a,b and 2a,b are included for the sake of completeness. However, the behavior of the cosmological model discussed changes at longer times, $x \sim 10^7$, when the density of matter exceeds the density of radition.

In the next section we shall calculate the relative concentrations of neutrons and protons as a function of the time in the universal expansion. This ratio, it will be recalled, plays a most important role in determining the relative abundances of the nuclear species as calculated according to the simple neutron-capture theory of element formation including the effects of thermo-nuclear reactions. As has been stated, this theory quite clearly requires that during the early epochs in the universal expansion the density of nucleons, and of the nuclear species formed, should be negligibly small compared with the radiation density. Consequently, the physical conditions in the expanding model as described in Tables 1a,b and 2a,b, in which nucleon density is taken to be negligible, are used as a basis for examining the various non-equilibrium reactions between neutrons and protons.

II. THE NEUTRON-PROTON RATIO

In this section we shall examine the reactions which may occur among neutrinos, electrons, positrons, and nucleons in the very early stages of the cosmological model described in the previous section. In particular we shall calculate the ratio of the concentrations of protons and neutrons as a function of time, a ratio upon which the results of the neutron-capture theory of element formation strongly depend.

The nuclear reactions which must be considered in determining the proton-neutron ratio are the following:[8]

$$n + e^{+} \rightleftarrows p + \nu \, , \tag{37a}$$

$$n + \nu \rightleftarrows p + e^{-} \, , \tag{37b}$$

$$n \rightleftarrows p + e^{-} + \nu \, . \tag{37c}$$

[8]The question of the existence of the antineutrino is of no concern in determining the individual reaction rates (see footnote 5). In the absence of charge, mass, and magnetic moment, the absorption of a neutrino from a negative energy state is in all respects equivalent to the emission of an antineutrino.

The probability per second w for these reactions may be obtained from the Fermi theory of β decay. For the reaction $n + e^+ \rightarrow p + \nu$, one has, per electron

$$w_{A'} = \frac{2\pi}{\hbar} \left(|\psi_p(0)| \cdot |\psi_\nu(0)| \cdot |\mathbf{M}| g \right)^2 \frac{dn}{dE} \sec^{-1} , \qquad (38)$$

where the expectation value at the origin for the product particles, proton, and neutrino, depends on $|\psi_p(0)|^2 |\psi_\nu(0)|^2$, the ψ are taken as plane wave states, $\mathbf{M}$, the matrix element, is taken as unity for lack of a better estimate, g is the Fermi coupling constant and the quantity dn/dE is the energy density of available final states. We can, therefore, write

$$w_{A'} = \frac{2\pi g^2}{\hbar\, \Omega^2} \frac{dn}{dE} \sec^{-1} , \qquad (39)$$

where Ω is any finite normalization volume. For the neutrino the number of available states per unit energy in the volume Ω is the difference between the total and occupied number of states, viz.,

$$\frac{4\pi (2i_\nu + 1) |\mathbf{p}_\nu|}{2\pi\hbar^3} \left\{ 1 - [1 + \exp(E_\nu/kT_\nu)]^{-1} \right\} \frac{d|\mathbf{p}_\nu|}{dE_\nu} , \qquad (40a)$$

while for each neutron the number of available final states in this particular reaction is $(2i_p + 1)$. Since all neutrons are equivalent, one may write for the total number of final states in Ω:

$$\Omega n_n (2i_p + 1) = 2\Omega n_n . \qquad (40b)$$

In Eqs. (40) the neutrino and proton spins are denoted by i_ν and i_p, $|\mathbf{p}_\nu|$ is the neutrino momentum and n_n the number of neutrons per unit volume. Since dn/dE in Eq. (39) is the product of terms given by Eq. (40a) and Eq. (40b) one can write

$$w_{A'} = \frac{4g^2 n_n |\mathbf{p}_\nu| \exp(E_\nu/kT_\nu)}{\pi\hbar^4 [1 + \exp(E_\nu/kT_\nu)]} \frac{d|\mathbf{p}_\nu|}{dE_\nu} \sec^{-1} . \qquad (41)$$

The number of such reactions, $n + e^+ \rightarrow p + \nu$, per second per unit volume, is given by

$$A' n_n n_{e^+} = \int_{m_e c^2}^{\infty} w_{A'} n_{e^+}(E_{e^+}) dE_{e^+} , \qquad (42)$$

where

$$E_\nu = E_{e^+} + Q \; ; \tag{42a}$$

$$Q = (m_n - m_p)c^2 \tag{42b}$$

is the neutron-proton energy difference and n_{e^+} is the concentration of positrons per unit energy at E_{e^+}. The lower limit of integration is the threshold energy, which in this case is the electron rest energy. Using the relation $E_\nu = |\mathbf{p}_\nu|c$ and replacing n_{e^+} by means of the Fermi-Dirac integral, viz.,

$$n_{e^+}(|\mathbf{p}_{e^+}|)d|\mathbf{p}_{e^+}| = \frac{8\pi|\mathbf{p}_{e^+}|^2 d|\mathbf{p}_{e^+}|}{(2\pi\hbar^3)[1 + \exp(E_{e^+}/kT)]} \; , \tag{43}$$

where $|\mathbf{p}_{e^+}|$ is given by Eq. (9c), one can write Eq. (42) in the form

$$A'n_n n_{e^+} =$$

$$a_0 n_n \int_{m_e c^2}^{\infty} \frac{E_{e^+}(E_{e^+} + Q)^2(E_{e^+}^2 - m_e^2 c^4)^{1/2} \exp[(E_{e^+} + Q)/kT_\nu]}{\{1 + \exp[(E_{e^+} + Q)/kT_\nu]\}\{1 + \exp[E_{e^+}/kT]\}} dE_{e^+} \; , \tag{44}$$

where Eq. (42a) has been used to eliminate E_ν, and

$$a_0 = (4g^2)/(\pi^3 c^6 \hbar^7) \; . \tag{44a}$$

It should be recalled that he neutrino temperature $T_\nu \neq T$, where the latter is the temperature of the remainder of the medium. Equation (44) may be rewritten by taking

$$\begin{aligned} x &= m_e c^2/(kT) \; , & x_\nu &= m_e c^2/(kT_\nu) \; , \\ q &= Q/(m_e c^2) \; , & \varepsilon &= E_e/(m_e c^2) \; , \end{aligned} \tag{45}$$

with the result that for the reaction $n + e^+ \rightarrow p + \nu$,

$$A'n_n n_{e^+} = m_e^5 c^{10} a_0 n_n I_{A'} \; \sec^{-1} \, cm^{-3} \; , \tag{46}$$

where

$$I_{A'} = \int_1^\infty \frac{\varepsilon(\varepsilon+q)^2(\varepsilon^2-1)^{1/2}\exp[(\varepsilon+q)x_\nu]d\varepsilon}{\{1+\exp[\varepsilon x]\}\{1+\exp[(\varepsilon+q)x_\nu)\}} . \qquad (46a)$$

Rates for the other reactions in Eq. (37a) and Eq. (37b) may be obtained from similar considerations, since they are all of second degree. One obtains the following:

for $p + \nu \rightarrow n + e^+$,

$$An_p n_\nu = m_e^5 c^{10} a_0 n_p I_A \ \sec^{-1}\ cm^{-3} , \qquad (47)$$

where

$$I_A = \int_1^\infty \frac{\varepsilon(\varepsilon+q)^2(\varepsilon^2-1)^{1/2}\ \exp(\varepsilon x)d\varepsilon}{\{1+\exp[\varepsilon x]\}\{1+\exp[(\varepsilon+q)x_\nu]\}} , \qquad (47a)$$

and where

$$E_{e^+} = E_\nu - Q ; \qquad (47b)$$

for $n + \nu \rightarrow p + e^+$,

$$B'n_n n_\nu = m_e^5 c^{10} a_0 n_n I_{B'} \ \sec^{-1}\ cm^{-3} , \qquad (48)$$

where

$$I_{B'} = \int_q^\infty \frac{\varepsilon(\varepsilon-q)^2(\varepsilon^2-1)^{1/2}\exp(\varepsilon x)d\varepsilon}{\{1+\exp[\varepsilon x]\}\{1+\exp[(\varepsilon-q)x_\nu]\}} , \qquad (48a)$$

and where

$$E_\nu = E_{e^-} - Q ; \qquad (48b)$$

for $p + e^- \rightarrow n + \nu$,

$$Bn_p n_{e^-} = m_e^5 c^{10} a_0 n_p I_B \ \sec^{-1}\ cm^{-3} , \qquad (49)$$

where

$$I_B = \int_q^\infty \frac{\varepsilon(\varepsilon-q)^2(\varepsilon^2-1)^{1/2}\exp[(\varepsilon-q)x_\nu]d\varepsilon}{\{1 + \exp[\varepsilon x]\}\{1 + \exp[(\varepsilon-q)x_\nu]\}}, \qquad (49a)$$

and where

$$E_\nu = E_{e^-} - Q \,. \qquad (49b)$$

The reaction rates for free neutron decay and the inverse process, Eq. (37c), require a slightly different calculation. Thus for the reaction $p + e^- + \nu \rightarrow b$, we note that the quantity dn/dE in Eq. (39) is given by just the number of protons present in the volume Ω, so that

$$w_C = (2\pi g^2/\hbar)n_p(2i_n + 1) \,,$$

and

$$Cn_p n_{e^-} n_\nu = \int_{m_e c^2}^{Q} w_C(E_{e^-})n_{e^-}(E_{e^-})n_\nu(Q - E_{e^-})dE_{e^-} \,, \qquad (50)$$

where

$$E_{e^-} = Q - E_\nu \,. \qquad (50a)$$

In Eq. (50) $n_{e^-}(E_{e^-})$ is the concentration of electrons per unit energy at E_{e^-} and $n_\nu(Q-E_{e^-})$ is the concentration of neutrinos per unit energy at $(Q-E_{e^-})$, the argument of n_ν being that required for energy balance in this reaction. Finally, one can rewrite Eq. (5) replacing n_{e^-} and n_ν by means of the Fermi-Dirac integral [see Eq. (42)] and using Eqs. (45), as

$$Cn_p n_{e^-} n_\nu = m_e^5 c^{10} a_0 n_p I_C \sec^{-1} \text{cm}^{-3} \,, \qquad (51)$$

where

$$I_C = \int_1^q \frac{\varepsilon(\varepsilon-q)^2(\varepsilon^2-1)^{\frac{1}{2}}d\varepsilon}{\{1 + \exp[\varepsilon x]\}\{1 + \exp[(q-\varepsilon)x\nu]\}} \,. \qquad (51a)$$

For the reaction $n \rightarrow p + e^- + \nu$, we note that dn/dE in Eq. (39) is the product of three quantities, viz., the available states per unit energy in the volume Ω for protons, electrons, and neutrinos. For protons the number of available states is the number of neutrons in the volume Ω, viz., $\Omega n_n(2i_n+1)$, while for electrons and neutrinos one can use the form of Eq. (40a) which gives the quantity for Fermi-Dirac particles. Since formally the reaction rate for free neutron decay is

$$C'n_n = \int_{m_e c^2}^{Q} w_{C'} dE \;, \tag{52}$$

it follows from Eq. (39) after some manipulation that

$$C'n_n = m_e^5 c^{10} a_0 n_n I_{C'} \; \sec^{-1} \; \mathrm{cm}^{-3} , \tag{53}$$

where

$$I_{C'} = \int_1^q \frac{\varepsilon(\varepsilon-q)^2(\varepsilon^2-1)^{\frac{1}{2}} \exp[(q-\varepsilon)x_\nu]\exp(\varepsilon x)d\varepsilon}{\{1+\exp[\varepsilon x]\}\{1+\exp[(q-\varepsilon)x_\nu]\}} \; . \tag{53a}$$

The foregoing reaction rates have been used in the equations developed below which describe the time dependence of neutron and proton concentrations. Let N_j be the number of nucleons of species j in the arbitrary finite volume V. The time derivative of N_j can be expressed formally for two- and three-body processes as

$$\frac{dN_j}{dt} = \sum_{\alpha,\beta} K_{\alpha\beta} n_\alpha N_\beta + \sum_{\alpha,\beta,\gamma} K_{\alpha\beta\gamma} n_\alpha n_\beta N_\gamma \; , \tag{54}$$

where

$$n_\alpha = N_\alpha / V \; . \tag{54a}$$

We note that

$$\frac{dn_j}{dt} = \frac{1}{V}\frac{dN_j}{dt} - \frac{n_j}{V}\frac{dV}{dt} \; , \tag{55}$$

where, since $V \propto \ell^3$,

$$\frac{1}{V}\frac{dV}{dt} = \frac{3}{\ell}\frac{d\ell}{dt} \; .$$

For the cosmological model described in Sec. II,

$$\frac{1}{\ell}\frac{d\ell}{dt} = \left(\frac{8\pi G}{3}\rho\right)^{\frac{1}{2}} ,$$

where ρ is the total density, so that

$$\frac{dn_j}{dt} = \sum_{\alpha,\beta} K_{\alpha\beta} n_\alpha n_\beta + \sum_{\alpha,\beta,\gamma} K_{\alpha\beta\gamma} n_\alpha n_\beta n_\gamma - 3n_j (8\pi G\rho/3)^{\frac{1}{2}} . \tag{56}$$

Consequently, we can write for neutrons the following rate equation:

$$dn_n/dt = An_p n_\nu - A'n_n n_{e^+} + Bn_p n_{e^-} - B'n_n n_\nu + Cn_p n_{e^-} n_\nu - C'n_n - 3n_n(8\pi G\rho/3)^{\frac{1}{2}} . \tag{57}$$

Equation (57) can be rewritten using some of Eqs. (46) - (53) as

$$dn_n/dt = m_e^5 c^{10} a_0 [n_p K_p - n_n K_n] - 3n_n (8\pi G\rho/3)^{\frac{1}{2}} , \tag{58}$$

where

$$K_p = I_A + I_B + I_C , \tag{58a}$$

and

$$K_n = I_{A'} + I_{B'} + I_{C'} . \tag{58b}$$

The limits of integration in the six integrals involved in K_p and K_n make it possible to combine certain pairs, with the result that

$$K_p = \int_1^\infty \frac{\varepsilon(\varepsilon^2-1)^{\frac{1}{2}}}{1 + \exp[\varepsilon x]} \left\{\frac{(\varepsilon+q)^2 \exp[\varepsilon x]}{1 + \exp[(\varepsilon+q)x_\nu} + \frac{(\varepsilon-q)^2 \exp[(\varepsilon-q)x_\nu]}{1 + \exp[(\varepsilon-q)x_\nu]}\right\} d\varepsilon , \tag{59}$$

and

$$K_n = \int_1^\infty \frac{\varepsilon(\varepsilon^2-1)^{\frac{1}{2}}}{1+\exp[\varepsilon x]} \left\{ \frac{(\varepsilon+q)^2 \exp[(\varepsilon+q)x_\nu]}{1+\exp[(\varepsilon+q)x_\nu]} + \frac{(\varepsilon-q)^2 \exp[\varepsilon x]}{1+\exp[(\varepsilon-q)x_\nu]} \right\} d\varepsilon \; . \tag{60}$$

The equation describing the time rate of change of proton concentration can be written in a manner analogous to Eq. (58) as follows:

$$dn_p/dt = m_e^5 c^{10} a_0 [n_n K_n - n_p K_p] - 3n_p (8\pi G\rho/3)^{\frac{1}{2}} \; . \tag{61}$$

As shall be seen below, Eqs. (58) and (61) can be combined to give a single equation for the proton-neutron ratio, with the effect of the universal expansion not appearing explicitly.

The rate coefficients K_p and K_n have been evaluated numerically using Eqs. (59) and (60) for the range of values of x of interest. The values of x_ν corresponding to x have been taken from Table 2a,b with[9]

$$q = 1+(m_n-m_p)/m_e = 2.53 \; .$$

It can be shown from Eqs. (59) and (60) that for small x, i.e., $x < 1$ where $x_\nu \to x$,

$$\lim_{x_\nu \to x} K_p = e^{-qx} K_n \; . \tag{62a}$$

Furthermore, for large values of x, one finds that

$$\lim_{x \to \infty} K_p = 0 \; , \tag{62b}$$

and

$$\lim_{x \to \infty} K_n = \int_1^\infty \varepsilon(\varepsilon-q)^2 (\varepsilon^2-1)^{\frac{1}{2}} d\varepsilon = 1.6318 \; . \tag{62c}$$

[9] For $(m_n-m_p)c^2$ we have used 0.782 MeV, as given by Van Patter (1952), while $m_e c^2$ has been taken as 0.5110 MeV, as given by DuMond and Cohen (1951).

The limit approached by K_n in Eq. (62c) for $x \to \infty$ is just the term $C'/(m_e^5 c^{10} a_0)$ where $C'n_n = dn_n/dt$ describes free neutron decay and $C'(=\lambda)$ is the neutron-decay constant. One can, therefore, select the Fermi coupling constant g, which is the only undetermined constant in a_0 [see Eq. (44a)], so that the value of C' is the observed neutron decay constant.[10] The neutron half-life measured recently by Robson(1951)[11] is $\tau_{1/2} = 12.8 \pm 2.5$ minutes. The values of $m_e^5 c^{10} a_0$ and of g corresponding to the measured half-life limits are given in Table 3.

In Tables 1a,b are given values of the dimensionless quantities K_n and K_p, evaluated numerically from Eqs. (59) and (60) in the range required for the present calculation. It may be noted that for x slightly greater than 3, K_n is close to the free decay value of 1.63 while K_p is negligibly small. Also fox $x < 1$ the relationship $K_p \cong K_n \exp(-qx)$ holds quite closely.

Equations (58) and (61) describing neutron and proton concentrations can be combined by defining

$$\phi(t) = n_n/(n_n+n_p) \quad . \tag{63}$$

Table 3

VALUES OF FERMI CONSTANT FOR VARIOUS NEUTRON HALF-LIVES.

Neutron half-life (min)	10.3	12.8	15.3
$m_e{}^5 c^{10} a_0 (\sec^{-1})$	4.627×10^{-4}	5.531×10^{-4}	6.876×10^{-4}
$g(\text{erg cm}^3)$	1.01×10^{-49}	1.11×10^{-49}	1.23×10^{-49}

[10] Although the numerical constants in a_0 which depend on spin, etc., have been carefully evaluated in building up the rate coefficients, it may be noted that the equality $C'n_n/(m_e{}^5 c^{10} a_0) = 1.6318$, with $C' = \lambda$ known from experiment, automatically yields a value for a_0, so that g, spin factors, etc., need not be separately specified.

[11] Recently Langer and Moffat (1952), obtained the value $\tau_{1/2} = 10.4 \pm 0.6$ min indirectly from studying tritium decay. This value and Robson's value agree within the probable errors.

Taking the time derivative of ϕ and employing Eqs. (58) and (61) as required, one can write

$$d\phi/dt = m_e^5 c^{10} a_0 [K_p(1-\phi) - K_n \phi] \;, \tag{64}$$

where a_0 is given by Eq. (44a). The actual integrations were done with lnx as independent variable where one writes

$$\frac{d\phi}{d\ln x} = \frac{d\phi}{dt} \frac{d\ln \ell}{d\ln x} \frac{dt}{d\ln \ell} \;. \tag{64a}$$

The quantity d $\ln\ell$/d lnx is given by Eq. (30) and calculated values are given in Table 1a,b while values of d $\ln\ell/dt$, determined from Eq. (23), are also given in Table 1a,b.

Equation (64) has been integrated numerically for the six cases of interest, viz., for $\nu \equiv \nu^*$ and $\nu \not\equiv \nu^*$ taking three values of the neutron half-life, namely, the mean value and the mean value plus-and-minus the probable error, as given by Robson (1951). The integration procedure was such as to give a final accuracy in ϕ of the order of the accuracy of the coefficients, i.e., ~1 percent. The solutions we have obtained are presented in Figures 2 and 3, where the proton-neutron ratio is plotted versus x and versus the time scale appropriate to the type of neutrinos involved. The integrations were carried from x = 0.1 toward larger x, the initial value of $\phi = n_n/(n_n+n_p)$ being taken as the equilibrium value, i.e., $\sim[1+\exp(qx)]^{-1}$, since at x = 0.1 the rate coefficients K_n and K_p are large and show negligible deviation from their respective equilibrium values. The integration interval was 0.1 in lnx, with the first approximation at each step being the equilibrium value which Eq. (64) would predict at the given value of x. It may be noted in Figs. 2 and 3 that by x = 4 further change in the proton-neutron ratio is almost entirely caused by free neutron decay.

A comparison of these results with those of Hayashi indicates that the major difference between our calculation and his may arise from the difference in neutron half-life used, viz., Hayashi used 20.8 min, while the remainder of the differences, amounting to perhaps 20 percent in the proton-neutron ratio, arise from the use of relativistic quantum statistics, a more detailed cosmological model, and different temperatures for neutrinos and the rest of the medium in the present calculation. The proton-neutron ratio obtained by Hayashi by the time free neutron decay predominated was ~4:1, whereas in the present calculation values from ~4.5:1 to ~6.0:1 are obtained depending on the half-life taken for the neutron and the type of neutrino considered.

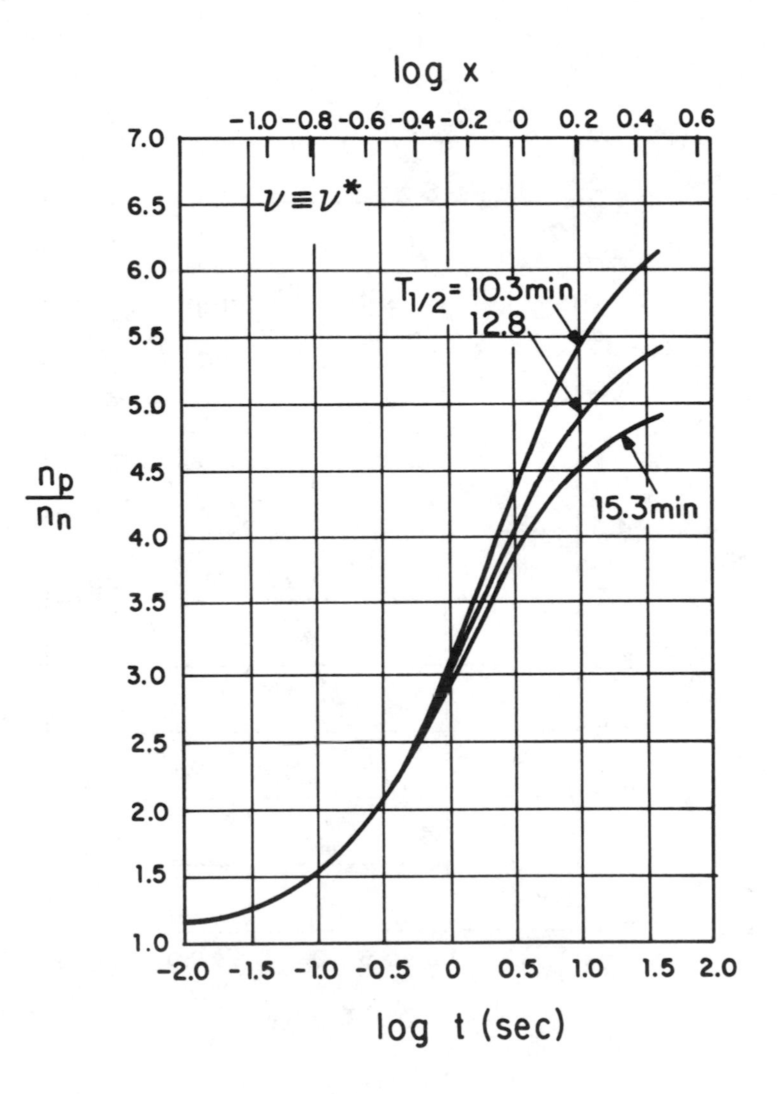

Figure 2

The proton-neutron concentration ratio versus time and temperature ($x = m_e c^2/kT$) in the case of the Dirac neutrino (distinguishable neutrino and antineutrino) for the Robson neutron-halflife value of 12.8 min, plus-and-minus the probable error.

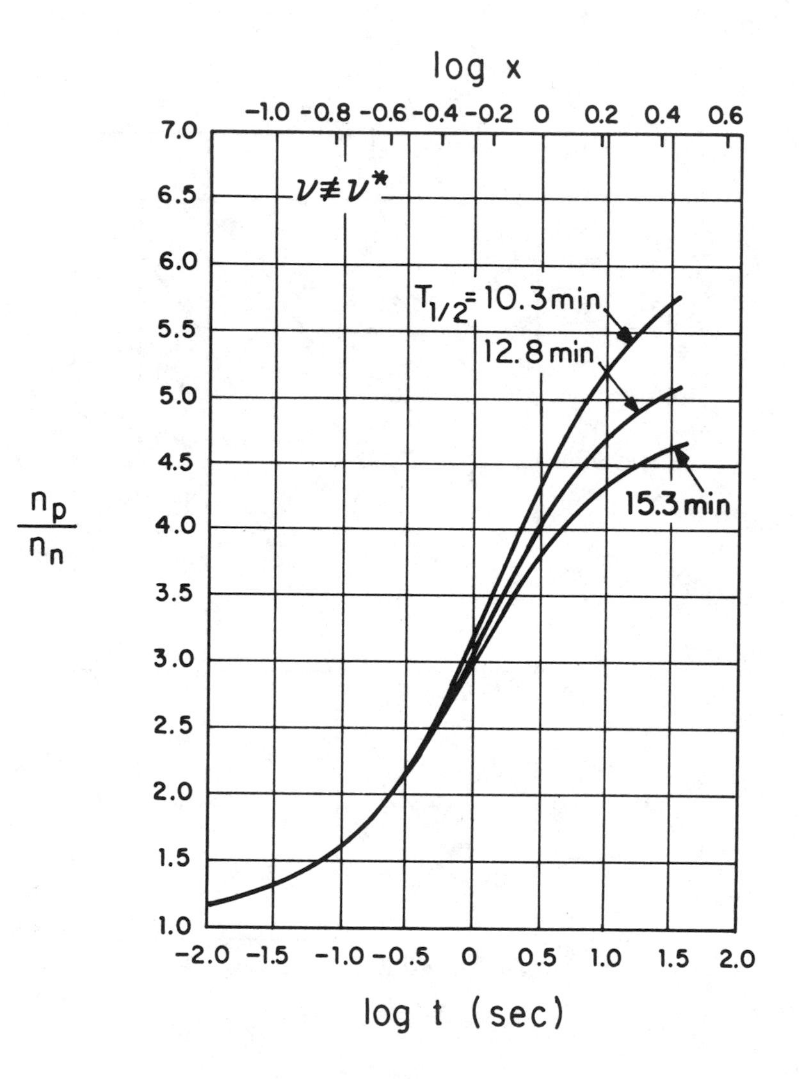

Figure 3

The proton-neutron concentration ratio versus time and temperature ($x = m_e c^2/kT$) in the case of the Majorana neutrino (indistinguishable neutrino and antineutrino) for the Robson neutron half-life value of 12.8 min, plus-and-minus the probable error.

It is interesting to note that if all the neutrons available at the start of element synthesis were used in making helium nuceli only, then the ratio of hydrogen to helium abundances corresponding to the range of proton-neutron ratios computed above would be from ~7:1 to ~10:1. Since some of the neutrons decay and some are involved in making the heavier elements, the above ratios would be minimum values of the initial universal H/He ratios. These values are consistent with the range of values obtained from astronomical data, viz., ~5:1 to ~30:1 as found in planetary nebulae, stellar atmospheres, and theoretical stellar models (Underhill 1952).

DISCUSSION

In the preceding sections we have discussed quantitatively the physical conditions in the initial stages of the universal expansion. It now seems pertinent to mention some of the small physical effects whose influence on the present calculations has been neglected and also to comment on some of the limitations of the cosmological model when extrapolated to very early epochs.

The first question to be considered is whether or not the various processes, such as pair production, Compton and Coulomb scattering, etc., occur at sufficiently rapid rates to maintain equilibrium. A qualitative criterion as first described by Hayashi (1950) is to compare the time required for the concentration of a constituent to change by about its own value with the time required for the universal temperature to change by about its own value. Characterizing these as relaxation times, τ, one finds from Eq. (6) and Table I that the relaxation time for temperature is given with sufficient accuracy for the present purpose by the following expression:

$$\tau_T \sim -\frac{dt}{d\ \ln T} = -\left(\frac{3c^2}{8\pi G a_\gamma}\right)^{1/2} T^{-2} = 2t \; . \qquad (65)$$

To take a specific example one can calculate from the equilibrium concentration of electron pairs and their rate of production by photon-photon collisions the relaxation times, τ_{pair}, for the pair production-annihilation. The result for $x \ll 1$ is

$$\tau_{pair} \sim -\frac{dt}{d\ \ln(n_{pair})} \sim \frac{144 x \hbar^3}{\pi^3 m_e e^4} \; , \qquad (66)$$

where e is the charge on the electron. Hence

$$\frac{\tau_{pair}}{\tau_T} \cong 10x^3\left(\frac{Gm^2}{\hbar c}\right)^{1/2}\left(\frac{e^2}{\hbar c}\right)^{-2} \sim 10^{-20}\ x^3 \ll 1 \; . \qquad (67)$$

A similar result is obtained for other reactions not involving neutrinos, for which the change in the coupling coefficient $e^2/\hbar c$ does not greatly change the order of magnitude of the ratio. This is not the case for neutrino interactions. Hence all processes not involving neutrinos proceed at sufficiently rapid rates to maintain equilibrium.

The question of electron degeneracy is most easily examined by considering the requirement of electrical neutrality (Hayashi 1950). If one integrates Eq. (43) with a degeneracy parameter, ζ, included, then for high temperatures the electron or positron concentration can be written as

$$n_e = \sim (2/\pi^2)(kT/\hbar c)^3 \, e^{\pm\zeta} \, , \tag{68}$$

and

$$n_{e^-} + n_{e^+} \sim (4/\pi^2)(kT/\hbar c)^3 \cosh\zeta \, . \tag{68a}$$

If the condition of electrical neutrality is imposed then $n_{e^-} - n_{e^+} = n_p$ and

$$\sinh\zeta = \frac{\pi^2 n_p x^3}{4(mc/\hbar)^3} \sim \frac{n_p}{n_{e^-} + n_{e^+}} \, . \tag{69}$$

As has been shown (Alpher and Herman 1950; 1950; 1951; also see footnote 3) the nucleon concentration during the early stages of expanding universe is very small compared with the density of radiation $(1:10^6)$ and, therefore, also small compared with the electron-positron pair concentration. It follows then from Eq. (69) that the parameter ζ is very small and therefore, the degeneracy of electrons or positrons properly has been neglected.

The charge on the electrons and positrons gives rise to a Coulomb interaction energy which contributes to the total energy of the medium. The reasonableness of neglecting this interaction energy can be seen from the following. The average distance between, say, electrons is found from Eq. (68), taking $\zeta \ll 1$ as

$$(1/n_{e^-})^{1/3} \sim h/|\mathbf{p}| \, , \tag{70}$$

i.e., the de Broglie wavelength. It follows tha the Coulomb energy, E_c, for two electrons is

$$E_c \sim -\frac{e^2}{(\hbar/|\mathbf{p}|)} = -\left(\frac{e^2}{\hbar c}\right)|\mathbf{p}|c \sim -\frac{1}{137}E_T \, , \tag{71}$$

where E_T is the mean thermal energy per electron. Because of this Coulomb interaction energy there will be a slight tendency for a given charge to have more nearest neighbors with charge of opposite sign. The fractional charge excess per nearest neighbor at the distance $\hbar|\mathbf{p}|$ may be expected to be of the order $\exp[-E_c/kT] - 1 \sim 1/137$. Therefore, the contribution of Coulomb energy due to nearest neighbors to the total energy of the medium is $\sim E_c/137 \cong E_T/(137)^2$ times the mean number of nearest neighbors. Assuming this number to be of the order of 10, the contribution of the Coulomb energy is $<10^{-3}E_T$, and can, therefore, be neglected.

The contribution of specifically nuclear forces is negligible because the nucleon density is very small compared with nuclear density. Furthermore, the energy evolution of nuclear reactions also can be neglected because it is itself small compared with the already small contribution of the low density of nucleons.

The foregoing small effects bear mainly on the cosmological model which has been discussed in Section II. There are also several questions of this kind which concern the calculation of the rates of the nuclear reactions in Eq. (37) which were determined in Section III. An examination of these reactions shows that of the six rates only $B'n_n n_\nu$ and $C'n_n$, Eqs. (48) and (53), involve two charged product particles. For these, one should more correctly include a factor in the reaction probability, w, to take into account the effect of Coulomb forces. In general this factor is given by[12]

$$2\pi\eta[1 - \exp(-2\pi\eta)]^{-1} , \qquad (72)$$

where

$$\eta = Ze^2 E_e (\hbar c^2 |\mathbf{p}_e|)^{-1} ,$$

Z is the nuclear charge, and E_e and $|\mathbf{p}_e|$ are electron energy and momentum, respectively. The effect of this correction on two integrals in $C'n_n$ and $B'n_n n_\nu$ has been estimated and found to be less than one percent. Thus the effect of the Coulomb forces can be completely neglected in these cases.

As has been mentioned the matrix elements for the nuclear reactions stated in Eqs. (37) have been taken equal to unity for lack of a better estimate (Fermi 1950). There seems to be little doubt that free neutron decay is a super-allowed transition since the decay rate is consistent with those of other light element β emitters. Furthermore, it would seem likely that the matrix elements for all the reactions considered here

[12] See for example Gamow and Critchfield (1949).

would remain about equal in the event that one included effects such as nucleon recoil.

It should also be pointed out that in calculating reaction rates we have considered that the nucleons, i.e., the heavy particles, are at rest. This approximation, which is customarily made, leads to a negligible error.

In addition to the above questions there are a number of more general points which may bear on the validity of the theory presented in this paper. One such question concerns the extrapolation of physical theories back to extremely high temperatures and densities. For example, some quantum field theories introduce a cutoff in, say, the electric field at the value it would have on the surface of the classical electron in order to avoid high-energy difficulties. This cutoff is introduced by appropriate modification of the field equations and, therefore, of the distribution of states in momentum space. When the mean electric field is equal to the foregoing cutoff, one has

$$\sim [e/(e^2/m_e c^2)^2]^2 \cong \rho_\gamma c^2 ,$$

which leads to a temperature of ~15 MeV. However, recent advances in quantum field theory obviate such a high-energy cutoff. In fact, if such a cutoff exists it is probably an order of magnitude higher. This is evidenced by the quantitative agreement between theory and the observed Lamb shift, for example. Cutoffs in momentum space must be larger than present day experimental energies, i.e., $kT > 100$ MeV, since observed bremsstrahlung and pair production, etc., agree with theory quite well.

Another pertinent question is the possible contribution of equilibrium concentrations of "gravitational quanta" to the total density. Although at equilibrium the "graviton" density would be expected to be equal to the photon density, one must consider at what temperatures such equilibria can be maintained. We may apply Eq. (67) to the present situation and replace the coupling coefficient $(e^2/\hbar c)^2$ by the product of $(Gm^2/\hbar c)$ with an electronic or mesonic coupling coefficient whose value will be in the range of $\sim 1 - \sim 10^{-2}$. Then since $(Gm^2/\hbar c) \cong 10^{-45}$ with $m = m_e$, one finds $\tau_{grav}/\tau_T \sim 10^{22}$ at ~ 1 MeV. In order for $\tau_{grav}/\tau_T \sim 1$, i.e., for the "gravitons" to maintain equilibrium, the temperature must be $\sim 10^4$ MeV.[13] It is difficult to see how the introduction of many-body processes would reduce this temperature drastically. One does not know how many different kinds of particles exist in the

[13]It should be noted that in the coupling coefficient the quantity m must be taken to be the relativistic mass of the interacting particles, i.e, $(Gm^2/\hbar c) = (Gm_e{}^2/\hbar c)x^{-2}$. Also note that the numerical results given here for extreme physical conditions are at best rather crude approximations.

range $\sim 10^2 - \sim 10^4$ MeV but on the basis of the presently known types of particles one can determine an upper limit to the graviton contribution. We can compute the ratio of graviton density to that of neutrinos down to the temperature at which neutrinos freeze-in, since beyond this temperature the ratio remains constant. From the analog to Eq. (35) one has, if F_i represents degrees of freedom for each constitutent present, i.e., $F_i \to \rho_i/\rho_\gamma$ as $T \to \infty$, the following relationship:

$$\frac{\rho_{grav}}{\rho_\gamma} = \left(\frac{\Sigma_i F_i \text{ at } T_\nu'}{\Sigma_i F_i \text{ at } T'_{grav}}\right)^{4/3} , \tag{73}$$

where T_ν' and T'_{grav} are the freeze-in temperatures of neutrinos and gravitons, respectively. From the presently known elementary particles which would exist at these temperatures one can estimate from Eq. (73) that $\rho_{grav}/\rho_\gamma < 0.1$ at T_ν'. During the subsequent expansion down to $T \sim 0.1$ MeV, ρ_{grav}/ρ_λ diminishes by a factor of ~ 4, just as ρ_ν/ρ_γ diminishes [see Table 2a,b]. At no time does the upper bound of the graviton contribution to the density exceed 2 or 3 percent, and the total contribution is probably much smaller.

Finally, it seems pertinent to comment on the question as to whether the density of nucleons relative to the density of radiation can be calculated at some very early time on the basis of theoretical considerations with complete symmetry between nucleons and anti-nucleons or whether it is a free initial condition. In particular can the nucleon density be the result of a statistical fluctuation in the competition between different processes of nucleon annihilation such as

$$\begin{aligned} p^+ + p^- &\rightleftarrows 2h\nu \\ &\rightleftarrows \pi^+ + \pi^-, \text{ etc.,} \end{aligned}$$

where p^+ and p^- are proton and antiproton, respectively, and as yet unknown high-energy processes such as

$$p^+ \rightleftarrows \text{mesons} + e^\pm, \text{ etc.?}$$

An examination of this question on statistical grounds yields a probable residual density of nucleons approximately equal to $\rho_\gamma/N^{1/2}$, where N is the number of nucleons initially present in any given finite volume under consideration in co-moving oordinates.[14] If we take an initial volume corresponding to the presently observable universe, the residual number of nucleons is found to be less than would be required to form the earth. It appears that the situation described above is untenable and that the inital nucleon concentration must be specified arbitrarily. This result is in agreement with present thinking in elementary particle physics which does not allow for single nucleon annihilation processes. In addition, it should be pointed out that no catalytic type of reaction (e.g., $2p^+ + p^- \rightarrow 2p^+ + \mu^-$) can vitiate the above statistical arguments because of the finite propagation velocity of disturbances noted in Section II.

V. CONCLUSION

The problem discussed in this paper has been concerned with the detailed nature of the general nonstatic homogeneous isotropic expanding cosmological model derived from general relativity as well as the elementary particle reactions which occur during early epochs. The study of the elementary particle reactions leads to a knowledge of the time dependence of the proton-neutron concentration ratio which is required in the problem of nucleogenesis. While the problem of element

[14]This can be seen from the following arguments. Let the number of protons, antiprotons, neutrons, and antineutrons in any finite co-moving volume V be equal to N. Let α be the probability per particle for any of these particles to transmute to mesons at high temperature. We shall suppose that such transmutations occur first in the expansion, and that annihilation occurs later. This situation yields the largest residual density. Then on the average $4\alpha N$ particles transmute to mesons. The standard deviation σ in the number transmuting is then

$$\sigma = [4\alpha N(1-\alpha)]^{1/2} ,$$

which is a maximum of $N^{1/2}$ for $\alpha = 1/2$. One may expect that in any volume V the excess of nucleons over antinucleons, or conversely, will be of the order of σ, i.e, of the order of $N^{1/2}$. The concentration of these residual nucleons at a later time when the initial volume V_0 has expanded to V_1 is given by $n_{nuc} = (N^{1/2}/V_0)(V_0/V_1)$. To a rough approximation the number of photons originally in V_0, a number approximately equal to N initially, has remained constant down to V_1, so that $V_0/V_1 \cong n_{\gamma_1}/n_{\gamma_0}$ where n_γ is photon concentration. Hence one can write

$$n_{nuc} N^{1/2} n_{\gamma_1}/(V_0 n_{\gamma_0}) \cong n\gamma_1/N^{1/2}, \text{ or } \rho_{nuc} \sim \rho_{\gamma_1}/N^{1/2} .$$

origin stimulated the present study, the results concerning the cosmological model are of interest in themselves. On the basis of the new physical conditions which have been discussed here, it would appear necessary to reexamine the specific re-

TABLE 4
TIMETABLE OF EVENTS IN THE EARLY EPOCHS OF THE EXPANDING UNIVERSE

Temperature (MeV)	Remarks: Neutrino ≡ antineutrino	Remarks: Neutrino ≢ antineutrino
>100	Region of doubtful validity of the field equations where ρ_γ exceeds nuclear density.	
~100	Thermodynamic equilibrium prevails.	
	$\rho_\gamma \cong 1.2\times10^{13}$ g/cm^3	Same as for $\nu \equiv \nu^*$ except
	$\rho_\mu=(7/4)\rho_\gamma$, $\rho_\pi=(3/2)\rho_\gamma$	$\rho_\nu=(7/4)\rho_\gamma$
	$\rho_\nu=(7/8)\rho_\gamma$, $\rho_e=(7/4)\rho_\gamma$	$t \cong 5.9\times10^{-5}$ sec
	$t \cong 6.3\times10^{-5}$ sec	
~100 - ~10	Mesons annihilate converting energy into photons, electrons, and neutrinos	
~10	Neutrinos are freezing-in during this period.	
	$\rho_\gamma \cong 1.2\times10^{9}$ g/cm^3	Same as for $\nu \equiv \nu^*$ except
	$\rho_\mu=10^{-6}\rho_\gamma$, $\rho_\pi \sim 10^{-6}\rho_\gamma$	$\rho_\nu=(7/4)\rho_\gamma$
	$\rho_\nu=(7/8)\rho_\gamma$, $\rho_e=(7/4)\rho_\gamma$	$t \cong 7.8\times10^{-3}$ sec
	$t \cong 8.7\times10^{-3}$ sec	
~10 - ~2	Continued adiabatic expansion of universe with $T_\nu \cong T$ despite negligible interaction of neutrinos with medium.	
~2	Start of electron-positron annihilation.	
	$\rho_\gamma \cong 1.9\times10^{6}$ g/cm^3	Same as for $\nu \equiv \nu^*$ except
	$\rho_\mu = \rho_\pi \sim 0$	$\rho_\nu \cong (7/4)\rho_\gamma$
	$\rho_\nu \cong (7/8)\rho_\gamma$, $\rho_e=(7/4)\rho_\gamma$	$t \cong 0.20$ sec
	$t \cong 0.22$ sec	

TABLE IV (continued)

TIMETABLE OF EVENTS IN THE EARLY EPOCHS OF THE EXPANDING UNIVERSE

Temperature (MeV)	Remarks: Neutrino ≡ antineutrino	Remarks: Neutrino ≢ antineutrino
~2 - ~0.05	Electron-positron annihilation, converting energy into photons. Neutrinos cool adiabatically relative to remaining particles, the latter maintaining thermodynamic equilibrium. [See Tables 1 and 2 for more details during this epoch.] The neutron-proton abundance ratio reaches the free decay value, 4.5:1-6.0:1, at T ~ 0.2 MeV. Nucleogenesis begins at T ~ 0.2 MeV.	
~0.05	Nucleogenesis is well under way.	
	$\rho_\gamma \cong 0.72$ g/cm^3	$\rho_\gamma \cong 0.72$ g/cm^3
	$\rho_\nu \cong 0.24\rho_\gamma$, $\rho_e \sim 0$	$\rho_\nu \cong 0.47\rho_\gamma$, $\rho_e \sim 0$
	$t \cong 600$ sec	$t \cong 550$ sec
~0.03	Nucleogenesis essentially complete except for charge adjustment by β decay.	
	t ~30 min	
~0.03 MeV - ~1 keV	Thermonuclear reactions among some of the light elements, viz., Li, Be, B, D with H, continue during this period.	
~0.015 eV	At t ~10^8 yr, T ~170° K and ρ ~ 10^{-26} g/cm^3, galaxies probably form.	

actions among the lighter nuclei, particularly as regards the missing species at A = 5.

In order to summarize, we have presented the above calculations in abbreviated form as a timetable of events in the very early stages of the expanding universe, through the period of residual thermonuclear reactions (Alpher, Herman, and Gamow 1948) and galaxy formation (Gamow 1953). In Table 4 are given for various temperatures the corresponding epochs according to the expanding cosmological model involving the interconversion of matter and radiation, the densities of the various constituents according to the appropriate relativistic quantum statistics, as well as remarks concerning some of the principal physical phenomena that occur during these various early stages. This tabulation, it will be noted, covers both distinguishable and indistinguishable neutrinos.

Finally, we should like to point out that all of the results presented in this paper follow uniquely from general relativity, relativistic quantum statistics, and β-decay theory without the introduction of any free parameters, so long as the density of matter is very small compared with the density of radiation.

VI. ACKNOWLEDGEMENTS

We wish to thank Mrs. Betty Grisamore, Mrs. Kathryn Stevenson, and Mr. Charles V. Bitterli for their assistance in some of the numerical work, Miss Shirley Thomas for typing this manuscript, and Miss Doris Rubenfeld for assistance with the illustrations.

REFERENCES

Alpher, R. A., and Herman, R. C. 1950, Rev. Modern Phys., **22**, 153.

Alpher, R. A., and Herman, R. C. 1951, Phys. Rev., **84**, 60.

Alpher, R. A., and Herman, R. C. 1953, Annual Review of Nuclear Science, (Stanford: Annual Review of Nuclear Science, Inc.), Vol. 2, p. 1.

Alpher, R. A., Herman, R. C., and Gamow, G. 1948, Phys. Rev., **74**, 1198.

Alpher, Ralph A., Follin, James W., Jr., and Herman, Robert C. 1953, Phys. Rev., **91**, 479.

Bearden, J. A., and Watts, H. M. 1951, Phys. Rev., **81**, 73.

Caianiello, E. R. 1952, Phys. Rev., **86**, 564.

DuMond, J. W. M., and Cohen, E. R. 1951, Phys. Rev., **82**, 555.

Einstein, A. 1945, The Meaning of Relativity (Princeton: Princeton University Press).

Fermi, E. 1950, Nuclear Physics (Chicago: University of Chicago Press).

Gamow, G. 1949, Revs. Modern Phys., **21**, 367.

Gamow, G. 1953, Kgl. Danske Videnskab. Selskab, Mat.-fys. Medd., **27**, No. 10.

Gamow, G., and Critchfield, C. L. 1949, Theory of Atomic Nucleus and Nuclear Sources (Oxford: Clarendon Press).

Hayashi, C. 1950, Progr. Theoret. Phys. (Japan), **5**, 224.

Langer, L. M., and Moffat, R. J. D. 1952, Phys. Rev., **88**, 689.

Robson, J. M. 1951, Phys. Rev., **83**, 349.

Tolman, R. C. 1934, Relativity, Thermodynamics and Cosmology (Oxford: Clarendon Press).

Van Patter, D. M. 1952, Massachusetts Institute of Technology, Technical Report Number 57, January (unpublished).

Underhill, A. 1952, Symposium on the Abundance of Elements, held at Yerkes Observatory, Williams Bay, Wisconsin, November 6-8 (unpublished).
Wu, C. S. 1952, Physica, **18**, 989.

PRIMORDIAL HELIUM ABUNDANCE AND THE PRIMORDIAL FIREBALL. II

P. J. E. Peebles
Palmer Physical Laboratory, Princeton University, Princeton, New Jersey

Received: May 23, 1966

ABSTRACT

This paper contains results of a calculation of the He^3, He^4, and deuterium abundances produced in the early stages of expansion of the Universe. It is assumed in the calculation that the present temperature of the primordial fireball is 3° K, and the abundances are computed for two values of the present mean density of matter in the Universe, and for a range of possible changes in the time scale for expansion of the early Universe.

INTRODUCTION

Subsequent to the suggestion by R. H. Dicke (1964, unpublished) that one ought to search for the residual, thermal radiation left over from the early stages of expansion of the universe (Dicke, Peebles, Roll and Wilkinson 1965) a new isotropic microwave background was discovered (Penzias and Wilson 1965). The available information on the spectrum of this radiation (Roll and Wilkinson 1966; Field and Hitchcock 1966; Thaddeus and Clauser 1966) is consistent with the assumption that it is thermal, blackbody radiation. If this thermal character is confirmed by further observations, it will be compelling evidence that the new background is the primordial fireball, for it is difficult to see how any process other than thermal relaxation could produce the characteristic thermal spectrum. If we assume that this is established it provides us with the very remarkable opportunity to bring observational evidence to bear on processes which occurred when the Universe was 10^{30} times more dense than it is now (Peebles 1966a), when according to general relativity an appreciable amount of helium should have formed.

The purpose of the present article is to describe in greater detail the calculation (Peebles 1966a) of the production of helium in the highly concentrated Universe. We also show how the amount of element production depends on the time scale for expansion of the Universe through the early, highly contracted phase. This is an important question, since the time scale for expansion could be reduced by the presence of gravitational radition, or of other new kinds of energy density (Dicke 1966), and since it has been emphasized also that the time scale could be appreciably modified by shear motion in the early Universe (Thorne 1966; Kantowski and Sachs 1966; Hawking and Tayler 1966).

The physical processes of interest here originally were considered in the "big-bang theory," the theory of the formation of the elements in the early, high contracted Universe (von Weizsäcker 1938; Chandrasekar and Henrich 1942; Gamow 1948; Hayashi 1950; Alpher, Follin, and Herman 1953). This theory seems to have been generally abandoned with the realization that elements much heavier than helium would not have been formed in any appreciable amount, and with the development of the theory of nucleosynthesis in stars (Burbidge, _et al._ 1957). The great importance here of the primordial fireball, if it exists, is that it provides us with tangible evidence that the universe did pass through a hot, highly contracted phase, so that it becomes of very direct concern to attempt to understand the observable consequences of physical processes that would have occurred in the very early Universe.

In this paper we will assume that the newly discovered microwave background is the primordial fireball - thermal blackbody radiation with a temperature of 3° K. The numerical results are presented for two assumed values of the mean mass density in the Universe, 7×10^{-31} gm/cm^3 and 1.8×10^{-29} gm/cm^3. The first value is the estimated mass density in ordinary galaxies (Oort 1958; van den Bergh 1961; Kiang 1961), and it is a reasonable lower limit to the total mass density. The second value is the mass density required to close the Universe, assuming a reciprocal Hubble constant $H^{-1} = 1\times10^{10}$ yr. This density is a factor of about 3 below the upper limit obtained from the measurements of the acceleration parameter (Sandage 1961). We shall confine attention in this paper to the production of nuclei up to helium.

TIME SCALE

In this section we list the equations used to compute the variation of temperature and density with time during the epoch of element formation. The equations are based on an isotropic homogeneous cosmological model, and are equivalent to those previously used by Alpher _et al._ (1953).

According to the simple isotropic homogeneous models the time scale for expansion is given by the equation

$$\left(\frac{1}{a}\frac{da}{dt}\right)^2 = \frac{8\pi G\varepsilon}{3c^2}, \tag{1}$$

where the energy density $\varepsilon(t)$ satisfies

$$\frac{d\varepsilon}{dt} = -\frac{3(\varepsilon+P)}{a}\frac{da}{dt}. \tag{2}$$

In these equations P is the pressure of the material and a(t) is the expansion parameter. Here and throughout, all lengths, times, and so on are proper quantities as measured by a comoving observer who uses ordinary instruments. The terms involving the curvature of the model, and the cosmological constant, which usually appear in equation (1) (Tolman 1934) must be negligible at the epoch of element formation, for we know that the Universe has been able to expand out to the present epoch with a general order-of-magnitude agreement between the present values of the left- and right-hand sides of equation (1).

We have assumed that the Universe expanded from an initial temperature of at least 10^{12} °K. This temperature is high enough to have assured thermal equilibrium, including thermal distributions of neutrino pairs of both kinds, in equilibrium with the radiation and the electron-positron pairs. As the Universe cooled below 10^{10} °K (~1 MeV), the electron-positron pairs would have recombined and fed their energy to the radiation. At a temperature of 10^{10} °K the most effective process for transferring energy to the neutrinos would be electron neutrino scattering. If this process exists, the cross-section would be of the order of 10^{-44} cm^2; and since at a temperature of 10^{10} °K the characteristic expansion time of the Universe is 1 sec, and the electron-pair density is somewhat less than 10^{32} cm^{-3}, the depth of the Universe for scattering a neutrino is only about 0.03. Thus it is an adequate approximation to suppose that at all temperatures the neutrinos expand according to the simple adiabatic law,

$$T_\nu(t) \propto 1/a(t) \ . \tag{3}$$

The total energy density in neutrinos of both types, electron and muon, is (Landau and Lifshitz 1958)

$$\varepsilon_\nu = \frac{7}{4} \sigma T_\nu^{\ 4} \ , \tag{4}$$

where σ is the electromagnetic radiation energy density constant, $\sigma = 7.6 \times 10^{-15}$ ergs/cm^3 °K^4.

Matter and electromagnetic radiation would remain in thermal equilibrium until long after nucleosynthesis is finished, the variation of the temperature T_e of radiation and matter being given by equation (2) (where we take account of the energy and pressure of the radiation and electron pairs only). We have adopted the free particle approximation for the energy density and pressure of the electron pairs,

$$\varepsilon_e = \frac{2}{\pi^2 \hbar^3} \int_0^\infty \frac{p^2 E dp}{e^{E/kT_e} + 1} \ , \tag{5}$$

$$P_e = \frac{2c^2}{3\pi^2 \hbar^3} \int_0^\infty \frac{p^4 dp}{E(e^{E/kT_e} + 1)} \ . \tag{6}$$

In these equations $E^2 = p^2c^2 + m_e{}^2c^4$, where m_e is the mass of an electron. We are assuming here, and will assume below, that the Fermi energies (chemical potentials) of the electrons and neutrinos may be neglected. This is almost implied by the assumption that the present radiation temperature is 3° K, for with this temperature the leptons would approach degeneracy only if the lepton number were a factor of 10^6 times larger than the nucleon number in the Universe. We believe that the most reasonable assumption would be that the lepton number of the Universe is of the same order as the nucleon number, so that the lepton number may be neglected.

The radiation energy density is $\varepsilon_r = \sigma T_e^4$. In the assumed model the significant contributions to the total energy density are the radiation density ε_r, the energy density ε_ν in the two kinds of neutrinos (eq. [4]), and the electron pair energy density ε_e (eq. [5]). When the temperature is 10^9 °K the contribution to the total density due to the nucleons is at most one part in 10^4 of the total density, and this contribution has been neglected.

In the next section we have listed the results (Table 1) of a numerical integration of equations (2), (5), and (6), to obtain the radiation temperature T_e as a function of the expansion parameter, and of equation (1) to obtain the time variation of the expansion parameter.

TABLE 1

NEUTRON PRODUCTION

T_e (10^{10}° K)	T_e/T_ν	λ(sec^{-1}) (Eq. 17)	$\hat{\lambda}$ (sec^{-1}) (Eq. 18)	Time (sec)	X_n (Eq. 16)
100	1.00	3.60×10^9	3.66×10^9	0	0.496
60	1.00	2.79×10^8	2.86×10^8	1.94×10^{-4}	.494
30	1.00	8.61×10^6	9.05×10^6	1.129×10^{-3}	.488
20	1.00	1.12×10^6	1.21×10^6	2.61×10^{-3}	.481
10	1.00	3.37×10^4	3.91×10^4	1.078×10^{-2}	.463
6	1.00	2.48×10^3	3.19×10^3	0.0301	.438
3	1.001	6.78×10^1	1.12×10^2	0.1209	.380
2	1.002	7.75	16.4	0.273	.331
1	1.008	0.151	0.682	1.103	.241
0.6	1.022	5.70×10^{-3}	7.12×10^{-2}	3.14	.1969
0.3	1.081	2.12×10^{-5}	3.42×10^{-3}	13.83	.1719
0.2	1.159	2.08×10^{-7}	5.28×10^{-4}	35.2	.1671
0.1	1.346	1.44×10^{-11}	1.95×10^{-5}	182.0	.1646
0.03	1.401	3.0×10^{-28}	3.3×10^{-7}	2.08×10^3	0.1640

To take account of the possibility that the time scale for expansion through this highly contracted phase was altered by the presence of gravitational radiation, or of other kinds of neutrinos, or of matter less compressible than radiation, or of shear motion, we have computed the element abundances with the assumption that the cosmic time has been scaled by a constant factor S. That is, the time $\hat{t}(T_e)$ for expansion of the model from infinite temperature to temperature T_e is given by the equation

$$\hat{t}(T_e) = St(T_e) , \tag{7}$$

where $t(T_e)$ is the corresponding time for the original simple model. The factor S is a constant. For more general models it may be noted that the amount of element production depends on two factors, the neutron abundance, which is frozen in at a temperature of $1 \times 10^{10\circ}$ K if S = 1, and the rate of reactions at the time element formation become possible, when $T_e \sim 1 \times 10^{9\circ}$ K. The neutron abundance is given in the next section (Table 2). It is seen that if the expansion time scale were substantially reduced ($S \ll 1$) the neutron abundance would approach a constant value, and element production would depend on the time scale for expansion through the epoch $T_e \sim 1 \times 10^{9\circ}$ K.

NEUTRON PRODUCTION

The production of elements in the early Universe depends on the formation of neutrons due to reactions with the thermal electron and neutrino pairs. The important reactions are

$$p + \bar{\nu} \rightarrow n + e^+ , \qquad p + e^- \rightarrow n + \nu . \tag{8}$$

We can neglect the reverse of free neutron decay, for we know that when $kT \gtrsim (m_n - m_p)c^2$, the rate of this reaction is that characteristic of free neutron decay, 10^{-3} sec^{-1}, and by the time free neutron decay becomes appreciable the lepton energies are too low to cause further reactions. For the same reason we have neglected the effect of the partial antineutrino degeneracy on the free-neutron decay rate.

The cross-sections for the reactions were computed using the interaction Hamiltonian

$$H_1 = \frac{1}{\sqrt{2}} [\bar{p}\gamma_\mu(C_V - C_A\gamma_5)n][\bar{e}\gamma_\mu(1 + \gamma_5)\nu] + \text{herm. conjugate} . \tag{9}$$

We are interested in reaction energies of the order of 1 MeV, so we can use the values of the coefficients C_V and C_A obtained from nuclear beta-decay (Källén 1964),

$$C_V = 1.418 \times 10^{-49} \text{ ergs cm}^3 , \qquad (C_A/C_V)^2 = 1.39 . \qquad (10)$$

The rate for the first reaction (8), from left to right, is

$$\lambda_a = \frac{(C_V^2 + 3C_A^2)}{2\pi^3\hbar^7c^3} \int \frac{v_e}{c} \frac{(E_\nu - Q)^2 p_\nu^2 dp_\nu}{(e^{E_\nu/kT_\nu} + 1)(1 + e^{-E_e/kT_e})} . \qquad (11)$$

This is the reaction rate per proton. In this equation v_e is the velocity of the positron produced in the reaction and E is its kinetic energy (including rest mass), while $E_\nu = p_\nu c$ is the kinetic energy of the incident neutrino. The second term in the denominator is the correction for the partial degeneracy of the positrons. In computing the cross-section and the electron-pair density we have neglected electromagnetic corrections. The threshold for the reaction is $E_\nu = Q + m_e c^2$, where

$$Q = (m_n - m_p)c^2 , \qquad (12)$$

and $m_n - m_p$ is the mass difference between a neutron and proton. The integral (11) extends over all neutrino energies greater than the threshold.

The rate of the second reaction (8), from left to right, per proton, is

$$\lambda_b = K \int \frac{(E_e - Q)^2 p_e^2 dp_e}{(e^{E_e/kT_e} + 1)(1 + e^{-E_\nu/kT_\nu})} , \qquad (13)$$

the threshold here being $E_e = Q$. The coefficient K is the same as the coefficient appearing in front of the integral in equation (11), p_e is the momentum of the incident electron, $E_e^2 = p_e^2c^2 + m_e^2c^4$, and E_ν is the energy of the neutrino produced in the reaction. The rate of the first reaction (8), from right to left, per neutron is

$$\hat{\lambda}_a = K \int_0^\infty \frac{(E_e + Q)^2 p_e^2 dp_e}{(1 + e^{E_e/kT_e})(1 + e^{-E_\nu/kT_\nu})} , \qquad (14)$$

where the symbols are the same as defined above, and the rate for the second reaction (8), from right to left, per neutron is

$$\hat{\lambda}_b = K \int_0^\infty \frac{v_e}{c} \frac{(E_\nu + Q)^2 p_\nu^{\ 2} dp_\nu}{\left(1 + e^{E_\nu/kT_\nu}\right)\left(1 + e^{-E_e/kT_e}\right)} . \tag{15}$$

These formulae are equivalent to the ones given previously by Alpher et al. (1953).

The neutron-proton abundance ratio is frozen in at a temperature of 1×10^{10}° K, when the characteristic expansion time of the Universe is 1 sec. This is well before free neutron decay becomes appreciable and well before element formation can commence (at $T_e \sim 1\times 10^{9}$° K). Since neutron production and burning thus can be separated in a first approximation, it is useful to compute the neutron abundance, neglecting for the moment neutron decay and element formation. In this case the neutron abundance X_n is given by the equation

$$\frac{dX_n}{dt} = \lambda - (\lambda + \hat{\lambda})X_n , \tag{16}$$

where

$$\lambda = \lambda_a + \lambda_b \tag{17}$$

is the total rate of neutron production per proton (eqs. [11] and [13]), and

$$\hat{\lambda} = \hat{\lambda}_a + \hat{\lambda}_b \tag{18}$$

is the total rate for destruction of neutrons per neutron (eqs. [14] and [15]). The ratio λ/λ would be equal to the Boltzmann factor $e^{Q/kT}$ if the electron and neutrino temperatures were equal.

In Table 1 we have listed some values of the total reaction rates λ and λ obtained by numerically integrating equations (11) and (13)-(15). Also listed is the ratio T_e/T_ν of the radiation temperature to the neutrino temperature. The neutrino temperature was given by equation (3), and the radiation temperature was found by numerically integrating equations (2), (5), and (6). It will be noted that the ratio T_e/T_ν approaches the value $(11/4)^{1/3} = 1.401$, as required by conservation of entropy. Finally, we have given the neutron abundance X_n as a function of time, according to equation (16), with the expansion time scale given by the simple cosmological model (S = 1 in eq. [7]). The neutron abundances are in good agreement with the previous results of Alpher et al. (1953).

We have also integrated equation (16) assuming other expansion time scales, as defined by the constant S in equation (7). The results are listed in Table 2. The neutron abundance X_n in the table is evaluated at the time the temperature has fallen to $T_e = 1\times10^{9\circ}$ K. This is the temperature at which element formation can begin, and at this temperature the value of X_n given by equation (16) is very nearly constant, independent of time.

NUCLEAR REACTIONS

The most important element-formation reactions are

$$n + p \rightarrow d + \gamma\ , \quad d + d \rightarrow H_e^3 + n\ , \quad d + d \rightarrow t + p\ , \tag{19}$$

$$H_e^3 + n \rightarrow t + p\ , \quad t + d \rightarrow H_e^4 + n\ , \tag{20}$$

At temperatures in excess of $10^{10\circ}$ K the rates of these reactions are very high, so that the element abundances relax to thermal equilibrium. The thermal equilibrium abundance ratios are given by the equations

$$\left(\frac{X_n X_p}{X_d}\right)_{th} \equiv G_{np} = \frac{4}{3}\,\frac{(2\pi m_n kT_e)^{3/2}}{(2\pi\hbar)^3 N}\left(\frac{m_p}{m_d}\right)^{3/2} e^{-B_1/kT_e}\ , \tag{21}$$

$$\left(\frac{X_d^3}{X_n X_3}\right)_{th} \equiv G_{n3} = \frac{9}{4}\left(\frac{m_d^2}{m_n m_3}\right)^{3/2} e^{-B_2/kT_e}\ , \tag{22}$$

$$\left(\frac{X_d^2}{X_t X_p}\right)_{th} \equiv G_{tp} = \frac{9}{4}\left(\frac{m_d^2}{m_t m_p}\right)^{3/2} e^{-B_3/T_e}\ , \tag{23}$$

$$\left(\frac{X_d X_t}{X_4 X_n}\right)_{th} \equiv G_{dt} = 3\left(\frac{m_d m_t}{m_4 m_n}\right)^{3/2} e^{-B_4/kT_e}\ . \tag{24}$$

In each of these equations the energy B in the exponential is the Q value of the reaction (König, Mattauch, and Wapstra 1962). On the left-hand sides of these equations in the abundances X represent the number densities of each nuclear species divided by the total nucleon number N per unit volume, so that we have

$$X_n + X_p + 2X_d + 3X_3 + 3X_t + 4X_4 = 1\ . \tag{25}$$

TABLE 2

NEUTRON ABUNDANCE AT $T_e = 1\times10^{9}$° K
ACCORDING TO EQ. (16)

Time-Scale Factor S (Eq. 7)	X_n (Eq. 16)	Time-Scale Factor S (Eq. 7)	X_n (Eq. 16)
3	0.0875	0.01	0.412
1	.1646	0.003	.441
0.3	.252	0.001	.459
0.1	.319	0.0001	0.481
0.03	0.376		

The masses of the nuclei on the right-hand sides of the equations are similarly labeled. We mean by the subscript on the abundance ratios that these are the thermal equilibrium values.

If the present radiation temperature is 3° K, the right-hand side of equation (21) is of the order of 10^{10} when $T_e = 10^{10}$° K. This means that at this temperature the deuterium abundance is very small, and by equations (22)-(24) the abundances of heavier nuclei are smaller still. As the Universe expands the temperature decreases and the Boltzmann factor becomes small enough to favor a high helium abundance. The amount of helium actually formed depends on how completely the abundance can relax to the thermal equilibrium value. It should be noted that in taking account only of the relaxation provided by the reactions (19)-(20) we have at least a minimum value of the helium abundance. If any other reaction were found to be important it could only increase the relaxation rate, and so increase the amount of helium formed.

The cross-section for the first reaction (19) varies inversely with relative velocity, so that the reaction rate is determined by the constant factor

$$(\sigma v)_{np} = 4.55\times10^{-20}\ \text{cm}^3/\text{sec}\ . \qquad (26)$$

This cross-section was obtained using detailed balance from the observed cross-section for photodissociation of the deuteron (Hulthen and Sugawara 1957). The cross-section for the first reaction (20) also varies approximately inversely as the relative velocity, and with sufficient accuracy we have for this reaction (Seagrave 1960)

$$(\sigma v)_{n3} = 8.3\times 10^{-16}\ \mathrm{cm}^3/\mathrm{sec}\ . \tag{27}$$

For the second and third reactions in (19) and the second reaction in (20) it was assumed that the cross-section varies with the reaction energy as

$$\sigma(E_d) = \frac{S}{E_d}\, e^{-A/\sqrt{(E_d)}}\ . \tag{28}$$

Here E_d is the energy of the incident deuteron when the other particle involved in the reaction at rest. This is the conventional Gamow form, but the constant S and A both were chosen to obtain the best fit to the observed cross-section near 100 keV ($\sim 10^{9\circ}$ K). For the second and third reactions in (19) the total cross-section including both reactions (Wenzel and Whaling 1952; Bransden 1960) is adequately represented by the constants

$$S_{dd} = 340\ \text{keV barns}, \qquad A_{dd} = 46.2\ \text{keV}^{1/2}\ , \tag{29}$$

with unity branching ratio to tritium and He^3. For the second reaction in (20) the values of the constants used are

$$S_{dt} = 7.75\times 10^4\ \text{keV barns} \qquad A_{dt} = 49.7\ \text{keV}^{1/2}\ . \tag{30}$$

This provides a satisfactory fit to the observed cross-section (Fowler and Brolley 1956) when E_d is less than or equal to 100 keV. At higher energies the cross-section given by equation (28) is larger than the observed cross-section, but this is not expected to affect the calculated abundances appreciably.

The cross-section formula (28) was numerically integrated with a Maxwell velocity distribution for the particles to obtain the total reaction rates as functions of the temperature T_e.

In terms of the nuclear abundances defined above (eq. [25]) the production of elements up to helium is determined by the following six equations:

$$\frac{dX_n}{dt} = \lambda X_p - (\lambda + \lambda_d)X_n - (\sigma v)_{np} N(X_n X_p - G_{np} X_d) +$$

$$+ \frac{1}{2} R_{dd}(X_d{}^2 - G_{n3} X_n X_3) - (\sigma v)_{n3} N(X_n X_3 - G_{tp}/G_{n3} X_t X_p) +$$

$$+ R_{dt}(X_d X_t - G_{dt} X_4 X_n)\ , \tag{31}$$

$$\frac{dX_p}{dt} = -\lambda X_p + (\hat{\lambda} + \lambda_d)X_n - (\sigma v)_{np}N(X_nX_p - G_{np}X_d) +$$

$$+ \frac{1}{2} R_{dd}(X_d^2 - G_{tp}X_tX_p) + (\sigma v)_{n3}N(X_nX_3 - G_{tp}/G_{n3}X_tX_p), \tag{32}$$

$$\frac{dX_d}{dt} = (\sigma v)_{np}N(X_nX_p - G_{np}X_d) - R_{dd}(2X_d^2 - G_{tp}X_tX_p - G_{n3}X_nX_3) -$$

$$- R_{dt}(X_dX_t - G_{dt}X_4X_n) , \tag{33}$$

$$\frac{dX_3}{dt} = \frac{1}{2} R_{dd}(X_d^2 - G_{n3}X_nX_3) - (\sigma v)_{n3}N(X_nX_3 - G_{tp}/G_{n3}X_tX_p) , \tag{34}$$

$$\frac{dX_t}{dt} = \frac{1}{2} R_{dd}(X_d^2 - G_{tp}X_tX_p) + (\sigma v)_{n3}N(X_nX_3 - G_{tp}/G_{n3}X_tX_p) -$$

$$- R_{dt}(X_dX_t - G_{dt}X_4X_n) , \tag{35}$$

$$\frac{dX_4}{dt} = R_{dt}(X_dX_t - G_{dt}X_4X_n) . \tag{36}$$

In these equations the total rate of the second and third reactions in (19) per unit volume is equal to $R_{dd}(T_e)$ multiplied by the square of the deuteron number density, and similarly $R_{dt}(T_e)$ yields the rate of the second reactions in (20). As mentioned above, these rates were obtained by numerically integrating the cross-section (28), using the constants (29) and (30), with a Maxwell velocity distribution for the particles. The free neutron decay rate is λ_d = log (2)/11.7 min (Sosnovsky, Spivak, Prokofiev, Kutikov, and Dobrinin 1959), and λ and $\hat{\lambda}$ were defined by equations (17) and (18).

The equations (31)-(36) were integrated numerically starting from a temperature of 10^{12}° K. The time variation of the abundances is given in Tables 3 and 4 for the two assumed values of the present mean mass density, in each case the present radiation temperature being 3° K. For these models the final helium abundance by mass was found to be 28 per cent if the present mass density is $\rho_0 = 1.8\times10^{-29}$ gm/cm^3, and 26 per cent if $\rho_0 = 7\times10^{-31}$ gm/cm^3. These abundances are slightly smaller than the values previously reported (Peebles 1966a), the difference being due to improved numerical accuracy of the integration.

TABLE 3

TIME VARIATION OF THE ABUNDANCES BY MASS OF NEUTRONS, DEUTERIUM, He^3, TRITIUM, AND He^4, ASSUMING $\rho_0 = 1.8\times10^{-29}$ gm/cm^3, $T_0 = 3°$ K, AND S = 1 (EQ. [7])*

T_e (10^{10}° K)	n	d	He^3	t	He^4
100	0.496	6.3×10^{-9}	1.8×10^{-17}	1.8×10^{-17}	3.3×10^{-27}
30	0.488	1.10×10^{-9}	6.1×10^{-19}	6.0×10^{-19}	3.2×10^{-28}
10	0.463	1.4×10^{-10}	4.0×10^{-20}	3.8×10^{-20}	2.0×10^{-29}
3	0.380	7.1×10^{-11}	1.0×10^{-20}	$.8\times10^{-20}$	1.7×10^{-29}
1	0.240	6.0×10^{-11}	1.5×10^{-19}	1.1×10^{-19}	2.0×10^{-21}
0.3	0.1698	2.7×10^{-9}	7.5×10^{-13}	2.9×10^{-12}	1.8×10^{-13}
0.16	0.1566	1.4×10^{-6}	1.3×10^{-10}	6.0×10^{-9}	5.8×10^{-7}
0.13	0.1500	3.5×10^{-5}	1.1×10^{-9}	1.6×10^{-7}	2.0×10^{-4}
0.11	0.1126	7.2×10^{-4}	4.3×10^{-8}	3.3×10^{-6}	0.0605
0.10	0.0119	5.7×10^{-4}	2.1×10^{-7}	2.7×10^{-6}	0.268
0.086	1.1×10^{-5}	2.7×10^{-5}	7.6×10^{-7}	1.3×10^{-7}	0.282

* The time is given in Table 1.

TABLE 4

TIME VARIATION OF THE ABUNDANCES BY MASS OF NEUTRONS, DEUTERIUM, He^3, TRITIUM, AND He^4, ASSUMING $\rho_0 = 7\times10^{-31}$ gm/cm^3, $T_0 = 3°$ K, AND S = 1 (EQ. [7])*

T_e (10^{10}° K)	n	d	He^3	t	He^4
100	0.496	2.4×10^{-10}	2.7×10^{-20}	2.7×10^{-20}	2.0×10^{-30}
30	0.488	4.3×10^{-11}	9×10^{-22}	9×10^{-22}	1.9×10^{-32}
10	0.462	1.0×10^{-11}	6.4×10^{-23}	6.0×10^{-23}	1.2×10^{-33}
3	0.380	2.8×10^{-12}	1.6×10^{-23}	1.3×10^{-23}	1.0×10^{-32}
1	0.241	2.3×10^{-12}	2.2×10^{-22}	1.7×10^{-22}	2.6×10^{-27}
0.3	0.170	1.1×10^{-10}	1.2×10^{-16}	4.5×10^{-16}	5.5×10^{-19}
0.12	0.147	5.6×10^{-6}	9×10^{-11}	2.3×10^{-8}	1.2×10^{-7}
0.11	0.143	3.4×10^{-5}	4.0×10^{-10}	1.6×10^{-7}	4.0×10^{-6}
0.10	0.137	3.2×10^{-4}	6.7×10^{-9}	1.5×10^{-6}	2.6×10^{-4}
0.09	0.116	3.6×10^{-3}	7.4×10^{-7}	1.7×10^{-5}	2.7×10^{-2}
0.08	5.3×10^{-3}	2.8×10^{-3}	8×10^{-6}	1.4×10^{-5}	0.243
0.07	1.5×10^{-5}	3.0×10^{-4}	2.7×10^{-5}	1.4×10^{-6}	0.256
0.03	7×10^{-8}	5.0×10^{-5}	4.9×10^{-5}	2.7×10^{-7}	0.256

* The time is given in Table 1.

The equations (31)-(36) were integrated also for a range of values of the time-scale factor S in equation (7). The abundances by mass of deuterium, He^3 (including the contribution from the beta-decay of tritium) and He^4 after the completion of nuclear burning are shown in Figures 1 and 2 as functions of S for the two assumed values of the present mass density.

CONCLUSIONS

With the assumption of the simple cosmological model we have concluded that the primordial helium abundance should be in the range 26-28 per cent by mass. Apparently a primordial abundance this high cannot yet be ruled out because we do not have a reliable measure of the helium abundance in very old Population II stars. The helium abundance in the Sun is thought to be 25-30 per cent by mass (Gaustad 1964), but we do not know how much of this helium was produced in earlier generations of stars.

It is important to ask what assumptions one might introduce to arrive at a lower primeval helium abundance. A reasonable upper limit to the possible amount of primordial gravitational radiation would be an energy density equivalent to 1.8×10^{-29} gm/cm^3. This is the density required to close the Universe, with the adopted value of the reciprocal Hubble constant ($H^{-1} = 1\times 10^{10}$ yr); although it will be noted that, if the Universe were closed by radiation of this amount, it would make the age of the Universe equal to 5×10^9 yr, in serious conflict even with the radioactive-decay ages of the meteorites, 4.5×10^9 yr. (We are here assuming that the cosmological constant is equal to zero.) Gravitational radiation of this maximum amount would decrease the time scale for expansion of the early Universe by a factor of 160. With this factor it is seen from Figure 2 that with the minimum acceptable nucleon density, 7×10^{-31} gm/cm^3, the helium abundance would be about 25 percent by mass, but the deuterium abundance would be unreasonably high. Thus, gravitational radiation apparently cannot appreciably reduce the primeval abundances, and in fact on the assumed model we must rule out the idea that the Universe could contain primordial gravitational radiation energy amounting to much more than the fireball radiation energy density. This conclusion applies to graviational radiation already present at the time of element formation.

Next, the basic assumption of a homogeneous, isotropic cosmological model might be questioned (Thorne 1966; Kantowski and Sachs 1966; Hawking and Tayler 1966). It should be noted first that, if the nucleon distribution were irregular but the radiation energy distributed in a homogeneous, isotropic way, it would not appreciably alter the computed element production. The important question is the time scale for cooling the radiation. It is seen from Figure 1 that element production would be effectively eliminated if irregularities could decrease the expansion time scale by a factor of 10^6 below that given by the simple models. A second possibility is that

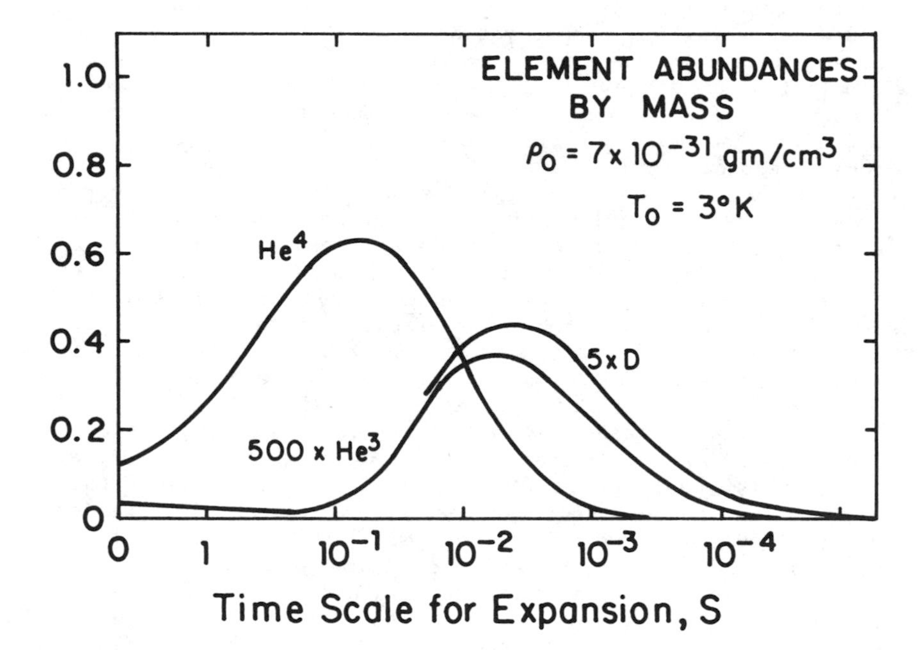

Figure 1

Element abundances by mass, produced in the early Universe, assuming present mean density 1.8×10^{-29} gm/cm^3 and fireball radiation temperature 3° K.

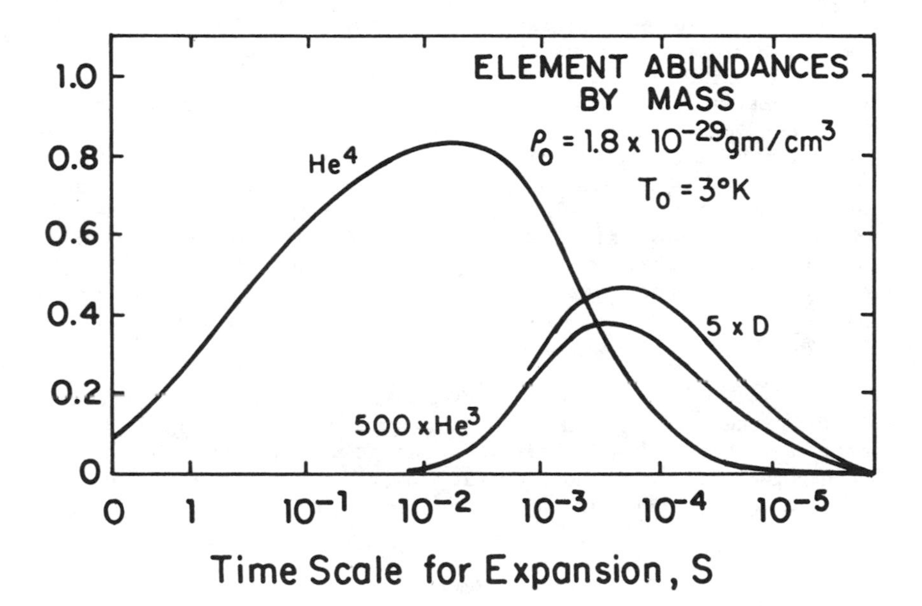

Figure 2

Element abundances by mass, produced in the early Universe, assuming present mean density 7×10^{-31} gm/cm^3, and fireball radiation temperature 3° K.

the time be increased by a factor of perhaps 10, thus eliminating the neutrons before any appreciable amount of deuterium can form. It is clearly important, therefore, to examine how one might reasonably alter the expansion time scale through the introduction of irregularities, although we have advanced general reasons (Peebles 1966a,b) for believing that the early Universe must have been quite isotropic and homogeneous.

Finally, it has been pointed out (Dicke 1966) that the time scale for expansion of the early Universe might be reduced very substantially if there existed a field which exerts a pressure in excess of one third of its energy density. The spatially uniform ϕ-field in the cosmology of Brans and Dicke (1961) has this property, and Dicke (1966) has shown that in this cosmology, assuming appropriate initial conditions, the time scale can be reduced by as much as the factor of 10^6 needed to eliminate element production.

Apparently a hot universe (T = 3° K now) with low primeval abundances of matter heavier than hydrogen might be achieved in two ways: by the introduction of very large irregularities in the early Universe or by the introduction of more incompressible matter, such as the scalar field in the Brans-Dicke theory. The first possibility is subject to the philosophical objection that the expanding Universe is in fact unstable, so that we would expect irregularities to grow larger rather than smaller. The difficulty with the second possibility is of course that the presence of such a field has not been experimentally established.

It is a pleasure to acknowledge useful discussions with R. H. Dicke and K. S. Thorne. This research was supported in part by the National Science Foundation and the Office of Naval Research, and made use of computer facilities supported in part by National Science Foundation grant NSF-GP-579.

REFERENCES

Alpher, R. A., Follin, J. W., Jr., and Herman, R. C. 1953, Phys. Rev., **92**, 1347.
Bergh, S. van den. 1961, Z. f. Ap., **53**, 219.
Brans, C., and Dicke, R. H. 1961, Phys. Rev., **124**, 925.
Bransden, B. H. 1960, Nuclear Forces and the Few-Nucleon Problem, T. C. Griffith and E. A. Power, eds. (New York: Pergamon Press), p. 527.
Burbidge, E. M., Burbidge, G. R., Fowler, W. A., and Hoyle, F. 1957, Rev. Mod. Phys., **29**, 547.
Chandrasekhar, S., and Henrich, L. R. 1942, Ap.J., **95**, 288.
Dicke, R. H. 1966, to be published.
Dicke, R. H., Peebles, P. J. E., Roll, P. G., and Wilkinson, D. T. 1965, Ap.J., **142**, 414.
Field, G. B., and Hitchcock. J. L. 1966, Phys. Rev. Lett., **16**, 817.
Fowler, J. L., and Brolley, J. E., Jr. 1956, Rev. Mod. Phys., **28**, 103.
Gamow, G. 1948, Nature, **162**, 680.

Gaustad, J. E. 1964, Ap.J., **139**, 406.
Hawking, S. W., and Tayler, R. J. 1966, Nature, **209**, 1278.
Hayashi, C. 1950, Progr. Theoret. Phys., **5**, 224.
Hulthén, L., and Sugawara, M. 1957, Hdb. d. Phys., S. Flügge, ed. (Berlin: Springer-Verlag), **39**, 6.
Källén, G. 1964, Elementary Particle Physics (Reading, MA: Addison-Wesley), p. 353.
Kantowski, R., and Sachs, R. K. 1966, J. Math. Phys., **7**, 443.
Kiang, T. 1961, M.N.R.A.S., **122**, 263.
König, L. A., Mattauch, J. H. E., and Wapstra, A. H. 1962, Nucl. Phys., **31**, 18.
Landau, L. D., and Lifshitz, E. M. 1958, Statistical Physics, trans. E. Peierls and R. F. Peierls (Reading, MA: Addison-Wesley), p. 325.
Oort, J. H. 1958, Structure and Evolution of the Universe, R. Stoops, ed. (76-78 Coudenberg, Bruxelles), p. 163.
Peebles, P. J. E., 1966a, Phys. Rev. Lett., **16**, 410.
Peebles, P. J. E., 1966b, to be published.
Penzias, A. A., and Wilson, R. W. 1965, Ap.J., **142**, 420.
Roll, P. G., and Wilkinson, D. T. 1966, Phys. Rev. Lett., **16**, 405.
Sandage, A. 1961, Ap.J., **133**, 355.
Seagrave, J. D. 1960, Nuclear Forces and the Few-Nucleon Problem, T. C. Griffith and E. A. Power, eds. (New York: Pergamon Press), p. 583.
Sosnovsky, A. N., Spivak, P. E., Prokofiev, Yu. A., Kutikov, I. E., and Dobrinin, Yu. P. 1959, Nucl. Phys., **10**, 395.
Thaddeus, P., and Clauser, J. F. 1966, Phys. Rev. Lett., **16**, 819.
Thorne, K. S. 1966, Bull. Am. Phys. Soc., **11**, 340.
Tolman, R. C. 1934, Relativity, Thermodynamics and Cosmology (Oxford: Oxford University Press), p. 377.
Weizsäcker, C. F. von. 1938, Phys. Z., **39**, 633.
Wenzel, W. A., and Wahling, W. 1952, Phys. Rev., **88**, 1149.

SECTION IV

THE VERY EARLY UNIVERSE

SECTION IV

THE VERY EARLY UNIVERSE

In the early 1970s, it became clear that the particle physicists had produced a theory which accurately described most of the phenomena that were observed in the laboratory. This theory had several distinct aspects. It recognized three types of subatomic particles; quarks, leptons and gauge particles. These particles interact in two distinct ways; a strong interaction, confined to quarks and gluons, and a combined weak and electromagnetic interaction, involving quarks, leptons, photons, and the W and Z particles, the weak interaction gauge particles. In some theories, a fourth type of particle with spin zero also exists, but these have not yet been observed. A summary of gauge theories is given by K. Huang (1982).

Soon after the success of these theories, it became clear that by applying them to conditions in the early universe, it might be possible to test some of their predictions under conditions of temperature and density that would not soon be available in our own laboratories. This is because, as we have already seen, as we go backwards in time, we reach conditions of ever higher temperature and density, ultimately, in the first fraction of a second after t = 0, reaching temperatures that correspond to kinetic energies much higher than those available in particle accelerators. Furthermore, it soon became apparent that the influence of particle physics on conditions in the early universe might help explain some puzzles about the present universe that were not previously understood, such as the prevalence of matter over antimatter.

The result of these developments has been a kind of symbiosis between cosmology and particle physics, which like most such relations has been beneficial to both disciplines. On the one hand, particle physics has gained a new arena for testing its predictions, allowing for a more effective choice among distinct models that are in accord with laboratory experiments. The cosmological context has also led particle physicists to consider processes, such as "vacuum tunneling" which would otherwise probably have not engaged their attention. Finally, cosmology has interested particle physicists in gravity, the first known of the forces of nature, which, however, plays a negligible role in the interactions of individual particles under familiar conditions. Again, this is something that might well have not happened otherwise.

For cosmology, the benefits have been of a different kind. Perhaps the most significant has been the prospect of finding explanations, through the ideas of particle physics, for some of the basic assumptions, such as the large scale homogeneity of the universe, which, as we have seen, are built into most cosmological models. Second, the extension of cosmological calculations to new classes of problems, such as how the observed asymmetry between the numbers of protons and antiprotons came about, puts constraints on allowable cosmological models that go beyond the usual astronomical data. Finally, the new symbiosis has attracted to the study of cosmological problems large numbers of particle theorists, whose way of thinking tends to be somewhat more concrete than that of the mathematical cosmologists of an earlier era. The result of this has been to make cosmology more like other branches of physics in that it is more closely tied to phenomena than has previously been the case.

Perhaps the most striking new idea that has emerged from the union of particle physics with cosmology has been the notion that the universe has gone through several "phase changes" during its evolution, especially during the period when its temperature corresponded to an average particle energy of 1 Gev or more. The possibility was first pointed out by the Soviet physicist D. Kirzhnits (1972) and then extended by others. The notion of a phase change is borrowed from statistical physics. A familiar phenomenon is that as the temperature of a substance, such as water, is changed by some outside agent, the physical state of the substance undergoes discontinuous change at certain values of the temperature. In these changes, such properties as density, entropy, specific heat, etc. may change suddenly from one set of values to another. It is these changes that we recognize as a change of phase, say from ice to water at 273° Kelvin. It has been known for a century that the reason for such phase changes is that a certain thermodynamic property of the substance, its free energy, can be different in one state than in another. Which state has the lower free energy, and is therefore stable, can vary with temperature. Thus as the substance is cooled to a specific temperature, which depends on the substance, its physical state may change from one form, stable at higher temperature, to another, stable at lower temperature. This is the phase change. The type described above is known as a first-order phase change. Other types have also been described, and some may be relevant to cosmology. A good discussion of the different types of phase change is given in the textbook of Landau and Lifshitz (1980).

In the present context, the "substance" undergoing the phase change as the universe expands and cools is more exotic. It has been known by particle physicists since the 1960s that otherwise empty space can contain varying levels of one or another quantum field. In its simplest version, this may be thought of as the expectation value, in the ground state, of a single scalar field, known as the Higgs field after the discoverer of the phenomenon. If the region of space involved is at zero temperature, then it often happens that the ground state expectation value of some Higgs field is non-zero. This leads to the effect known as spontaneous symmetry breaking. The equations describing the Higgs field, and other fields interacting with it, have solutions which do not have all the symmetry of the equations themselves. Again, this behavior has analogues in condensed matter physics, where, for example, the crystalline ground state of a solid body does not possess the full translation invariance of the equations describing the interacting atoms that make up the body. The idea of spontaneous symmetry breaking played an essential role in the late 1960s and early 1970s in the unification of weak and electromagnetic interactions (see the review article by Bernstein 1974).

When the temperature of space is sufficiently high, the solutions to the equations describing the Higgs field may differ from the zero temperature solution and take on the more symmetric form, corresponding to zero expectation value for the Higgs field. This is again somewhat analogous to the restoration of translation symmetry when a solid body is heated and melts into the liquid form. Just as the transition from solid to liquid is a phase change that takes place at a definite temperature, so it is expected that the transition from the situation in which one expectation value of the Higgs field is the most stable to that in which another is most stable would take place at a definite temperature. But in the latter case, the temperatures are incomparably higher, so much so that it has not been possible to produce them in laboratory situations. Furthermore, it is believed that as the universe cooled from the highest temperatures, several different phase changes could have taken place, just as happens for many ordinary substances.

The idea that in its earliest stages the universe was much more symmetric than it now appears, and that this symmetry was lost through a series of phase changes, has a number of interesting cosmological consequences, many of which are spelled out in detail in the papers of this section.

1. There is usually a difference in energy between the two phases. In this case, the energy difference is released in the form of heat when the phase transition occurs, as, for example, steam gives up some 540 calories/gm when it condenses into water. During the cosmological expansion, this extra energy can have two effects. It can result in the creation of new particles, or in the increase of the kinetic energy of particles already present. Also, we have seen that the energy content of the universe is what determines its rate of expansion. But energy in the form of an expectation value of a scalar field has a different effect than does energy in the form of particles. Indeed, the effect of "vacuum energy" is the same as that of non-zero cosmological constant in Einstein's equations, so that for this purpose we may think of the phase change as inducing a corresponding change in the cosmological constant. Significant use of this possibility is made in the "inflationary" models of the very early universe, which are discussed in the papers of section IVc.

2. The masses of subatomic particles are strongly influenced by the expectation value of the Higgs field, which varies from phase to phase. Therefore, as phase transitions took place in the early universe, the mass of any particles present changed. For example, the W and Z particles, whose mass in the present

universe is around 100 Gev, are believed to have had zero mass before the temperature dropped below about 1000 Gev. Under those conditions, these particles are stable. Furthermore, at that time there were about as many of them present as photons. After a phase transition generated an expectation value for the Higgs field that interacts with W and Z particles, the latter became massive and unstable, and quickly disappeared, with their energy going into the kinetic energy of their decay products.

Beginning in the 1970s, physicists devised "grand unified theories" that describe strong, electromagnetic and weak interactions as arising from the exchange of a single multi-component gauge field. A summary of the ideas involved in these theories is given by Zee (1982). In such theories there are also other components of this field corresponding to so-called X-particles, whose mass, because of spontaneous symmetry breaking in the present universe, is thought to be 10^{14} Gev or more. These X-particles would therefore be extremely unstable even if they could be produced now. Again, however, when the temperature was high enough, their mass would become zero, and they would be as common as other particles, The interactions responsible for the production and decay of X-particles do not conserve the number of baryons or leptons. In the present universe, virtual processes involving these particles are thought to make the proton unstable, with an expected lifetime of 10^{31} years or more.

In the early universe both real and virtual processes involving X-particles can occur. Because of the expansion of the universe, after a short time the processes involving decay of massive X-particles take place much more quickly than those in which X-particles are reconstituted from their decay products. Furthermore, because of the asymmetry between the rates of processes involving particles and those involving the corresponding antiparticles, known from laboratory experiments and called CP non-invariance, the decay rate of X-particles into baryons and antibaryons need not be equal. By a complex chain of events, these decay processes can therefore result in the generation of a difference in the number of baryons and antibaryons, even if the initial numbers were equal. This analysis was made qualitatively by A. Sakharov in 1967, in paper 18. More quantitative analyses have been given in papers 19-21 of section IVa, and in a paper by Toussaint, Treiman and Wilczek (1979) that we do not reprint because of its length.

The result that is being sought in the calculation is the ratio of the number of protons in the present universe to the number of photons. This ratio is a measure of the number of protons that escaped annihilation with antiprotons in the early universe, when the temperature dropped below a few Gev, that temperature necessary to reconstitute protons and antiprotons from their annihilation products. The observational situation is that there is essentially no antimatter present. Furthermore, as we have seen, the ratio of the number of protons to that of photons is very small, being 10^{-8} or less. In other words, we must account for the fact that there was a tiny but non-zero surplus of matter over antimatter in the early universe, and that this surplus survived to give most of the observable features of the present universe. In an earlier stage of physics, it was thought that the numerical value of this matter-antimatter asymmetry had to be imposed as an initial condition on the universe. But according to the scenario just described, any initial value for such an asymmetry would have disappeared during the time that X-particles were present in large numbers. So in this picture, the present asymmetry must have been generated slightly later, and should be calculable. The calculations have not yet succeeded in giving quantitatively accurate results for the proton to photon ratio, perhaps because the correct theory of X-particles and of CP violation is not known. Furthermore, it must be noted that as of 1985, there was no direct evidence for proton decay in the laboratory, at the rate required by some of the theories that predict the existence of X-particles. It should also be pointed out that in doing such calculations, we are extrapolating over 12 orders of magnitude in energy from the physics that is known from laboratory experiments, and it is entirely possible that unexpected effects that change the results may arise in the intervening energy region. Therefore, the explanation for the matter-antimatter asymmetry must be regarded as tentative. Nevertheless, the possibility of accounting for one of the striking features of the present universe, the surplus of what we call matter over what we call antimatter, is one of the significant accomplishments of the symbiosis of particle physics and cosmology.

3. Perhaps an even more striking consequence of phase changes in the early universe is the possible formation of objects that are distinct from subatomic particles but are nevertheless aspects of the underlying quantum fields. This possibility arises because the symmetry breaking that occurs as a phase transition takes place need not occur everywhere in the same way. Just as a ferromagnet can develop a domain structure, in which different regions are magnetized in different directions, so space in the early universe might have become divided into many regions, differing by the expectation value of the Higgs field. The form that these regions of differing Higgs field can take is constrained by the quantum field theory whose solutions they represent. Depending on how the regions of different Higgs field intersect one another, objects of various types and dimensionality can be formed. A detailed discussion of the different types is given in by T. W. B. Kibble in 1976 in the paper "Topology of Cosmic Domains and Strings," reprinted as paper 22.

Pointlike objects of this kind are magnetic monopoles, whose existence within the context of gauge theories had been pointed out by 't Hooft (1974) and Polyakov (1974). These authors have shown that in some gauge theories whose symmetry is broken by a Higgs expectation value, one should expect monopoles to exist whose mass is much larger than that of the gauge particles of the group. Work on the production and annihilation of magnetic monopoles in the early universe was done by Zeldovich and Khlopov (1979). More detailed calculations were carried out by J. Preskill in his paper "Cosmological Production of Superheavy Monopoles," reprinted as 23. The gist of Preskill's conclusion is that one would expect many more monopoles to have produced and survive from the early universe than can possibly exist in the present universe, on the basis of the total mass density.

This conclusion led to a good deal of soul searching among those concerned with the cosmology-particle physics link. Of the various attempts to circumvent it, the most profound, as well as the most popular, has been the idea that the universe, soon after its expansion began, went through an "inflationary" period of extremely rapid expansion. This idea in some respects involves a return to one of the earliest versions of relativistic cosmology, that due to de Sitter (paper 2). Recall from the discussion in the introduction to section I that the de Sitter universe can be written in a form in which the scale factor R expands exponentially with time rather than the power dependence that characterizes a typical Friedmann universe. This exponential increase of the scale factor is what is referred to as inflation.

In the present version, however, there is a more precise reason for the inflation than the existence of a cosmological constant in the Einstein equations, as assumed by de Sitter. The inflation is attibuted to the occurrence of a large energy density and negative pressure during a stage of the universe that began when the temperature of the universe dropped below a critical value T_c, thought to be around 10^{14} Gev. At this temperature, the phase transition responsible for the symmetry breaking that distinguishes the strong from the electroweak interaction is expected to occur. However, as in any first-order phase transition, it is not sufficient for the temperature to reach its critical value for the transition to actually take place. Instead, what will often happen is that the temperature will drop below T_c until random fluctuations produce nuclei of the new phase within the old, and these nuclei grow until they eventually encompass the whole of the substance. The phenomenon is known as supercooling in the context of ordinary phase transitions, and the same term is now used in the cosmological context.

During the period of supercooling, the energy density of the universe is largely that of the Higgs field expectation value, which is far greater than that of any particles present. This expectation value can be seen, by using considerations of Lorentz invariance, to enter into Einstein's equations in the same way as a cosmological constant term would, except that here the value of the constant is in principle calculable from particle physics. Although this has not been precisely done, it appears from qualitative arguments that the value of the constant would be very large, of the order of ${T_c}^4$. Therefore, during the period of supercooling, the universe expands exponentially, as it would if it contained no matter but had a large cosmological constant.

If the supercooling persists to a low enough temperature, the increase in the scale factor R during the exponential expansion is quite large, perhaps of the order of exp(100) or more. This expansion would have several important cosmological consequences. One is that any particles present in the universe before the expansion begins would become extremely diluted in density, by the cube of the scale factor increase, and so become essentially unobservable. Hence if the inflation occurs after the production of magnetic monopoles or other exotic topological structure is ended, the problem that these are not observed in the present universe is avoided. The particle content of the present universe then originates when the supercooling finally ends, and the phase transition is actually consummated, releasing as a large amount of latent heat the difference in energy density between the Higgs fields in the symmetric and unsymmetric states. This energy is not enough to reheat the universe to a temperature sufficient to produce monopoles, but it is enough to reproduce the conditions needed for generation of the matter-antimatter asymmetry, and the subsequent expansion of the universe occurs as in the standard model.

The inflationary model also sheds light on two basic assumptions of standard cosmology. One is the apparent causal connectivity of the observable universe. The essence of this problem is that observations show that conditions in widely separated parts of the universe are similar enough that they satisfy the usual assumption of homogeneity and isotropy that is built into the Robertson-Walker models. For example, the cosmological microwave radiation arriving from different directions in the sky shows about the same temperature, implying a similar state of the matter in the regions of space where these photons originated. However, in the standard picture, these regions are causally unconnected. That is, there is no way in which anything that occurred in one such region could have influenced the other region, even if its effect traveled at the speed of light over the whole lifetime of the universe until now. The homogeneity must then be attributed either to some Leibnizian "preestablished harmony" of nature or to some other mysterious agency. This is sometimes described by the statement that the two regions have been beyond each other's horizon in the past universe, and the associated problem is referred to as the "horizon" problem.

Another problem that is adumbrated by inflation is that of the extreme smallness of the curvature dependent term in the Einstein-Friedmann equations in the early universe. We have seen that in the present universe, the actual density is close to the critical density, certainly within a factor of ten of it. On the other hand, the deviation from 1 of the ratio of the two changes with time, and in early stages of the universe, say when the temperature was 1 Gev, the ratio was within one part in 10^{20} of 1. Some physicists have felt that this close agreement requires explanation, since, if the density in the early universe had not been so close to the critical density, the subsequent evolution of the universe would have been quite different. This is sometimes referred to as the "flatness" problem, since if the density is exactly equal to the critical density, the universe is spatially flat.

A staightforward explanation of these problems is implied by the inflationary model. One can imagine that all of the presently observable universe (and a good deal of space beyond it) originated in a tiny region, which <u>was</u> causally connected. If the scale factor has expanded sufficiently there is no difficulty in satisfying this assumption. The initial region, being causally connected, could easily become homogeneous before inflation, and would remain so during the subsequent periods of inflation and Friedmann type expansion. Furthermore, at the end of the inflationary stage, the scale factor has immensely increased and the temperature has hardly changed. The result of this is to drive the effective value of the curvature term in the Friedmann equation to a value very close to zero, whatever its value before the inflation began. Thus the subsequent Friedmann-type expansion begins with a density almost equal to the critical density. Indeed, this is a prediction of inflationary models. We have seen that observations are not clearly in agreement with this prediction, but a better understanding of possible "dark matter" content of the universe is necessary before we can decide the question.

We have described the original version of the inflationary model, as presented by Guth in paper 24. This version was soon found to have problems connected with ending the inflation. The rate of nucleation of bubbles of the asymmetric phase within the symmetric phase, and their subsequent coalescence, is too small to produce a universe that is all in the observed asymmetric phase (see Guth and Weinberg 1983). A new type of inflationary model was soon invented by Albrecht and Steinhardt (paper 25) and by Linde (1982). This version differs from Guth's in the gauge theory mechanism that produces the phase transition. In the cosmological context, its main difference is that the inflation is thought to take place within a single bubble, which slowly converts from the symmetric to the asymmetric phase, eliminating the problem of bubble coalescence that exists in Guth's version. While there are still some technical problems connected with the Linde-Albrecht-Steinhardt inflationary model, it commands a good deal of support among theoretical cosmologists and has been used as a starting point for the investigation of a number of open questions, such as the formation of galaxies.

Some current work in cosmology goes in the direction of pushing back our understanding to the time when the effects of quantum theory on gravitational phenomena played an important role. On dimensional grounds, this is believed to have been the case only up to some 10^{-43} seconds after the expansion began. It is hoped that such understanding might shed light on questions such as why the present cosmological constant is so small, as well as eliminating the initial singularity that exists in nonquantized cosmologies such as we have discussed. See Hartle (1983) and Hawking (1983) for examples of work in this direction. This area of work is somewhat hampered by the absence of a theory of quantum gravity that provides unambiguous answers to the questions put to it. Perhaps this problem may be alleviated by one or another proposed new theory of quantum gravity. Examples are theories based on "superstrings" such as that described by Green and Schwarz (1984) and Gross et al. (1985).

Another direction of research in contemporary cosmology is the analysis of how the initial expansion itself began, i.e., what happened before the Big Bang? Several papers have appeared attempting to show how an expanding universe of our type might have originated through some quantum process that took place in a preexisting universe of another type. See articles by Atkatz and Pagels (1982) and by Vilenkin (1983) for examples of such attempts. This approach could have several important consequences. It may allow for the elimination of the initial singularity in the Robertson-Walker universes, replacing it by a smooth continuation from a previous universe. Also, it could extend the horizons of cosmology to much earlier times in the past, and, at least in the mind's eye, allow for the study of new worlds, never before even imagined by scientists. However, more work is necessary before we will know whether efforts in this direction will succeed.

Finally, we should mention efforts in the direction of considering cosmology in more spacetime dimensions than the usual four. This idea arises from work of Kaluza and Klein many years ago (Kaluza 1921; Klein 1926). In the modern versions of this work, general relativity is formulated in a higher dimensional spacetime, and the extra degrees of freedom of the metric tensor are identified with other gauge fields, such as those associated with photons, W and Z particles. The particle physics implications of such theories have been vigorously pursued in the last few years, but it is the cosmological aspects that are of interest here. The obvious question is if spacetime is really higher dimensional, why are we only aware of the usual four dimensions? The answer that has been suggested is that very early in the evolution of the multidimensional universe, all the extra dimensions "compactified," that is, were rolled into a tiny "sphere," whose radius is 10^{-33} cm or so, and have remained in that situation ever since. Direct evidence for the existence of the extra dimensions is therefore unavailable, since motion then can only take place within the tiny sphere. Only the four familiar dimensions have taken part in the cosmic expansion. The interested reader will find a discussion of modern Kaluza-Klein theories in the article by Duff, Nilsson and Pope (1984). Whether this idea will prove useful and whether any indirect evidence for the multidimensional character of spacetime can be found are questions that cannot be answered at present.

REFERENCES

Atkatz, D., and Pagels, H. 1982, Phys. Rev., **D25**, 2065.
Bernstein, J. 1974, Rev. Mod. Phys., **46**, 7.
Duff, M. J., Nilsson, B. E. W., and Pope, C. N. 1984, Nucl. Phys., **B233**, 433.
Guth, A., and Weinberg, E. 1983, Nuc. Phys., **212**, 321.
Green, M. B., and Schwarz, J H. 1984, Phys. Lett., **149B**, 117.
Gross, D., Harvey, J., Martinec, E., and Rohm, E. 1985, Phys. Rev. Lett., **54**, 502.
Hartle, J. B. 1983, "Quantum Cosmology and the Early Universe," in Proc. of the Nuffield Workshop on the Very Early Universe, ed. G. Gibbons and S. W. Hawking (Cambridge: Cambridge University Press).
Hawking, S. 1983, Phil. Trans. Roy. Soc. Lon., **310**, 311.
Huang, K. 1982, Quarks, Leptons and Gauge Fields (Singapore: World Scientific Publishing).
Kaluza, T. 1921, Sitz. Preuss. Akad. Wiss. Phys. Math., **K1**, 966.
Kirzhnits, D. 1972, JETP Letters, **15**, 529.
Klein, O. 1926, Z. Phys., **37**, 875.
Landau, L., and Lifshitz, E. M. 1980, Statistical Physics (Oxford: Pergamon Press).
Linde, A. 1982, Phys. Lett., **108B**, 389.
Polyakov, A.M. 1974, JETP Letters, **20**, 194.
't Hooft, G. 1974, Nucl. Phys., **B79**, 276.
Toussaint, D., Treiman, S. B., and Wilczek, F. 1979, Phys. Rev. D, **19**, 1036.
Vilenkin, A. 1983, Phys. Rev., **D27**, 2848.
Zee, A. 1982, Unity of Forces in the Universe (Singapore: World Publishing Co.).
Zeldovich, Ya. B., and Khlopov, M. Y. 1979, Phys. Lett., **79B**, 239.

VIOLATION OF CP INVARIANCE, C ASYMMETRY, AND BARYON ASYMMETRY OF THE UNIVERSE

A. D. Sakharov

Received: September 23, 1966

The theory of the expanding Universe, which presupposes a superdense initial state of matter, apparently excludes the possibility of macroscopic separation of matter from antimatter; it must therefore be assumed that there are no antimatter bodies in nature, i.e., the Universe is asymmetrical with respect to the number of particles and antiparticles (C asymmetry). In particular, the absence of antibaryons and the proposed absence of baryonic neutrinos implies a non-zero baryon charge (baryonic asymmetry). We wish to point out a possible explanation of C asymmetry in the hot model of the expanding Universe (Zel'dovich 1966; 1967) by making use of effects of CP invariance violation (Okun' 1966; 1967). To explain baryon asymmetry, we propose in addition an approximate character for the baryon conservation law.

We assume that the baryon and muon conservation laws are not absolute and should be unified into a "combined" baryon-muon charge $n_c = 3n_B - n_\mu$. We put:

$$n_\mu = -1,\ n_K = +1 \text{ for antimuons } \mu_+ \text{ and } \nu_\mu = \mu_0 \ ,$$

$$n_\mu = +1,\ n_K = -1 \text{ for muons } \mu_- \text{ and } \nu_\mu = \mu_0 \ ,$$

$$n_B = +1,\ n_K = +3 \text{ for baryons P and N} \ ,$$

$$n_B = -1,\ n_K = -3 \text{ for antibaryons P and N}$$

This form of notation is connected with the quark concept; we ascribe to the p, n, and λ quarks $n_c = +1$, and to antiquarks $n_c = -1$. The theory proposes that under laboratory conditions processes involving violation of n_B and n_μ play a negligible role, but they were very important during the earlier stage of the expansion of the Universe.

We assume that the Universe is neutral with respect to the conserved charges (lepton, electric, and combined), but C-asymmetrical during the given instant of its development (the positive lepton charge is concentrated in the electrons and the negative lepton charge in the excess of antineutrinos over the neutrinos; the positive electric charge is concentrated in the protons and the negative in the electrons; the positive combined charge is concentrated in the baryons, and the negative in the excess of μ-neutrinos over μ-antineutrinos).

According to our hypothesis, the occurrence of C asymmetry is the consequence of violation of CP invariance in the nonstationary expansion of the hot universe during the superdense stage, as manifest in the difference between the partial probabilities of the charge-conjugate reactions. This effect has not yet been observed experimentally, but its existence is theoretically undisputed (the first concrete example, Σ_+ and Σ_- decay, was pointed out by S. Okubo as early as in 1958) and should, in our opinion, have an important cosmological significance.

We assume that the asymmetry has occurred in an earlier stage of the expansion, in which the particle, energy, and entropy densities, the Hubble constant, and the temperatures were of the order of unity in gravitational units (in conventional units the particle and energy densities were $n \sim 10^{98}$ cm^{-3} and $\varepsilon \sim 10^{114}$ erg/cm^3).

M. A. Markov (Zel'dovich and Gershtein 1966) proposed that during the early stages there existed particles with maximum mass on the order of one gravitational unit ($M_0 = 2\times10^{-5}$ g in ordinary units), and called them maximons. The presence of such particles leads unavoidably to strong violation of thermodynamic equilibrium. We can visualize that neutral spinless maximons (or photons) are produced at $t < 0$ from contracting matter having an excess of antiquarks, that they pass "one through the other" at the instant $t = 0$ when the density is infinite, and decay with an excess of quarks when $t > 0$, realizing total CPT symmetry of the Universe. All the phenomena at $t < 0$ are assumed in this hypothesis to be CPT reflections of the phenomena at $t > 0$. We note that in the cold model CPT reflection is impossible and only T and TP reflections are kinematically possible. TP reflection was considered by Milne, and T reflection by the author; according to modern notions, such a reflection is dynamically impossible because of violation of TP and T invariance.

We regard maximons as particles whose energy per particle ε/n depends implicitly on the average particle density n. If we assume that $\varepsilon/n \sim n^{-1/3}$, then ε/n is proportional to the interaction energy of two "neighboring" maximons $(\varepsilon/n)^2 n^{1/3}$. Then $\varepsilon \sim n^{2/3}$ and $R_0^0 \sim (\varepsilon + 3p) = 0$, i.e., the average distance between maximons is $n^{-1/3} \sim t$. Such dynamics are in good agreement with the concept of CPT reflection at the point $t = 0$.

We are unable at present to estimate theoretically the magnitude of the C asymmetry, which apparently (for the neutrino) amounts to about $[(\bar{\nu} - \nu)/(\bar{\nu} + \nu)] \sim 10^{-8} - 10^{-10}$.

The strong violation of the baryon charge during the superdense state and the fact that the baryons are stable in practice do not contradict each other. Let us consider a concrete model. We introduce interactions of two types.

1. An interaction between the quark-muon transformation current and the vector boson field $a_{i\alpha}$, to which we ascribe a fractional electric charge $\alpha = \pm1/3$, $\pm2/3$, $\pm4/3$ and a mass $m_a \sim (10 - 10^3)m_p$. This interaction produces reactions $q \rightarrow a +$

$\bar{\mu}$, $q + \mu \to a$, etc. The interaction of the first type conserves the fractional part of the electric charge and therefore the actual number of quarks minus the number of antiquarks ($= 3n_B$) is conserved in processes that include the a-boson only virtually.

We estimate the constant of this interaction as $g_a = 137^{-3/2}$, from the following considerations: The vector interaction of the a-boson with the μ-neutrino leads to the presence of a certain rest mass in the latter. The upper bound of the mass μ_0 is estimated on the basis of cosmological considerations. If we assume a flat cosmological model of the Universe and assume that the greater part of its density $\rho \sim 1.2\times10^{-29}$ g/cm^3 should be ascribed to μ_0, then the rest mass of μ_0 is close to 30 eV. The given value of g_a follows then from the hypothetical formula

$$\frac{m_{\mu_0}}{m_e} = \frac{g_a^2}{e^2} \sim (137)^{-2} .$$

We note that the presence in the Universe of a large number of μ_0 with finite rest mass should lead to a number of very important cosmological consequences.

2. The baryon charge is violated if the interaction described in Item 1 is supplemented with a three-boson interaction leading to virtual processes of the type $a_{\alpha 1} + a_{\alpha 2} + a_{\alpha 3} \to 0$. At the advice of B. L. Ioffe, I. Yu. Kobzarev, and L. B. Okun', the Lagrangian of this interaction is assumed to be dependent on the derivatives of the a-field, for example,

$$L_2 = g_2(\sum_\alpha f_k^i f_j^k f_i^j + \text{h.c.}),\ f_{ik} = R_0 t a_i .$$

Inasmuch as L_2 vanishes when two tensors coincide, in this concrete form of the theory we should assume the presence of several types of a-fields. Assuming $g_2 = 1/M_0^2$ and $M_0 = 2\times10^{-5}$ g, we have strong interaction at $n \sim 10^{98}$ cm^{-3} and very weak interaction under laboratory conditions. The figure shows a proton-decay diagram including three vertices of the first type, one vertex of the second, and the vertex of proton decay into quarks, which we assume to contain the factor $1/p_q^2$ (due, for example, to the propagator of the "diquark" boson binding the quarks in the baryon). Cutting off the logarithmic divergence at $p_q = M_0$, we find the decay probability

$$\omega \sim \frac{m_p^5\, g_a^6\, [\ln (M_0/m_a)]^2}{M_0^4} .$$

The lifetime of the proton turns out to be very large (more than 10^{50} years), albeit finite.

The author is grateful to Ya. B. Zel'dovich, B. Ya. Zel'dovich, B. L. Ioffe, I. Yu Kobzarev, L. B. Okun', and I. E. Tamm for discussions and advice.

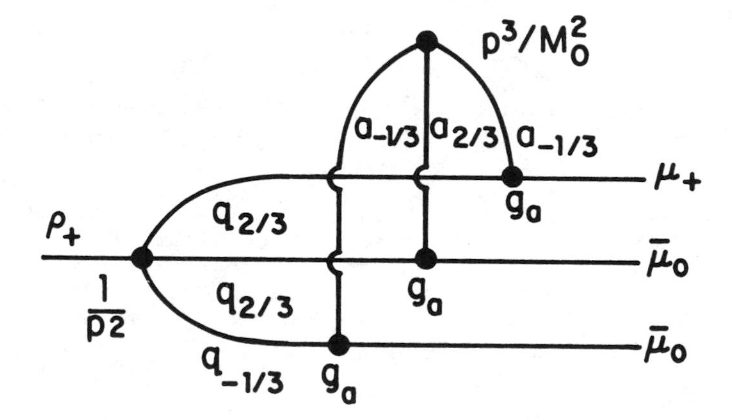

REFERENCES

Markov, M. A. 1966, JETP, **51**, 878. Sov. Phys. JETP, **24**, to be published.

Okun', L. B. 1966, UFN, **89**, 603. Soviet Phys. Uspekhi, **9**, to be published.

Sakharov, A. D. 1966, JETP Letters, **3**, 439 (translation p. 288).

Zel'dovich, Ya. B. 1966, UFN, **89**, 647. Soviet Phys. Uspekhi, **9**, to be published.

Zel'dovich, Ya. B., and Gershtein, S. S. 1966, JETP Lett., **4**, 174 (translation p. 120)

UNIFIED GAUGE THEORIES AND THE BARYON NUMBER OF THE UNIVERSE

Motohiko Yoshimura
Department of Physics, Tohoku University
Sendai, Japan

Received: April 27, 1978

I suggest that the dominance of matter over antimatter in the present universe is a consequence of baryon-number-nonconserving reactions in the very early fireball. Unified gauge theories of weak, electromagnetic, and strong interactions provide a basis for such a conjecture and a computation in specific SU(5) models gives a small ratio of baryon- to photon-number density in rough agreement with observation.

It is known that the present universe is predominantly made of matter, at least in the local region around our galaxy, and there has been no indication observed[1] that antimatter may exist even in the entire universe. I assume here that in our universe matter indeed dominates over antimatter, and I ask within the framework of the standard big-bang cosmology[2] how this evolved from an initially symmetric configuration, namely an equal mixture of baryons and antibaryons. Since the baryon number is not associated with any fundamental principle of physics,[3] such an initial value seems highly desirable. I find in this paper that generation of the required baryon number is provided by grand unified gauge theories (Georgi and Glashow 1974; Fritzsch and Minkowski 1975; Gürsey and Sikivie 1976; Inoue, Kakuto, and Nakano 1977) of weak, electromagnetic, and strong interactions, which predict simultaneous violation of baryon-number conservation and CP invariance. More interestingly, my mechanism can explain why the ratio of the baryon- to the photon-number density in the present universe is so small, roughly of the order (see footnote 1) of 10^{-8} - 10^{-10}.

The essential point of my observation is that in the very early, hot universe the reaction rate of baryon-number-nonconserving processes, if they exist, may be enhanced by extremely high temperature and high density. In gauge models discussed below, the relevant scale of temperature is given by the grand unification mass around 10^{16} GeV where fundamental constituents, leptons and quarks, begin to become indistinguishable.

[1] See e.g., Weekes (1969).
[2] For a review, see S. Weinberg (1972), chapter 15.
[3] However, for a possibility of elevating the baryon number to an absolute conservation law in grand unified models, see Yoshimura (1977); Abud, Buccella, Ruegg, and Savoy (1977); Langacker, Segre, and Weldon (1978). Also Gell-Mann, Ramond, and Slansky (to be published).

This mass is high enough to make futile virtually all attempts to observe proton decay in the present universe: proton lifetime >> 10^{30} year (Reines and Crouch 1974). Instead, if my mechanism works, we may say that a fossil of early grand unification has remained in the form of the present composition of the universe.

The laws obeyed by the hot universe at temperatures much above a typical hadron mass (~ 1 GeV) might, at first sight, appear hopelessly complicated because of many unknown aspects of hadron dynamics. Recent developments of high-energy physics, however, tell that perhaps the opposite is the case. At such high temperatures and densities hadrons largely overlap and an appropriate description of the system is given in terms of pointlike objects - quarks, gluons, leptons, and any other fundamentals. The asymptotic freedom (Politzer 1973; Gross and Wilczek 1973) of the strong interaction and weakness of the other interactions further asssure (Collins and Perry 1975; Kislinger and Morely 1977) that this hot universe is essentially in a thermal equilibrium state made of almost freely moving objects. I shall assume that this simple picture of the universe is correct up to a temperature close to the Planck mass, $G_N^{-1/2} \sim 10^{19}$ GeV, except possibly around the two transitional regions where spontaneously broken weak-electromagnetic and grand-unified symmetries become restored (Kirzhnits and Linde 1972; Weinberg 1974; Bernard 1974; Dolan and Jackiw 1974).

In such a hot universe the time development of the baryon-number density $N_B(t)$ is given by

$$\frac{dN_B}{dt} = - 3 \frac{\dot{R}}{R} N_B + \sum_{a,b} (\Delta n_B)\langle\sigma v\rangle N_a N_b \; . \qquad (1)$$

Here R is the cosmic scale factor; $\langle\sigma v\rangle$ is the thermal average of the reaction cross section for a + b → anything times the relative velocity of a and b; Δn_B is the difference of baryon number between the final and initial states of this elementary process; N_a is the number density of a in thermal equilibrium. At the temperatures I am considering here, the energy density is dominated by highly relativistic particles and the following relations holds (see footnote 2):

$$\dot{R}/R = -\dot{T}/T = (8\pi\rho G_N/3)^{1/2} \qquad (2)$$

where ρ is the energy density

$$\rho = d_F \pi^2 T^4/15 \; . \qquad (3)$$

I use units such that Boltzmann constant k = 1. The effective number of degrees of freedom d_F is counted as usual, 1/2 and

7/16, respectively, for each boson or fermion species and spin state. To solve Eq. (1) with (2) and (3) given, it is convenient to rewrite (1) in the form

$$\frac{dF_B}{dT} = -(8\pi^3 G_N d_F/45)^{-1/2} \left(\frac{3}{8}\right)^2 F_\gamma^{\ 2} \delta \ , \tag{4}$$

where $\delta = \Sigma(\Delta n_B)\langle\sigma v\rangle$; $F_B = N_B/T^3$; $N_a = 3N_\gamma/8$; $F_\gamma = N_\gamma/T^3$ with N_γ the photon-number density. I used the fact, valid in the following example, that only massless fermions participate in bayron-number-nonconserving reactions.

To obtain a nonvanishing baryon number one must break the microscopic detailed balance (more precisely, reciprocity), because otherwise the inverse reaction would cancel the baryon number gained. This necessity of simultaneous violation of baryon-number conservation and CP or T invariance has a further consequence that the amount of generated baryon number may be severely limited. This is because the detailed balance is known[4] not to be broken by Born terms, and one must deal with higher-order diagrams.

As an illustration of the ideas presented above, I shall work in grand unified models based on the group SU(5), which are direct generalizations of the original Georgi-Glashow model (Georgi and Glashow 1974; Fritzsch and Minkowski 1975; Gürsey and Sikivie 1976; Inoue, Kakuto, and Nakano 1977) to allow more than six flavors of quarks and heavy leptons sequentially in accord with recent experimental observations (Herb *et al.* 1977; Innes *et al.* 1977; Perl 1977). Fundamental fermions are thus classified as follows:

$$\underline{5},\ (\phi)_{iR} = \begin{vmatrix} \ell_1 & \\ & q_3^{\ a} \\ \ell_2 & \end{vmatrix}_{iR} \tag{5a}$$

$$\underline{10},\ (\psi)_{iL} = \begin{vmatrix} & q_1^{\ a} & \\ \ell_3 & & C\bar{q}_4^{\ a} \\ & q_2^{\ a} & \end{vmatrix}_{iL} , \tag{5b}$$

with several sequences ($i = 1-n_s$). Here $(\ell_1\ell_2)$ and (q_1q_2) form $SU(2)_W$ doublets and the other singlets; a = 1, 2, 3 are three colors. The baryon number is assigned as 1/3 to any $SU(3)_C$ triplet and -1/3 to any antitriplet.

[4]See, e.g., Blatt and Weisskopf (1952).

There are two characteristic mass scales, $m(W)$ and $m(\tilde{W})$, in this class of models with W a representative of ordinary weak bosons and with $\tilde{W}$ that of colored weak bosons. Existence of the two extremely different mass scales reflects two Higgs systems, $\tilde{H}(\underline{24})$ and (presumably several of) $\tilde{H}(\underline{5})$, being responsible for the breaking of SU(5) and $[SU(2)\otimes U(1)]_W$ (Weinberg 1967; Salam 1968), respectively. At the temperatures that most concern us, $m(W) \ll T \lesssim m(\tilde{W})$, the universe is effectively $[SU(2)\otimes U(1)]_W \otimes SU(3)_C$ symmetric and all fermions remain massless, which makes subsequent computations easier.

The baryon nonconservation is caused by exchange of $\tilde{W}$ coupled to fermions,

$$(g/2\sqrt{2})\Sigma_i (\bar{\ell}_1\bar{\ell}_2)\gamma_\alpha (q_3{}^c)_R + \bar{\ell}_3\gamma_\alpha (q_2{}^c, - q_1{}^c)_L + \\ + \varepsilon_{abc}(\bar{q}_1{}^a \bar{q}_2{}^a)\gamma_\alpha (C\bar{q}_4{}^b)_L]_i \left|\begin{matrix}\tilde{W}_1{}^c \\ \tilde{W}_2{}^c\end{matrix}\right|_\alpha + (\text{H.c.}) \ , \tag{6}$$

and by exchange of colored Higgs $H_i(\underline{5})$ contained in the full Yukawa coupling,

$$L_Y = \frac{1}{2} f_{ij}\bar{\phi}_i{}^{\alpha\beta}[H_1{}^\alpha(\phi_j{}^\beta)_R - H_1{}^\beta(\phi_j{}^\alpha)_R] + \\ + \frac{1}{4} h_{ij}\varepsilon_{\alpha\beta\gamma\delta\eta}\phi_i{}^{\alpha\beta} C(\phi_j{}^{\gamma\delta})_L H_2{}^\eta + (\text{H.c.}) \ , \tag{7}$$

where the two Higgs bosons of $\underline{5}$, H_1 and H_2, may or may not coincide. The Yukawa coupling constants in (7) are related to fermion masses when $SU(2)\otimes U(1)$ is broken,

$$(f_{ij}) = (2\sqrt{2}G_F)^{1/2} \cos\zeta M_1 \ , \tag{8a}$$

$$(h_{ij}) = (2\sqrt{2}G_F)^{1/2} \sin\zeta U^T M_2 U \ , \tag{8b}$$

where G_F is the Fermi constant, $G_F m_N{}^2 \simeq 10^{-5}$; M_1 and M_2 are diagonalized mass matrices for quarks of charge 2/3 and −1/3, respectively [masses of charged leptons are equal to those of q(−1/3) except for renormalization effects (Georgi, Quinn, and Weinberg 1974; Buras, Ellis, Gaillard, and Nanopoulos to be published)]; U is a unitary matrix that generalizes the Cabibbo rotation for more than three flavors (Kobayashi and Maskawa 1973). To allow CP conservation also in the Higgs sector as in the Weinberg (1976) model of CP nonconservation, I introduce the complex dimensionless parameters, α and β, in propagators by

$$i\langle T(H_1{}^{\dagger a} H_2{}^{a})\rangle_{q=0} = \begin{cases} \alpha/m^2(W) \text{ for } a = 1,2 \\ \beta/m^2(W) \text{ for } a = 3\text{-}5, \end{cases} \tag{9}$$

where q is the momentum involved in propagation. This is possible only if more than three Higgs bosons $H_i(\underline{5})$ exist with complex, trilinear and quartic couplings to $\tilde{H}(\underline{24})$.

I now calculate the quantity δ in Eq. (4) by keeping only two-body reactions. Here it is reasonable to suppose that masses of Higgs bosons are of the order of m(W) for uncolored and of $m(\tilde{W})$ for colored ones and fermion masses are $<< m(W)$, and hence α, β = O(1). Remarkably, I found after summing over all sequences that δ vanishes if α and β are real. I can actually prove that this is true to any order of perturbation. The result of this computation is particularly simple when $m(W) << T << m(\tilde{W})$. The dominant contribution to δ of (4) comes from interference of the diagrams of Figs. 1(a) and 1(b) and, after the thermal average is taken, leads to

$$\frac{\delta}{T^2} \simeq \frac{3g^2}{8\pi^2 m^4(\tilde{W})} \, \mathrm{Im}\beta\alpha^* [\mathrm{tr}hh^\dagger \mathrm{tr}f^2 - \mathrm{tr}hh^\dagger f^2] \, . \tag{10}$$

In this computation I ignored the Pauli exclusion effect in the final state caused by occupied thermal fermion states. Correct inclusion of this effect would not affect the result (10) drastically. Under the reasonable assumption $F_B << F_\gamma$ or $N_B << N_\gamma$, F_γ on the right-hand side of (4) may be approximated by the value at an initial temperature T_i where $N_B(T_i) = 0$, and the rate equation (4) is integrated to

$$\frac{N_B(T)}{N_\gamma(T)} \simeq - (8\pi^3 G_N d_F/5)^{-1/2} \left(\frac{3}{8}\right)^2 \frac{\delta}{T^2} N_\gamma(T_i) \, . \tag{11}$$

During the evolution of the universe from this high temperature T down to the recombination temperature (~4000 °K) all fundamental constituents in the initial universe annihilate each other or go out of thermal contact, and the ratio (11) is roughly conserved to give finally the present value $(N_B/N_\gamma)_0$.

To obtain a rough quantitative idea of this ratio, I shall make a drastic extrapolation of formula (11) up to $T_i = m(\tilde{W})$. In the case of six flavors (n_s = 3) the best guess (Georgi, Quinn, and Weinberg 1974; Buras, Ellis, Gaillard, and Nanopoulos to be published) for the parameters of the model is $g^2/4\pi$ = 0.022, $m(\tilde{W}) = 2\times10^{16}$ GeV. I also use the large difference of quark mass scales, $m(b) >> m(s)$, m(d) and $m(t) >> m(c)$, m(u), and the small mixing parameters (Ellis, Gaillard and Nanopoulos 1976) ($\lesssim 0.1$) in U of (8b). Combining (8), (10), and (11) and putting in some numerical factors, I find

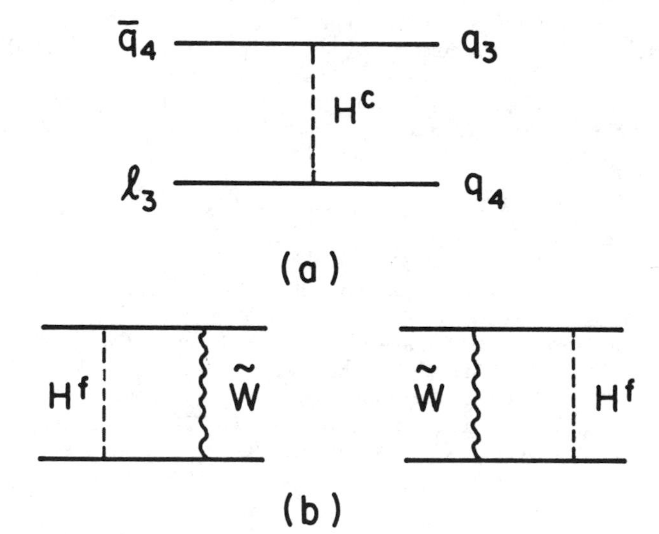

Figure 1

Diagrams leading to generation of the baryon number. $H^c(H^f)$ means a colored (uncolored) Higgs boson.

$$\left(\frac{N_B}{N_\gamma}\right)_0 \approx 0.12 \frac{(d_F G_N)^{-1/2}}{m(\tilde{W})} \varepsilon \,, \tag{12a}$$

$$\varepsilon \simeq \frac{g^2}{\pi^2} G_F^{\,2} m^2(t) m^2(b) (\sin\theta_2 + \sin\theta_3)^2 A \,, \tag{12b}$$

where $A = \mathrm{Im}\beta\alpha^* \sin^2\zeta \cos^2\zeta \approx O(1)$ in general. For definiteness, mixing angles are set by $\sin\theta_2 = \sin\theta_3 = 0.1$, which is consistent with present data (Ellis, Gaillard, and Nanopoulos 1976). Furthermore, $d_F = 64$ (69) with three H_i of 5 and real (complex) $\tilde{H}$ of 24, and hence

$$\left(\frac{N_B}{N_\gamma}\right)_0 \approx 2\times 10^{-9}\left[\frac{m(t)}{10\text{ GeV}}\right]^2\left[\frac{m(b)}{5\text{ GeV}}\right]^2 A \,. \tag{13}$$

An estimate of this quantity (see footnote 2) deduced from data ranges from 10^{-8} to 10^{-10}, which agrees with (13). The numerical value of (13) should not be taken too seriously because of the very crude approximations assumed, but it is hard to imagine that neglected corrections would alter (13) by more than 100.

Although my result (13) depends on the specific SU(5) model taken, the formula (12a) is presumably more general than this example provided that $\varepsilon \approx$ (CP-invariance violation parameter)/137. To this extent the small ratio of the number densities appears to be a general consequence of grand unification in the earliest history of our universe. The possibility that this fundamental parameter of cosmology is related to those of elementary-particle physics seems intriguing and deserves much investigation.

REFERENCES

Abud, M., Buccella, F., Ruegg, H., and Savoy, C. A. 1977, Phys. Lett., **67B**, 313.

Bernard, C. 1974, Phys. Rev. D, **9**, 3312.

Blatt, J. M., and Weisskopf, V. F. 1952, Theoretical Nuclear Physics (New York: Wiley), p. 530.

Buras, A. J., Ellis, J., Gaillard, M. K., and Nanopoulos, D. V. to be published.

Collins, J. C., and Perry, M. J. 1975, Phys. Rev. Lett., **34**, 1353.

Dolan, L., and Jackiw, R. 1974, Phys. Rev. D, **9**, 3320.

Ellis, J., Gaillard, M. K., and Nanopoulos, D. V. 1976, Nucl. Phys., **B109**, 213.

Fritzsch, H., and Minkowski, P. 1975, Ann. Phys. (N.Y.), **93**, 193.

Gell-Mann, M., Ramond, P., and Slansky, R., to be published.

Georgi, H., and Glashow, S. L. 1974, Phys. Rev. Lett., **32**, 438.

Georgi, H., Quinn, H. R., and Weinberg, S. 1974, Phys. Rev. Lett., **33**, 451.
Gross, D. J., and Wilczek, F. 1973, Phys. Rev. Lett., **30**, 1343.
Gürsey, F., and Sikivie, P. 1976, Phys. Rev. Lett., **36**, 775.
Herb, S., et al. 1977, Phys. Rev. Lett., **39**, 252.
Innes, W. R., et al. 1977, Phys. Rev. Lett., **39**, 1240, 1640(E).
Inoue, K., Kakuto, A., and Nakano, Y. 1977, Progr. Theor. Phys., **58**, 630.
Kirzhnits, D. A., and Linde, A. D. 1972, Phys. Lett., **42B**, 471
Kislinger, M. B., and Morley, P. D. 1977, Phys. Lett., **67B**, 371.
Kobayashi, M., and Maskawa, T. 1973, Progr. Theor. Phys., **49**, 652.
Langacker, P., Segre, G., and Weldon, A. 1978, Phys. Lett., **73B**, 87.
Perl, M. L. 1977, in Proceedings of the International Symposium on Lepton and Photon Interactions at High Energies, Hamburg 1977, F. Gutbrod, ed. (Hamburg: DESY).
Politzer, H. D. 1973, Phys. Rev. Lett., **30**, 1346.
Reines, F., and Crouch, M. F. 1974, Phys. Rev. Lett., **32**, 493.
Salam, A. 1968, in Elementary Particle Physics, N. Svartholm, ed. (Stockholm: Almquist and Wiksels), p. 367.
Weeks, T. 1969, High Energy Astrophysics (London: Chapman and Hall).
Weinberg, S. 1967, Phys. Rev. Lett., **19**, 1264.
Weinberg, S. 1972, Gravitation and Cosmology (New York: Wiley), Chapter 15.
Weinberg, S. 1974, Phys. Rev. D, **9**, 3357.
Weinberg, S. 1976, Phys. Rev. Lett., **37**, 657.
Yoshimura, M. 1977, Prog. Theor. Phys., **58**, 972.

ERRATUM

UNIFIED GAUGE THEORIES AND THE BARYON NUMBER OF THE UNIVERSE

Motohiko Yoshimura

[Phys. Rev. Lett. 1978, **41**, 281]

In the computation of δ of Eq. (10), Feynman diagrams containing triangle loops were omitted. A mistake was made by incorrectly ignoring a finite discontinuity that remains after the vertex renormalization. This correction leads to a conclusion that contributions of massless fermions to the baryon asymmetry vanish to the order of $g^2hh^\dagger f^2$ in the approximation of this paper. This agrees with a recent result of D. Toussaint et al. (to be published) based on two-body unitarity. More satisfactory computation, including effects near the unification temperature, will be dealt with in a separate paper. The author should like to thank Dr. J. Arafune, Dr. S. M. Barr, and Dr. S. Weinberg for critical comments.

Toussaint, et al. to be published.

BARYON NUMBER OF THE UNIVERSE

Savas Dimopoulos[1]
Enrico Fermi Institute, University of Chicago,
Chicago, Illinois

Leonard Susskind
Stanford Linear Accelerator Center, Stanford University,
Stanford, California

Received: June 9, 1978

We consider the possibility that the observed particle-antiparticle imbalance in the universe is due to baryon-number, C, and CP nonconservation. We make general observations and describe a framework for making quantitative estimates.

I. INTRODUCTION

Evidence exists (Steigman 1976) that the universe contains many more particles than antiparticles. A quantitative measure of this particle excess is given by the number of baryons within a unit thermal cell of size $R = T^{-1}$. Such a cell contains a single black-body photon (Weinberg 1972). In current cosmological theories it is a box, expanding according to (Weinberg 1972)

$$R^{-1} \frac{d}{dt} R(t) \approx \left(\frac{8\pi}{3} G\rho\right)^{1/2} . \qquad (1.1)$$

In the very early universe there was approximately 1 of every species of particle within a unit cell. However, the unit cell today contains only 10^{-9} baryons and essentially no antibaryons. If baryon number is conserved then the unit cell has always contained a baryon number of order 10^{-9}.

One cannot rule out the possibility that the universe was created with net baryon number and no explanation is needed. However, to quote Einstein: "If that's the way God made the world then I don't want to have anything to do with Him." In fact modern theories of particle interactions suggest that baryon number is not strictly conserved (Georgi and Glashow 1974; Pati and Salam 1973; 1974; Gürsey and Sikivie 1976; 't Hooft 1976a; 1976b; Belavin *et al.* 1975). If this is true then today's baryon number is as much dependent on dynamical processes as on initial conditions. Indeed Yoshimura (1978) has made the exciting suggestion that baryon-number violation can combine with CP noninvariance to produce a calculable net

[1]Present address: Physics Department, Columbia University, New York, New York 10027.

baryon number even though the universe was initially baryon neutral. Yoshimura has also made estimates (Yoshimura 1978) which indicate that this may be quantitatively plausible.

There are three interesting reasons to believe that baryon number is not exactly conserved:

(1) Black holes can swallow baryons (Hawking 1975; Wald 1975; Parker 1977).

(2) Quantum-mechanical baryon-number violations have been discovered by 't Hooft in the standard Weinberg-Salam theory ('t Hooft 1976a; 1976b; Belavin *et al.* 1975).

(3) Superunified theories of strong, electromagnetic, and weak interactions naturally violate baryon number at superhigh energy (Georgi and Glashow 1974; Pati and Salam 1973; 1974; Gürsey and Sikivie 1976).

Although baryon-number violations are minute at ordinary energy, in cases (2) and (3) they may become significant at sufficiently high temperature.

Baryon-number violation is not enough to create an excess of baryons. The process itself must be particle-antiparticle asymmetric (Hawking 1975; Wald 1975; Parker 1977). Otherwise the sign of the effect will be random and cancel in different cells. In this case the total baryon excess would be of the order of the square root of the total number of photons. However, the total number of photons in the observed universe is $\sim 10^{88}$ and the baryon number is $\sim 10^{79}$.

The required particle-antiparticle asymmetry is known to exist. Indeed charge conjugation is maximally violated in ordinary weak interactions. Were this the only asymmetry, CP invariance would destroy any possible effect because total baryon number changes sign under CP as well as C. Luckily CP violations are known to exist (Lee, Oehme, and Yang 1957; Lee and Wu 1966).

CPT invariance also imposes a very interesting constraint on the expansion rate of the universe. As we shall see, CPT invariance ensures vanishing baryon density in thermal equilibrium. Therefore the expansion rate must remain rapid enough to prevent the baryon-number-violating forces from coming to equilibrium.

In this paper we will discuss how baryon-number, C, and CP nonconservation can conspire with the early Hubble expansion to produce an observable baryon excess.

As we shall see, the baryon excess may originate at or close to the very earliest times, $\sim 10^{-42}$ sec. At that time the temperature, energy density, and local space-time curvature are assumed to be of order unity in units of the Planck mass. The metric in Planck units is of the Robertson-Walker type (Steigman 1976)

$$(ds)^2 = dt^2 - R(t)^2 dx_i dx_i ,$$

where $R(t) \sim 1$ at the Planck time $t = 1$.

Let us follow the evolution of a single unit coordinate cell of dimensions $\Delta x_i = 1$. At the earliest of times it is a cube of unit volume (10^{-100} cm^3) in Plank units. We will assume that quantum fluctuations and gravitational interactions between gravitons and matter rapidly bring the universe to equilibrium at a temperature of unity. It follows that our unit cell initially contains about one elementary particle of each species. In current unified theories this means ~100 particles (photons, leptons, gravitons, intermediate bosons, quarks, vector gluons, Higgs bosons, superheavy bosons, ...).

As the unit cell evolves it expands and cools. The process is not too different from the slow expansion of a box containing radiation. As in this case, the entropy within the cell is not significantly changed during the expansion. Roughly speaking this implies that the number of particles within that cell is the same today as it was at creation. Of course, by now, the only particles left are photons, neutrinos, and any excess protons and electrons. The others all annihilated or decayed when the temperature decreased below their mass.

The excess, expressed as a baryon number in the unit coordinate cell, is a number of order

$$n_B = \frac{N_B}{N_\gamma} n_\gamma ,$$

where $N_B/N_\gamma \approx 10^{-9}$ and n_γ is the number of photons in the unit cell today. Assuming it is of the order of the number of elementary particle types, we must account for 10^{-7} baryons per box.

The estimates made in later sections for the baryon excess are too uncertain to be taken seriously. In addition to particle physics uncertainties, the properties of the initial condition at creation are unknown and can influence the result. Our estimates are made for the most pessimistic case which we call "chaotic initial conditions." Such an initial conditions is described by a density matrix ρ which is diagonal in baryon number and symmetric under the interchange of baryons and antibaryons. It is the sort of initial condition which would describe equilibrium if the earliest interactions respected baryon-number, C and CP invariance.

II. CPT AND EQUILIBRIUM

It is self-evident that if C or CP are symmetries of the equations of motion then no global baryon excess can result from baryon-number-violating processes. To illustrate the constraints imposed by CPT in an expanding universe we discuss some examples.

Consider a complex scalar field $\phi(x)$ in an expanding universe described by the metric

$$(ds)^2 = (dt)^2 - R(t)^2(d\vec{x})^2 . \tag{2.1}$$

The action for this model is taken to be

$$S = \int d^4x \sqrt{-g} \left[g^{\mu\nu} \partial_\mu \phi \partial_\nu \phi^* - V(\phi) \right] , \tag{2.2}$$

where

$$V(\phi) = \lambda(\phi\phi^*)^n(\phi + \phi^*)(\alpha\phi^3 + \alpha^*\phi^{*3}) \tag{2.3}$$

and α is a complex phase. The baryon current density is

$$B_\mu = \sqrt{-g}\ i\phi\overleftrightarrow{\partial}_\mu\phi^* . \tag{2.4}$$

Note that $V(\phi)$ violates baryon-number conservation, C invariance ($\phi \leftrightarrow \phi^*$), and CP invariance.

The Hamiltonian for this model is

$$H(t) = \int d^3x \left[\frac{\pi\pi^*}{R^3(t)} + R(t)\left|\nabla\phi\right|^2 + R^3(t)V(\phi) \right] . \tag{2.5}$$

This Hamiltonian is invariant under the following CPT transformation:[2]

$$\phi(x) \to \phi(-x) , \tag{2.6}$$

$$\pi(x) \to -\pi(-x) . \tag{2.7}$$

The baryon number

$$B_\mu(x) = \begin{cases} i(\phi\pi - \pi^*\phi^*), & \mu = 0 \\ \sqrt{-g}\ i\phi\overleftrightarrow{\nabla}\phi^*, & \mu = i \end{cases} \tag{2.8}$$

[2] We use Schwinger's definition of CPT transformation which differs from the standard (Wigner) definition by a Hermitian conjugation.

changes sign under (2.6) and (2.7).

The CPT transformation is a symmetry of the spectrum of the instantaneous Hamiltonian but not of the equation of motion because of the explicit time dependence of H.

Now consider the case where the universe expands so slowly that at every instant it is in thermal equilibrium with respect to the instantaneous Hamiltonian H(t). The density matrix at time t is

$$\rho(t) = \exp[-\beta(t)H(t)] \quad . \qquad (2.9)$$

Since CPT conjugate states carry equal energy but opposite baryon charge B the expectation value of B vanishes,

$$\langle B \rangle = \mathrm{Tr}(e^{-\beta(t)H(t)}\hat{B}) = 0 \quad . \qquad (2.10)$$

Therefore the only hope of generating baryon excess is for the baryon-number-violating interactions to remain out of thermal equilibrium. This implies that the rate of expansion of the universe has to be faster than the baryon-number-violating reaction rates.

Now we will discuss a second model to illustrate the possibility of baryon-number generation if we are out of equilibrium.

Consider a time-independent Hamiltonian $H = H_0 + V$. H_0 is baryon-number, C, and CP conserving and V is a small perturbation which violates these quantum numbers. Suppose that at time $t = 0$ the system is in thermal equilibrium with respect to the Hamiltonian H_0,

$$\rho(0) = e^{-\beta H_0} \quad . \qquad (2.11)$$

Under the action of the full Hamiltonian the density matrix at time t has evolved to

$$\rho(t) = e^{-iHt}\, e^{-\beta H_0}\, e^{+iHt} \quad .$$

The mean baryon number is

$$\langle B(t) \rangle = \mathrm{Tr}[e^{-\beta H_0}\hat{B}(t)]/\mathrm{Tr}\rho \quad , \qquad (2.12)$$

where

$$\hat{B}(t) = e^{iHt}\,\hat{B}e^{-iHt} . \tag{2.13}$$

The CPT invariance of H and H_0 implies that $\langle B(t)\rangle$ is an odd function of time

$$\langle B(t)\rangle = \mathrm{Tr}\left(e^{-\beta H_0}\,\hat{B}(t)\right) \tag{2.14}$$

$$= \mathrm{Tr}\left[\theta e^{-\beta H_0}\,\theta^{-1}\,\theta\hat{B}(t)\theta^{-1}\right]$$

$$= \mathrm{Tr}\left\{e^{-\beta H_0}\left[-\hat{B}(-t)\right]\right\}$$

$$= -\,\mathrm{Tr}\left[e^{-\beta H_0}\,\hat{B}(-t)\right]$$

$$= -\,\langle B(-t)\rangle , \tag{2.15}$$

where θ = CPT. This asymmetry of $\langle B\rangle$ with time is the only constraint implied by CPT.

Interesting information can also be extracted by looking at the rate of change of B,

$$\langle \dot{B}(t)\rangle = i\mathrm{Tr}\left\{e^{-iHt}\left[e^{-\beta H_0}, V\right]e^{iHt}\,\hat{B}\right\} . \tag{2.16}$$

If we approximate e^{-iHt} by e^{-iH_0t} then $\langle\dot{B}\rangle$ must vanish since $[B,H_0] = 0$. This implies $\langle\dot{B}\rangle$ is at least second order in V and first order in time, $\langle\dot{B}\rangle \sim t$. But since $\langle B\rangle$ is an odd function of t, $\langle\dot{B}\rangle$ must be even and cannot be of order t. It follows that $\langle\dot{B}\rangle$ is at least second order in t and $\langle B\rangle$ is third order:

$$\langle B(t)\rangle \sim t^3 . \tag{2.17}$$

That baryon-number excess vanishes to first order in V is to be expected. The nontrivial part of the time translation operator U is anti-Hermitian to first order. Therefore amplitudes changing B by opposite amounts have equal magnitude and cancel. The relation $\langle B\rangle \sim t^3$ shows that baryon excess builds up slowly in the beginning.

In this example, a period of time will elapse during which $\langle B\rangle$ is not zero. Eventually the interactions in V will restore the system to true thermal equilibrium with vanishing $\langle B\rangle$. If, however, the baryon-number-violating force is switched off after a finite time the system will retain a finite net baryon excess.

The process of early expansion can disturb thermal equilibrium and lead to a temporary excess. If the universe expands and cools sufficiently rapidly the baryon-number-violating forces may not have time to come back to equilibrium. This is especially true if the reaction rates for these processes are rapidly falling with decreasing temperature. In order to estimate if this is so we consider the quantity $\dot{R}/R$ which measures the rate of expansion of the universe. The condition for equilibrium is

$$\frac{\dot{R}}{R} < \text{reaction rate} . \qquad (2.18)$$

The expansion rate in the radiation-dominated epoch is given by

$$\frac{\dot{R}}{R} \approx T^2 , \qquad (2.19)$$

where the temperature T and time are in units $c = \hbar = G = 1$.

The dependence of the reaction rate on temperature can be obtained from dimensional considerations. For example, in a renormalizable theory with all mass scales much lower than T the reaction rate must be proportional to T. This is because coupling constants in renormalizable theories are dimensionless. Accordingly the condition for equilibrium is

$$T^2 < T \qquad (2.20)$$

or

$$T < 1 . \qquad (2.21)$$

Therefore the condition for thermal equilibrium in renormalizable theories is increasingly satisfied as the universe cools. This continues as long as explicit masses can be ignored. From these arguments it is easy to see that ordinary strong electromagnetic and weak interactions are in thermal equilibrium from superhigh temperatures ($\sim 10^{15}$ GeV) down to ordinary temperature (~ 1 GeV).

In superunified theories baryon-number-violating processes are effectively nonrenormalizable Fermi interactions below energies $\sim 10^{18}$ GeV. This energy corresponds to the mass $\tilde{M}$ of the superheavy bosons which mediate the process. The effective Fermi coupling constant is

$$\tilde{G} \approx \frac{\alpha}{\tilde{M}^2} \sim 10^{-38} \text{ GeV}^{-2} . \qquad (2.22)$$

The reaction rate is proportional to $\tilde{G}^2$ and by dimensional arguments is

$$(\text{reaction rate}) \approx \tilde{G}^2\, T^5 .$$

The condition for equilibrium becomes

$$T^2 < \tilde{G}^2 T^5 \text{ (in Planck units)}$$

or

$$T > \left(\frac{\tilde{M}^4}{\alpha^2}\right) . \tag{2.23}$$

For $\tilde{M} \sim M_{Planck}$ it is unlikely that the baryon-number-violating forces were ever in equilibrium.

Note that the baryon-number violations are of order α at temperatures $\sim 10^{18}$ GeV. Effectively we are in a situation where these interactions are switched on for a brief time interval and are then switched off. These considerations indicate that the possibility of generating baryon excess is viable.

III. MODELS WITH BARYON-NUMBER VIOLATION

By a unified theory (Weinberg 1972) we mean a theory in which the strong, weak, and electromagnetic gauge invariance are embedded in a simple unifying group. Such theories involve a single coupling constant of the order of the electric charge. Both leptons and quarks appear in the same multiplets. Therefore quarks can turn into leptons by the emission of vector bosons called $\tilde{W}$. For example, in the SU_5 theory of Georgi and Glashow the process shown in Figure 1 is possible. This process implies that a proton can decay into a positron and photons.

In order to suppress the decay of the proton, the mass of the $\tilde{W}$ must be made large. Consistency with the empirical bounds on the lifetime of the proton requires

$$\tilde{M} > 10^{15} \text{ GeV} . \tag{3.1}$$

We will assume $\tilde{M}$ is approximately the Planck mass and set it equal to unity. This assumption simplifies our discussion.

At energies below $\tilde{M}$ the baryon-number-violating processes are effectively described by four-Fermi interactions. The coupling constant is approximately

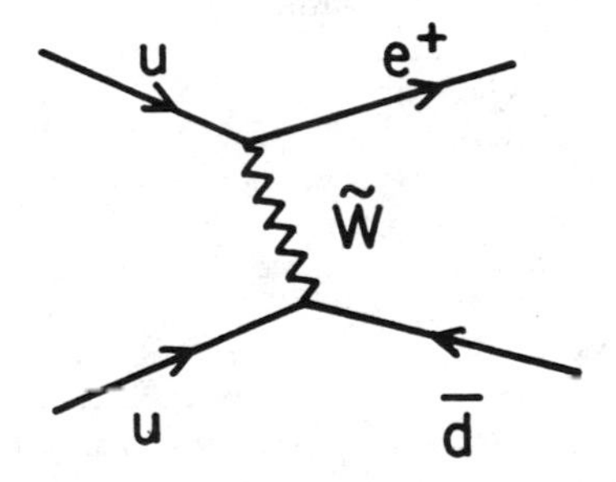

Figure 1

Baryon-number-violating process occurring in the SU_5's unified theory.

$$G = \frac{\alpha}{\tilde{M}^2} = \alpha \qquad (3.2)$$

in Planck's units. The baryon-number-changing interactions obviously are unimportant for temperatures very much smaller than $\tilde{M}$.

The other ingredient needed for baryon excess is CP violation (Lee, Oehme, and Yang 1957; Lee and Wu 1966). In principle the observed CP violation could arise spontaneously (Lee 1973; 1974) or from explicit asymmetry of the Lagrangian (Weinberg 1976). If it arises spontaneously then it disappears at temperatures well above 1 TeV. In this case the CP and baryon processes cannot combine to yield an excess.

We will assume that a CP violation, perhaps unrelated to observed CP violation, exists at the superheavy scale. We might suppose that this breaking is also spontaneous. However, in this case it could not be effective in producing an excess. The reason is because the radius of an event horizon is very small at the time when the baryon excess is produced. This means that uncorrelated domains of different CP directions must occur with small spatial extent. Within these domains the baryon excess will have opposite sign and therefore cancel. Thus we must have an explicit CP violation in the part of the Lagrangian which is relevant at superheavy scales. This does not exclude the idea (Lee 1973; 1974) that the observed CP violation is spontaneous.

For definiteness we will assume explicit four-Fermi vertices which break both CP and baryon-number conservation.

A second source of baryon-number violation has been discovered in the standard Weinberg-Salam theory. In this model the baryon-number violation is of purely quantum-mechanical origin ('t Hooft 1976a; 1976b; Belavin *et al.* 1975). There exists a discrete infinity of classical degenerate vacuums (Jackiw and Rebbi 1976; Callan, Dashen, and Gross 1976) labelled by the "winding number" n. Quantum-mechanical transitions between these classical vacuums can occur by tunneling through an energy barrier. These events are called instantons. The physics is analogous to tunneling between the minima of a periodic potential. As 't Hooft first noted ('t Hooft 1976a; 1976), each instanton event is accompanied by a change in baryon number. A change in lepton number also occurs in order to compensate the electric charge. The tunneling amplituded at zero temperature is proportional to ('t Hooft 1976a; 1976b; Belavin *et al.* 1975)

$$e^{-8\pi^2/g^2} ,$$

which is of the order of 10^{-93}. At very high temperatures $T >$ 250 GeV two quantitatively new things happen. First, the Higgs vacuum expectation values goes away (Kirzhnits and Linde 1972; Dolan and Jackiw 1974; Weinberg 1974). Second, there exists a lot of thermal energy available. This can be used to overcome the potential barrier.

To estimate the importance of this effect we must compare the barrier height with the available thermal energy. Consider an instanton of space-time radius ρ. For temperatures >> 250 GeV the expectation value of the Higgs potential vanishes and the action of an instanton is roughly what it would be for pure Yang-Mills theory:

$$\text{action} = 8\pi^2/g^2 \ . \tag{3.3}$$

The tunneling barrier is estimated by dividing this action by the duration of the event ρ,

$$V = 8\pi^2/g^2\rho \ . \tag{3.4}$$

[We remind the reader that Equation (3.4) only applies above the transition temperature for the Higgs field to disappear.]

Equation (3.4) suggests that we can always lower the barrier as small as we like by considering arbitrarily large instantons. This is not so. The reason is that a tunneling event is a coherent process in which the instanton density $F_{\mu\nu}\tilde{F}_{\mu\nu}$ is of a definite sign over the size of the tunneling region. Thus ρ cannot exceed the coherence length which is given by the Debye screening length in the gauge-field plamsa (Kislinger and Morely 1976a; 1976b). This is given by the plasmon Compton wavelength which for pure Yang-Mills theory is

$$\lambda_{\text{plasma}} \approx \left(\frac{gT}{\sqrt{6}}\right)^{-1} = \rho_{\max} \ . \tag{3.5}$$

The thermal energy within such a volume is $\sim(\pi^2/2)\rho_{\max}{}^3T^4 \sim (\pi^2/2)T^4\lambda_P{}^3$. The condition that this thermal energy overcomes the barrier B is

$$\frac{\pi^2}{2}T^4\lambda_P{}^3 > \frac{8\pi^2}{g^2}\frac{1}{\lambda_P}$$

or (3.6)

$$(18 - 8g^2)\frac{\pi^2}{g^4} \geqslant 0 \ .$$

This appears to be satisfied for the coupling constants characteristic of weak-electromagnetic theories.

These crude estimates only suggest the possibility that baryon-number-violating interactions are not suppressed at T >

250 GeV. Quantitative calculations are needed to decide the importance of this effect. In particular, the effects of fermions will probably suppress the tunneling. For the remainder of this paper we will ignore this quantum-mechanical source of baryon-number violation, although it is possible for it to seriously alter the results of this paper.

IV. BARYON GENERATION MECHANISM IN FIELD THEORY

In this section we will describe field-theoretic methods for computing the baryon-number excess in an expanding universe. For definiteness we will consider a model in which both baryon and CP violation are mediated by superheavy bosons of mass $\sim M_{Planck}$. In practice this means that these interactions are described as four-Fermi couplings.

We are going to consider a field theory in an expanding universe described by the metric

$$ds^2 = (dt)^2 - R(t)^2(d\vec{x})^2 \tag{4.1}$$

$$= (dt)^2 - t(d\vec{x})^2 \ . \tag{4.2}$$

The choice $R = \sqrt{t}$ is appropriate to a radiation-dominated epoch. We will illustrate such a system by considering a scalar field with action

$$S = \int d^4x \sqrt{-g} \left[g^{\mu\nu} \frac{\partial\phi}{\partial x^\mu} \frac{\partial\phi^*}{\partial x^\nu} + V(\phi) \right] \ . \tag{4.3}$$

Now the metric in Eq. (4.2) is of the conformally flat type meaning that by a change of variables it can be brought to the form

$$ds^2 = \rho^2(x)[(dx_0)^2 - (d\vec{x})^2] \ . \tag{4.4}$$

In particular, if we change variables from t to $\tau = (2t)^{1/2}$ then

$$ds^2 = \tau^2(d\tau^2 - d\vec{x}^2) \ . \tag{4.5}$$

Now the reader can verify that if the field ϕ is replaced by

$$s = \rho^{-1}\phi \ , \tag{4.6}$$

then the free part of the Lagrangian becomes

$$S = \int d^3x d\tau \left[\left(\frac{ds}{d\tau}\right)^2 - (\nabla s)^2 \right]$$

$$+ \text{ pure divergence} . \qquad (4.7)$$

Furthermore, if a renormalizable ϕ^4 interaction is present in V then it is replaced by s^4. If on the other hand nonrenormalizable terms such as ϕ^{4+2n} are present they are replaced by

$$V(s) = \frac{s^{4+2n}}{\tau^{2n}} . \qquad (4.8)$$

Thus, in the new time coordinate, the free and renormalizable terms in the action take their flat-space form and appear to be τ indepedndent. The nonrenormalizable terms appear time dependent with rapidly falling coefficients.

Similar results hold for more general theories. If we consider the usual type of theory containing scalar spinor and vector fields ϕ, ψ, A_μ and define conformal fields by

$$\phi \rightarrow \rho^{-1}\phi ,$$

$$\psi \rightarrow \rho^{-3/2}\psi , \qquad (4.9)$$

$$A_\mu \rightarrow A_\mu ,$$

then the free and renormalizable terms take their flat-space form. The nonrenormalizable Fermi couplings are replaced by their flat-space counterparts times the factor $1/\tau^2$. Thus the form that the action for our model takes is

$$S = \int d^3x d\tau \left(L_0 + \frac{1}{\tau^2} L_1 \right) , \qquad (4.10)$$

where L_0 is a renormalizable τ-independent Lagrangian containing all the usual interactions and L_1 is a four-Fermi coupling containing the superheavy mediated effects.

We will make two cautionary remarks before proceeding to study baryon-number excess generation. The first is that the flat-space form for renormalizable theories ignores mass effects. Since we only use it for very high temperatures this is no problem. The second remark concerns ultraviolet divergences. The above analysis was purely classical and fails when renormalization is accounted for. However, because the unified coupling is small at the Planck length, the failure only involves very weakly varying logarithms. In fact, these effects would show up as logarithms of τ multiplying the renormalizable interactions. They are completely unimportant for our problem.

Let us now return to the baryon excess problem. We write the Hamiltonian resulting from Equation (4.10) as

$$H = H_0 + V(\tau) , \tag{4.11}$$

where H_0 is baryon-number and CP conserving. $V(\tau)$ contains the violating terms and scales like τ^{-2}.

Suppose the initial density matrix at the Planck time $\tau = 1$ is given by $\rho(1)$. The expectation value of the baryon number at this time is

$$\langle B(1)\rangle = \mathrm{Tr}\big(\rho(1)\hat{B}\big) . \tag{4.12}$$

At a later time τ the value of $\langle B\rangle$ is

$$\langle B(\tau)\rangle = \mathrm{Tr}\big(\rho(1)U^{\dagger}(\tau)\hat{B}U(\tau)\big)$$

$$= \mathrm{Tr}\big(U(\tau)\rho(1)U^{\dagger}(\tau)\hat{B}\big) , \tag{4.13}$$

where $U(\tau)$ is the time translation operator from $\tau = 1$ to τ.

For the case that $V(\tau)$ is τ independent (renormalizable interactions) we may immediately conclude that as $\tau \to \infty$ $\langle B\rangle \to 0$. This is because a field theory with time-independent Hamiltonian will eventually come to thermal equilibrium and we have seen that CPT ensures $B = 0$ in this case.

On the other hand, if $V(\tau) \to 0$ fast enough we can use ordinary perturbation theory in V to compute the baryon-number excess as $\tau \to \infty$. To do this we use the standard interaction-picture formalism to obtain

$$U(\tau) = U_0(\tau)U_V(\tau), \quad U_0(\tau) = e^{-iH_0(\tau-1)} ,$$

$$U_V(\tau) = T \exp\left[-i \int_1^{\tau} V_1(\tau')d\tau'\right] , \tag{4.14}$$

$$V_1(\tau) = U_0^{\dagger}(\tau)V(\tau)U_0(\tau) .$$

Thus using $[B,U_0] = 0$,

$$\langle B(\tau)\rangle = \mathrm{Tr}\left(\rho(1)U_V^\dagger(\tau)\hat{B}U_V(\tau)\right) \quad . \qquad (4.15)$$

Graphical rules are derived in Appendix A for the evaluation of (4.15). The following features emerge from analysis of these rules:

(1) For the case $V(\tau) \sim 1/\tau^2$ each order has a finite limit as $\tau \to \infty$. These limits give an order-by-order expansion of the final baryon-number excess.

(2) The first order in which a nonvanishing excess occurs depends on certain features of $\rho(1)$. In particular, if $[\rho(1), B] = 0$ then the first order vanishes.

(3) If in addition to $\rho(1)$ being diagonal in baryon number it is CP symmetric then the second order also vanishes. Thus in the case of initially chaotic conditions, baryon-number excess is a third-order effect. Thus, since we suppose [see Equation (3.2)] that $V \sim \alpha$, baryon-number excess will be $\sim \alpha^3$ for an initially chaotic ρ.

We are currently constructing Feynman rules for the evaluation of Equation (4.15). These rules will be applied to some unified models in a future paper.

V. SCALAR TOY MODEL

Consider the model introduced in Section II [Eq. (2.2)]. In conformal coordinates the action becomes

$$S = \int d^3x d\tau \Big[\left|\frac{\partial\phi}{\partial\tau}\right|^2 - \left|\nabla\phi\right|^2 - \frac{\lambda}{\tau^{2n}} (\phi\phi^*)^n(\phi+\phi^*)$$
$$\times\ (\alpha\phi^3 + \alpha^*\phi^{*3}) - g(\phi\phi^*)^2\Big] \ , \qquad (5.1)$$

where we have added the renormalizable term $g\phi^4$ to represent all the renormalizable interactions. In this section we will make some very crude approximations which reduce the system to a single degree of freedom.

First we shall assume that the initial density matrix is in thermal equilibrium at a temperature ~1. If we ignore the small ($\sim\alpha$) nonrenormalizable couplings then the system will remain in equilibrium at this temperature for all τ. (Note that in transforming to the original coordinates the temperature becomes $1/\tau$ since it scales like energy.) Thus the average value of $|\phi|$ will remain constant of order unity. Indeed the first simplification will be to replace $|\phi|$ by unity.

The other drastic simplification will be to focus on a single unit coordinate cell over which ϕ will be assumed spatially constant. Setting $\phi = e^{i\theta}$ we obtain a system described by the Lagrangian.

$$\mathcal{L} = \left(\frac{d\theta}{d\tau}\right)^2 - \tau^{-2n} V(\theta) . \tag{5.2}$$

The baryon number of a unit cell is given by Eq. (2.4):

$$B = R^3(t) i\phi \overleftrightarrow{\partial}_t \phi^*$$

$$= 2 \frac{d\theta}{d\tau} . \tag{5.3}$$

Equation (5.2) describes a pendulum in a time-dependent unsymmetric potential and Eq. (5.3) says that the baryon number of a single cell is given by its angular velocity. The CPT invariance of the original instantaneous Hamiltonian corresponds to the time-reversal invariance of the pendulum.

The approximation of ignoring the interaction of neighboring cells is surely too severe to correctly describe the high-temperature nonequilibrium properties of the subsystem. In particular, it is impossible for the single pendulum to relax to thermal equilibrium if it is disturbed. For example, if the pendulum is given a hard "clock-wise" swing it will forever continue to rotate so that $\dot{\theta} \neq 0$. But in thermal equilibrium $\langle\dot{\theta}\rangle = 0$ by the same arguments which we used to prove $\langle B\rangle = 0$.

By ignoring the surrounding heat bath we have eliminated the possibility of dissipation. A simple method for incorporating it is to introduce a dissipative damping term into the equation of motion. Thus we write the equation of motion

$$\frac{d^2\theta}{d\tau^2} + \tau^{-2n} \frac{\partial V}{\partial \theta} + f(\tau) \frac{d\theta}{d\tau} = 0. \tag{5.4}$$

The computation of friction coefficients in nonequilibrium statistical mechanics typically involves the computation of the absorptive (imaginary) part of some thermal Green's function (Fetter and Walecka 1971). This is to say, we calculate the width of some excitation which propagates in the medium.

In the case of electical resistance we calculate the absorptive part of the plasmon propagator (Fetter and Walecka 1971). In our case, a nonzero baryon charge must dissipate as equilibrium is restored. Accordingly we must compute the width of the charge-carrying excitation described by the field ϕ due to baryon-number-violating processes. In the model field theory with interaction $V(\phi) = \lambda(\phi^*\phi)^n(\phi+\phi^*)(\alpha\phi^3+\alpha^*\phi^{*3})$

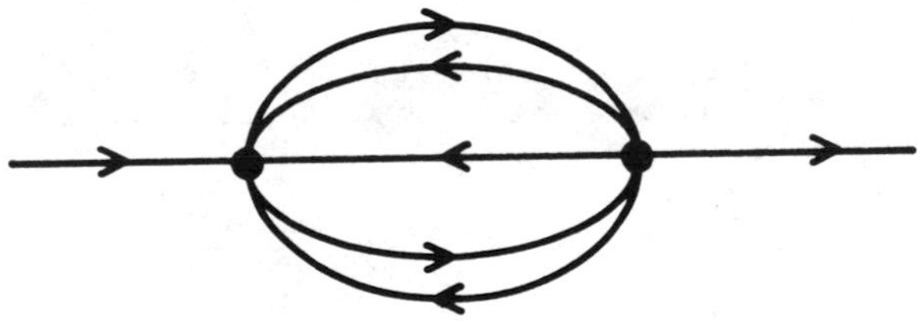

Figure 2

Graph contributing to baryon dissipation.

the relevant width is described by graphs shown in Figure 2. Dimensional arguments require the temperature-dependent width to be

$$\gamma(T) \approx \lambda^2 T^{4n+1} . \tag{5.5}$$

Thus if the number of baryons in the unit cell is B, the number lost by dissipation is

$$\left(\frac{d}{dt} B\right)_{dis} = - B\gamma = - B\lambda^2 T^{4n+1} \tag{5.6}$$

or

$$\left(\frac{dB}{dt}\right)_{dis} = - \frac{\lambda^2 B}{\tau^{4n}} . \tag{5.7}$$

Recalling that B is identified with $d\theta/d\tau$ we interpret Equation (5.7) to mean that the coefficient f in Equation (5.4) is λ^2/τ^{4n}:

$$\frac{d^2\theta}{d\tau^2} + \tau^{-2n} \frac{\partial V}{\partial \theta} + \frac{\lambda^2}{\tau^{4n}} \frac{d\theta}{d\tau} = 0 . \tag{5.8}$$

Equation (5.8) defines the toy model.

To see how the toy model can lead to an asymmetric distribution of baryons and antibaryons consider a $V(\theta)$ which looks like Figure 3, i.e., it has no point of reflection symmetry. Now suppose the initial probability density in θ and $\dot{\theta}$ is uniform in θ and symmetric under $\dot{\theta} \to -\dot{\theta}$. We observe that a particle has a large probability to get a small kick to the left and a small probability for a large kick to the right. Thus the probability distribution becomes asymmetric. However, to first order in time no average change in $\dot{\theta}$ occurs. This is because the average force $\partial V/\partial \theta$ vanishes for a uniform distribution in θ. In fact $\langle\dot{\theta}\rangle$ only becomes nonzero in order τ^5. Furthermore, the first nonvanishing order in V is third order.

If the universe were a nonexpanding box at fixed temperature then no net baryon excess could be maintained at long times. Indeed the toy model is consistent with this. In a nonexpanding universe the form of the toy model is

$$\frac{d^2\theta}{dt^2} + \lambda T^{2n+2} \frac{\partial V}{\partial \theta} + \lambda^2 T^{4n+1} \frac{d\theta}{dt} = 0 . \tag{5.9}$$

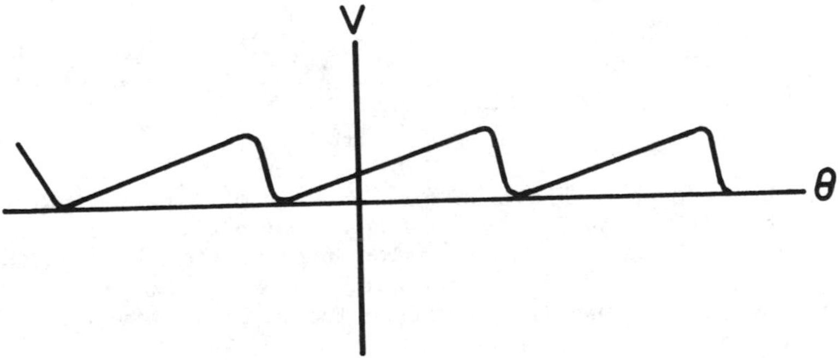

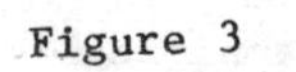

Figure 3

A potential which violates $V(\theta)$.

Let us suppose after a long time that the baryon number $T^2 d\theta/dt$ is constant. Then

$$\lambda T^{2n+2} \frac{\partial V}{\partial \theta} + \lambda^2 T^{2+n+1} \dot{\theta} = 0 . \tag{5.10}$$

Integrating this over a period and using the periodicity of $V(\theta)$ we see that the baryon number has to vanish. Note that both the periodicity of the potential and the existence of the friction term are important in reaching this conclusion.

Now we find the conditions that will allow a non-vanishing baryon number at large times. Multiplying Eq. (5.8) by τ^{2n} and integrating over a period we obtain

$$\oint \tau^{2n} \frac{dK}{d\tau} d\tau = -2\lambda^2 \oint \frac{K}{\tau^{2n}} d\tau , \tag{5.11}$$

where $K = 1/2(d\theta/d\tau)^2$. After a long time, this equation effectively becomes

$$\frac{dK}{d\tau} = - 2\lambda^2 \frac{K}{\tau^{4n}} . \tag{5.12}$$

For the renormalizable case $n = 0$ we see that baryon number is exponentially damped. This agrees with our previous expectations. For $n > 1/4$ this equation has solutions for which the baryon number tends to a constant. Thus we see that for non-renormalizable theories the friction term can be neglected, and baryon-number excess occurs as $\tau \to \infty$.

VI. CONCLUDING REMARKS

In this paper we have argued that a baryon-number excess may be produced in an expanding universe even though the initial conditions are symmetric. For the case of unified theories the excess is developed at times of order 10^{-40} sec while the temperature is comparable to the Planck mass. An admittedly oversimplified model yields a small number of baryons per unit cell of the order α^3.

This conclusion that the effect is $\sim \alpha^3$ does not appear to be general. It is a consequence of replacing the superheavy interactions by four-Fermi interactions. While this helps us visualize the process, it is not entirely consistent. This is because the main action occurs at energies of order $\tilde{M}$ and not much lower energies. Therefore it is important to open up the "black box" hiding the superheavy-boson exchange. As far as we can tell there are then order-α^2 effects. This is somewhat too large empirically but we must keep in mind that there are effects which we ignored which decrease N_B/N_γ. We have treated the universe expansion as if it were a reversible process with respect to the ordinary interactions. In fact there are possible sources of irreversibility which can heat up the system (Liang 1977). Eventually this heat must appear as photons.

Unfortunately this optimistic picture which emerges in unified theories may be drastically changed if the baryon-number-violating tunneling events are really important. The point is that the rates for these processes are of the renormalizable type for T > 250 GeV. Thus they can allow the system to return to equilibrium and may wash out any excess which developed at superhigh temperature.

Of course as the temperature goes below 250 GeV the tunneling processes also go out of equilibrium. In principle the observed baryon-number excess could be attributed to this final stage of baryon-number violation. In this case the number of baryons in the universe is independent of the initial conditions and the details of the particular unified model.

ACKNOWLEDGMENTS

We would like to thank R. Wagoner for many critical insights without which we could not have written this paper. We are also indebted to E. Liang, D. Sciama, and G. Steigman for interesting comments. One of us (S. D.) would like to thank Sid Drell and Dirk Walecka for hospitality at SLAC and Stanford University. This work was supported in part by the National Science Foundation under Contract No. 76-16992.

APPENDIX A

Graphical rules for computing $\langle B(\tau)\rangle$. Consider a theory of fermions interacting with baryon-number-, C- and CP-violating four-Fermi forces. The Hamiltonian of this theory in the expanding universe in terms of the conformal coordinates is of the form

$$H = H_0 + V(\tau) .$$

The baryon-number-violating piece $V(\tau)$ is of the form

$$V(\tau) = \frac{\alpha}{\tau^2} \int d^3x(\bar{\psi}\Gamma\psi)^2 \equiv \frac{v}{\tau^2} .$$

The graphical rules for the evaluation of $\langle B(\tau)\rangle$ can be deduced from the expression

$$\langle B(\tau)\rangle = \mathrm{Tr}\left(\rho(1) U^{+}_{V_1}(\tau) U^{\dagger}_{H_0}(\tau) \hat{B} U_{H_0}(\tau) U_{V_1}(\tau)\right) , \qquad \text{(A1)}$$

where

$$U_{H_0}(\tau) = T \exp\left[-i \int_1^\tau H_0(\tau')d\tau'\right]$$
$$= e^{-iH_0(\tau-1)} ,$$

$$U_{V_1}(\tau) = T\exp\left[-i \int_1^\tau V_1(\tau')d\tau'\right] , \qquad (A2)$$

$$V_1(\tau) = U^\dagger_{H_0}(\tau)V(\tau)U_{H_0}(\tau) .$$

Since H_0 conserves the baryon number, expression (A1) simplifies to

$$\langle B(\tau)\rangle = \mathrm{Tr}\left(\rho(1)U^\dagger_{V_1}(\tau)\hat{B}U_{V_1}(\tau)\right) . \qquad (A3)$$

The graphical rules for the evaluation of this quantity are the following:

(1) Draw the closed loop shown in Figure 4 in order $\ell + r$.

(2) For each cross on the right write $ie^{iH_0(\tau'-1)} \times ve^{-iH_0(\tau'-1)}$. For each cross on the left write $-ie^{iH_0(\tau-1)}\, ve^{-iH_0(\tau-1)}$.

(3) Write down the terms indicated in Figure 4 in anticlockwise order and take the trace.

(4) Carry out the time integrations with weight $1/\tau^2$. Respect time ordering.

Do the same for the $\ell + r + 1$ graphs appearing in order $\ell + r$. Note that the lines in Figure 4 are not particle lines. They represent propagation of states.

APPENDIX B

Here we will show explicitly that for the model discussed in Appendix A the second-order contributions to $\langle B(\tau)\rangle$ vanish. We shall label each state solely by its baryon number $|n\rangle$. The CPT conjugate state will be denoted by $|-n\rangle$, and by CPT invariance, $\rho_n(1) = \rho_{-n}(1)$. Since $[B, \rho(1)] = 0$, $\rho_{nm}(1) \equiv \rho_{nm} = \rho_n\delta_{nm}$. Since B is CPT odd, $B_{-n} = -B_n$. The second-order contributions to $\langle B(\tau)\rangle$ arise from the graphs of Figure 5. The contribution of graph (a) is

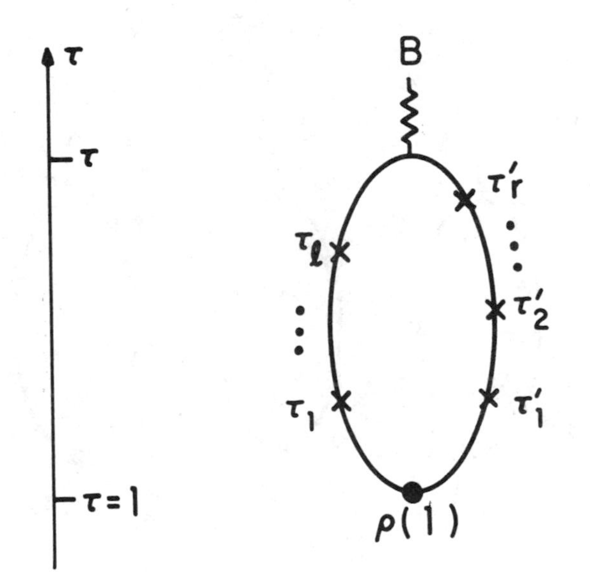

Figure 4

Graphical notations. Solid lines represent propagating state vectors. Crosses represent the action of V. The black dot represents the initial density matrix and the wavy line represents the measurement of baryon number.

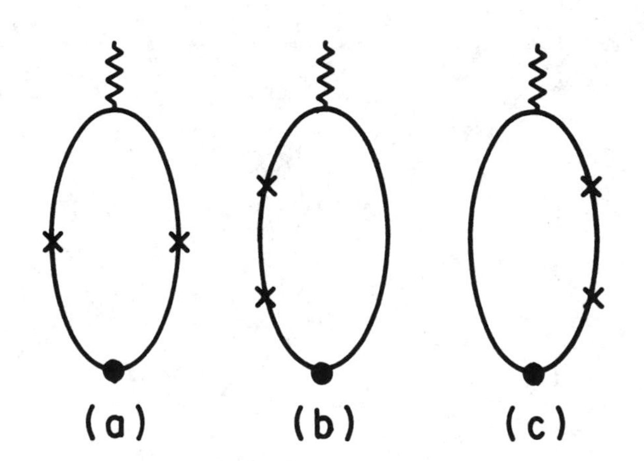

Figure 5

The second-order contributions to $\langle B(\tau) \rangle$.

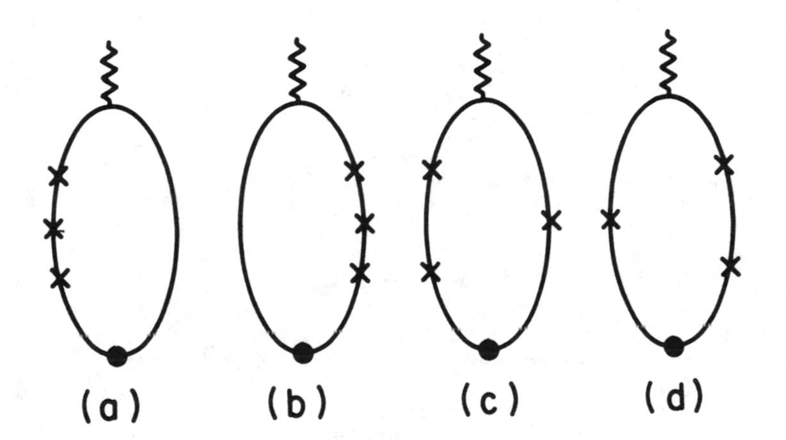

Figure 6

The third-order contributions to $\langle B(\tau) \rangle$

$$i(-i)\int_1^{\tau}\frac{d\tau_1}{\tau_1^2}\int_1^{\tau}\frac{d\tau_2}{\tau_2^2}\,\rho_n e^{i\varepsilon_n\tau_1}(v_{nm})\,e^{-i\varepsilon_m\tau_1}(B_m)\,e^{i\varepsilon_m\tau_2}$$

$$\times\,(v_{mn})e^{-i\varepsilon_n\tau_2} = 0 .$$

In deriving this we used the CPT invariance of the Hamiltonian H,

$$\varepsilon_n = +\,\varepsilon_{-n} \text{ and } |v_{nm}|^2 = |v_{-m,-n}|^2 .$$

The contribution of graph (b) is

$$(-i)^2\int_1^{\tau}\frac{d\tau_2}{\tau_2^2}\int_1^{\tau_2}\frac{d\tau_1}{\tau_1^2}\,\rho_n B_n\,e^{i\varepsilon_n\tau_2}(v_{nm})\,e^{-i\varepsilon_m\tau_2}\,e^{i\varepsilon_m\tau_1}$$

$$\times\,(v_{nm})\,e^{-i\varepsilon_n\tau_1} = 0 .$$

This vanishes for the same reason with graph (a). The vanishing of the second-order contribution to $\langle B(\tau)\rangle$ is not a general feature of all models. It only happens because the explicit time dependence of $V(\tau)$ can be factored out.

APPENDIX C

In this appendix we write down the third-order contributions to $\langle B(\tau)\rangle$ for the model of Appendix A. The graphs contributing are those of Figure 6.

Graph (a) contributes

$$(-i)^3\int_1^{\tau}\frac{d\tau_3}{\tau_3^2}\int_1^{\tau_3}\frac{d\tau_2}{\tau_2^2}\int_1^{\tau_2}\frac{d\tau_1}{\tau_1^2}$$

$$\rho_n B_n\,e^{i\varepsilon_n\tau_3}\,v_{nm}\,e^{-i\varepsilon_m(\tau_3-\tau_2)}\,v_{me}\,e^{-i\varepsilon_e(\tau_2-\tau_1)}\,v_{en}\,e^{-i\varepsilon_n\tau_1}$$

$$= (-i)^3\int_1^{\tau}\frac{d\tau_3}{\tau_3^2}\,e^{i\tau_3(\varepsilon_n-\varepsilon_m)}\int_1^{\tau_3}\frac{d\tau_2}{\tau_2^2}\,e^{i\tau_2(\varepsilon_m-\varepsilon_e)}$$

$$\int_1^{\tau_2}\frac{d\tau_1}{\tau_1^2}\,e^{i\tau_1(\varepsilon_e-\varepsilon_n)}\,\rho_n B_n v_{nm} v_{me} v_{en} .$$

Graph (b) contributes

$$i^3 \int_1^{\tau} \frac{d\tau_3}{\tau_2^2} \int_1^{\tau_3} \frac{d\tau_2}{\tau_2^2} \int_1^{\tau_2} \frac{d\tau_1}{\tau_1^2} \rho_n B_n$$

$$e^{i\varepsilon_n \tau_1} v_{nm} e^{-i\varepsilon_m(\tau_1 - \tau_2)} v_{me} e^{-\varepsilon_e(\tau_2 - \tau_3)} v_{en} e^{-i\varepsilon_n \tau_3}$$

$$= i \int_1^{\tau} \frac{d\tau_3}{\tau_3^2} e^{i\tau_3(\varepsilon_e - \varepsilon_n)} \int_1^{\tau_3} \frac{d\tau_2}{\tau_2^3} e^{i\tau_2(\varepsilon_m - \varepsilon_e)}$$

$$\int_1^{\tau_2} \frac{d\tau_1}{\tau_1^2} e^{i\tau_1(\varepsilon_n - \varepsilon_m)} \rho_n B_n v_{nm} v_{me} v_{en} \, .$$

Graph (b), of course, is just the complex conjugate of graph (a).

Graph (c) yields

$$(-i)^2 i \int_1^{\tau} \frac{d\tau'_1}{\tau_1'^2} e^{i(\varepsilon_n - \varepsilon_m)\tau_1'} \int_1^{\tau} \frac{d\tau_2}{\tau_2^2} e^{i(\varepsilon_m - \varepsilon_e)\tau_2}$$

$$\int_1^{\tau_2} \frac{d\tau_1}{\tau_1^2} e^{i(\varepsilon_e - \varepsilon_n)\tau} \rho_n v_{nm} B_m v_{me} v_{en} \, .$$

Graph (d) yields the complex conjugate of (c),

$$i^2(-i) \int_1^{\tau} \frac{d\tau'_2}{\tau_2'^2} e^{i\tau'_2(\varepsilon_m - \varepsilon_e)} \int_1^{\tau_2'} \frac{d\tau'_1}{\tau_1'^2} e^{i\tau'_1(\varepsilon_n - \varepsilon_m)}$$

$$\int_1^{\tau} \frac{d\tau_1}{\tau_1^2} e^{i\tau_1(\varepsilon_e - \varepsilon_n)} \rho_n v_{nm} v_m B_e v_{en} \, .$$

These expressions do not vanish in general. They, of course, vanish if we assume C- or CP-invariant matrix elements for v.

REFERENCES

Belavin, A. A., et al. 1975, Phys. Lett., **59B**, 85.

Callan, C., Dashen, R., and Gross, D. 1976, Phys. Lett., **63B**, 334.

Dolan, L., and Jackiw, R. 1974, Phys. Rev. D, **9**, 3320.

Fetter, A. L., and Walecka, J. D. 1971, Quantum Theory of Many Particle Systems (New York: McGraw Hill).

Georgi, H., and Glashow, S. L. 1974, Phys. Rev. Lett., **32**, 438.

Gürsey, F., and Sikivie, P. 1976, Phys. Rev. Lett., **36**, 775.

Hawking, S. 1975, Commun. Math. Phys., **43**, 199.

Jackiw, R., and Rebbi, C. 1976, Phys. Rev. Lett., **37**, 172.

Kirzhnits, D. A., and Linde, A. D. 1972, Phys. Lett., **42B**, 471.

Kislinger, M. B., and Morley, P. D. 1976a, Phys. Rev. D, **13**, 2765.

Kislinger, M. B., and Morley, P. D. 1976b, Phys. Rev. D, **13**, 2771.

Lee, T. D. 1973, Phys. Rev. D, **8**, 1226.

Lee, T. D. 1974, Phys. Rep., **3C**, 143.

Lee, T. D., Oehme, R., and Yang, C. N. 1957, Phys. Rev., **106**, 340.

Lee, T. D., and Wu, C. S. 1966, Ann. Rev. Nucl. Sci., **16**, 511.

Liang, E. P. T. 1977, Phys. Rev. D, **16**, 3369.

Parker, L. 1977, in Strong Gravitational Fields, F. P. Esposito and L. Witten, eds. (New York: Plenum), pp. 107-226.

Pati, J. C., and Salam, A. 1973, Phys. Rev. D, **8**, 1240.

Pati, J. C., and Salam, A. 1974, Phys. Rev. D, **10**, 275.

Steigman, G. 1976, Ann. Rev. Astron. Astrophys., **14**, 339.

't Hooft, G. 1976a, Phys. Rev. Lett., **37**, 8.

't Hooft, G. 1976b, Phys. Rev. D, **14**, 3432.

Wald, R. M. 1975, Comm. Math. Phys., **45**, 199.

Weinberg, S. 1972, Gravitation and Cosmology (New York: Wiley).

Weinberg, S. 1974, Phys. Rev. D, **9**, 3357.

Weinberg, S. 1976, Phys. Rev. Lett., **37**, 657.

Yoshimura, M. 1978, Phys. Rev. Lett., **41**, 281.

Yoshimura, M. 1978, unpublished.

COSMOLOGICAL PRODUCTION OF BARYONS

Steven Weinberg
Lyman Laboratory of Physics, Harvard University,
and Harvard-Smithsonian Center for Astrophysics,
Cambridge, Massachusetts

Received: October 27, 1978

Departures from thermal equilibrium which are likely to occur in an expanding universe allow the production of an appreciable net baryon density by processes which violate baryon-number conservation. It is shown that the resulting baryon to entropy ratio can be calculated in terms of purely microscopic quantities.

It is an old idea[1] that the observed excess matter over antimatter in our universe may have arisen from physical processes which violate the conservation of baryon number. Of course, the rate of baryon-nonconserving processes like proton decay are very small at ordinary energies, but if the slowness of these processes is due to the large mass of intermediate vector or scalar "X bosons" which mediate baryon nonconservation, then at very high temperatures with $kT \simeq m_X$, the baryon-nonconserving processes would have rates comparable with those of other processes. However, even if there are reactions which do not conserve C, CP, T, and baryon number, and even if these reactions proceed faster than the expansion of the universe, there can be no cosmological baryon production once the cosmic distribution functions take their equilibrium form,

[1]For one example, see Weinberg (1964). The subject has been considered in recent papers by Yoshimura (1978); Dimopoulos and Susskind (1978); Toussaint, Treiman, Wilczek, and Zee (1979). After this work was completed I also became aware of a discussion by Ignatiev, Krosnikov, Kuzmin, and Tavkhelidze (1978); and new reports have appeared by Dimopoulos and Susskind (to be published), and Ellis, Gaillard, and Nanopoulos (to be published). The approach followed and the conclusions reached here differ from those of Yoshimura, for reasons indicated below; from Ignatiev _et al._ because they adopt a different picture of baryon nonconservation (without superheavy X bosons); and from Ellis, Gaillard, and Nanopoulos, for reasons indicated in their erratum (to be published). The assumptions and general approach followed here is similar in many respects to that of Dimopoulos and Susskind and Section 2 of Toussaint _et al._ A major difference is that by following the baryon production scenario in detail, a formula is obtained here, Eq. (9), which gives the ratio of baryon number to entropy in terms of purely microscopic quantities.

until the expansion of the universe had a chance to pull these distribution functions out of equilibrium. This can be seen from the generalized Uehling-Uhlenbeck equation (Uehling and Uhlenbeck 1933) for a homogeneous isotropic gas,

$$dn(p_1)/dt = \Sigma_{k1} \int dp_2 \ldots dp_k dp_1' \ldots dp_\ell'$$

$$\times \{\Gamma(p_1' \ldots p_\ell' \to p_1 \ldots p_k) n(p_1') \ldots n(p_\ell')[1 \pm n(p_1)] \ldots [1 \pm n(p_k)]$$

$$- \Gamma(p_1 \ldots p_k \to p_1' \ldots p_\ell') n(p_1) \ldots n(p_k)[1 \pm n(p_1')] \ldots [1 \pm n(p_\ell')]\} , \tag{1}$$

where n is the single-particle density in phase space; p labels the three-momentum and any other particle quantum numbers, including baryon number; and Γ is a rate constant, equal, for $k = \ell = 2$, to the cross section times the initial relative velocity. The factors $1 \pm n(p)$ represent the effect of stimulated emission or Pauli suppression for bosons or fermions, respectively. If at any instant, n(p) takes its equilibrium form, then $n(p)/[1 \pm n(p)]$ is an exponential of a linear combination of the energy and any other conserved quantities; so for any allowed reaction with $\Gamma \neq 0$, we have

$$n(p_1') \ldots n(p_\ell')[1 \pm n(p_1)] \ldots [1 \pm n(p_k)] =$$

$$= n(p_1) \ldots n(p_k)[1 \pm n(p_1')] \ldots [1 \pm n(p_\ell')] . \tag{2}$$

Under T invariance, Γ would be symmetric, and the two terms in the integrand of Eq. (1) would cancel. But even without T invariance, unitarity always gives

$$0 = \Sigma_\ell \int dp_1' \ldots dp_\ell' [1 \pm n(p_1')] \ldots [1 \pm n(p_\ell')]$$

$$[\Gamma(p_1 \ldots p_k \to p_1' \ldots p_\ell') - \Gamma(p_1' \ldots p_\ell' \to p_1 \ldots p_k)] , \tag{3}$$

so that the p' integrals in (1) still cancel.[2] For an expand-

[2]A very general version of this argument in the context of the "master" equation was given about a decade ago in an unpublished work of C. N. Yang and C. P. Yang. Also see Aharony (1973), and references cited therein. I first learned of this argument for the special case of massless distinguish-

ing gas there are also terms in Eq. (1) which represent the effects of dilution and red shift, and these terms can produce departures from equilibrium, but of course they have no direct effect on the baryon number per co-moving volume.

This note will describe a mechanism for production of a cosmic baryon excess, based on the departures from thermal equilibrium which are likely to have occurred in the early universe. It is assumed here that all particles have masses below (though not necessarily far below) the Planck mass $m_P \equiv G^{-1/2} = 1.22 \times 10^{19}$ GeV. For simplicity, it will be assumed that the only superheavy particles with masses above 1 TeV or so are the X bosons which mediate baryon nonconservation; however, it would not be difficult to incorporate superheavy fermions with masses $m \simeq m_X$ in these considerations. Aside from gravitation itself, all interactions are supposed to have dimensionless coupling constants. For the interaction of X bosons with fermions, this coupling is denoted g_X. Finally, it will also be assumed that $\alpha_X^2 N \ll 1$, where $\alpha_X = g_X^2/4\pi$, and N is the number of helicity states of all particle species. Under these assumptions, we can trace the following chain of events:[3]

(1) At very early times, when $kT \approx m_P$, the interactions of gravitons were so strong that thermal equilibrium distributions would have been established at least approximately for all particle species; for instance, by graviton-graviton collisions.[4] (Of course, we do not know how to calculate detailed reaction rates at these times, but we can be confident that gravitational interactions were strong, because this is indicated by lowest-order calculations, and it is only the strength of the interactions that invalidates such calcula-

able particles from the original version of the paper of Toussaint _et al._ (1979). For indistinguishable particles, the factors $1 \pm n(p')$ in Eq. (3) arise from the effects of the ambient bosons or fermions on identical virtual particles in these reactions; in old-fashioned perturbation theory, the ambient particles generate a product of $1 \pm n$ factors for each intermediate state. These factors were omitted in the unitarity relation as given by Aharony, so that it was not possible in his paper to see how the Uehling-Uhlenbeck equation yields a vanishing rate of change for equilibrium distributions. Equation (3) shows that the physical processes considered by Yoshimura (1978) cannot produce an appreciable net baryon density if all relevant channels are taken into account, as pointed out by Toussaint _et al._ (1979).

[3]This scenario was developed in the course of conversations with F. Wilczek, and is also discussed in Section 2 of Toussaint _et al._ (1979). I am very grateful to F. Wilczek for numerous discussions of these ideas.

[4]Horizon effects may prevent complete establishment of thermal equilibrium at $kT \simeq m_P$; G. Steigman, private communication.

tions.) If gravitational interactions conserved baryon number at $kT \gtrsim m_P$, then the universe could have begun with a nonvanishing value for the baryonic chemical potential; I assume here that this is not the case.

(2) As kT fell below m_P, gravitational interactions became ineffective. The rates for X-boson decay, baryon-nonconserving collisions (or, for $kT \gtrsim m_X$, all collisions) and cosmic expansion may be estimated as[5]

$$\Gamma_X \simeq \alpha_X m_X{}^2 N/[(kT)^2 + m_X{}^2]^{1/2} \ , \tag{4}$$

$$\Gamma_C \simeq \alpha_X{}^2 (kT)^5 N/[(kT)^2 + m_X{}^2]^2 \ , \tag{5}$$

$$\dot{R}/R \equiv H = 1.66(kT)^2 N^{1/2}/m_P \ . \tag{6}$$

With $\alpha_X{}^2 N << 1$ and $m_X < m_P$, both Γ_X and Γ_C were much less than H at $kT \simeq m_P$. However, as long as kT remained above all particle masses, the expansion preserved the equilibrium form of all particle distributions, with red-shifted temperature $T \propto 1/R$.

(3) The X bosons began to decay when $\Gamma_X \simeq H$. If at this time $kT > m_X$, the collisions of the decay products with each other or with ambient particles would have rapidly recreated the X bosons through the inverse of the decay process, thus reestablishing equilibrium distributions. In order to produce any appreciable baryon excess, it is necessary that $kT \lesssim m_X$ when $\Gamma_X \simeq H$, so that the Boltzmann factor $\exp(-m_X/kT)$ could block inverse decay. Equation (4) then gives $\Gamma_X \simeq H$ at a temperature

$$kT_D \simeq (N^{1/2}\alpha_X m_X m_P)^{1/2} \ , \tag{7}$$

so that the condition $m_X \gtrsim kT_D$ yields a lower bound on m_X

$$m_X \gtrsim N^{1/2}\alpha_X m_P \ . \tag{8}$$

[5]If we keep track of all factors of 2π from Fourier integrals and 4π from solid-angle integrals, but set all other numerical constants equal to unity, then factors 4 and $8/\pi$ would appear in the right-hand sides of Eqs. (4) and (5), respectively. The powers of $m_X{}^2/[(kT)^2 + m_X{}^2]$ in Eqs. (4) and (5) are inserted to take account of time dilation and the virtual X-boson propagator, respectively.

(For gauge bosons we expect $\alpha_X \simeq \alpha$, so (8) requires $M_X \gtrsim 10^{17}N^{1/2}$ GeV, while for Higgs bosons α_X is presumably in the range of 10^{-4} to 10^{-6}, and the lower bound on m_X would be of order 10^{13} to $10^{15}N^{1/2}$ GeV.) Note also that (5), (6), and (8) give $\Gamma_C \ll H$ for all temperatures. This justifies the neglect of X-boson production or annihilation in reactions other than X decay and its inverse, and insures that any baryon excess produced when the X bosons decayed would have survived to the present time.

Before the X bosons decayed, at temperatures just above T_D, their number density was $n_{XD} = \zeta(3)(kT_D)^3 N_X/\pi^2$, where N_X is the total number of X (and $\bar{X}$) spin states. Also, the total entropy density of all other particles was $S_D = 4\pi^2 k(kT_D)^3 \times N/45$, with N now understood to include factors of 7/8 for fermion spin states. If the mean net baryon number produced in X or $\bar{X}$ decay is ΔB per decay, and if one can ignore the entropy released in X-boson decay, then the ratio of baryon number to entropy after the X bosons decayed was

$$kn_B/s = kn_{XD}\Delta B/S_D = 45\zeta(3)(N_X/N)\Delta B/4\pi^4 \ . \tag{9}$$

If one assumes the subsequent expansion to be adiabatic, both n_B and s would have scaled as R^{-3}, so that Eq. (9) would give the ratio of baryon number to entropy of the present universe.

Strictly speaking, one should take into account the entropy contributed by the X-boson decay products when they finally thermalize. This increases the energy density by a factor

$$\lambda = 1 + \frac{m_X n_{XD}}{\pi^2 (kT_D)^4 N/30} = 1 + \frac{30\zeta(3)N_X m_X}{\pi^4 NkT_D}$$

$$\simeq 1 + (N_X/N)(m_X/N^{1/2}\alpha_X m_P)^{1/2} \ , \tag{10}$$

and so decreases the ratio of baryon number to entropy by a factor of $\lambda^{-3/4}$. However, this effect can be ignored if $N \gg N_X$.

The crucial quantity ΔB in Eq. (9) can be determined from the branching ratios for X-boson decay. For instance, suppose that an X boson decays into two channels with baryon numbers B_1 and B_2 and branching ratios r and 1-r. The antiparticle will then decay into channels with baryon number $-B_1$ and $-B_2$, with the same total rate, but with different branching ratios $\bar{r}$ and $1 - \bar{r}$. The mean net baryon number produced when X or $\bar{X}$ decays is then

$$\Delta B = \frac{1}{2}[rB_1 + (1 - r)B_2 - \bar{r}B_1 - (1 - \bar{r})B_2]$$
$$= \frac{1}{2}(r - \bar{r})(B_1 - B_2) \ . \qquad (11)$$

CPT invariance gives $r = \bar{r}$ in the Born approximation. If the leading contribution to $r - \bar{r}$ arises from an interference of graphs with a total of ℓ loops, then one expects $r - \bar{r}$ to be of order $\varepsilon(\alpha_X/2\pi)^\ell$, where ε is whatever small angle characterizes CP violation. Of course, to be more definite, a detailed model of baryon nonconservation is needed. Hoewever, in any given model, Eqs. (9) and (1) give a precise prediction for the ratio of baryon number to entropy kn_B/s, which may be compared with the observed value[6] 10^{-8} to 10^{-10}.

The above discussion has assumed a homogeneous isotropic expansion, in which the entropy stays fixed except for the small effects of bulk viscosity (see footnote 6). However. it is also possible to deal with gross departures from thermal equilibrium that might be produced by cosmic inhomogeneities. As any part of the universe relaxes toward equilibrium, the rate at which its entropy increases will be proportional to the difference between the entropy and its maximum value S_{max}. Baryon production vanishes in the equilibrium configuration with $S = S_{max}$, so the rate of increase of baryon number will also be proportional to $S - S_{max}$. Thus, the ratio of the baryon-number production to the entropy production will be given by the ratio of the coefficients of $S - S_{max}$ in dB/dt and dS/dt, and independent of the amount of the initial departure from thermal equilibrium. If most of the entropy and baryon number of the universe were created in this way, then it is this ratio that would have to be compared with the experimental value of 10^{-8} to 10^{-10}.

Note added - (1) Any X bosons which can mediate baryon-nonconserving reactions are necessarily much heavier than the Z° or $W^\pm$; so their interactions can be analyzed using the weak and electromagnetic gauge group $SU(2)\otimes U(1)$ as well as the strong gauge group SU(3) as if they were all unbroken symmetries. In this way one finds in general there are just three kinds of bosons which can couple to channels consisting of a pair of ordinary fermions, with these channels not all having equal baryon numbers: They are an SU(3) triplet SU(2) singlet X_S of scalar bosons with charge -1/3; an SU(3) triplet SU(2) doublet X_V of vector bosons with charges -1/3, -4/3; and an SU (3) triplet SU(2) doublet X_V' of vector bosons with charges 2/3, -1/3; plus their corresponding SU(3)-$\bar{3}$ antibosons. For all these bosons, the decay channels are $X \to q\ell$, $\bar{q}\bar{q}$ and $\bar{X} \to \bar{q}\bar{\ell}$, qq, with q and ℓ denoting general quarks and leptons. Hence $B_1 = +1/3$ and $B_2 = -2/3$ in Eq. (11). This analysis incidentally shows that lowest-order baryon-number-nonconserving interactions always conserve baryon number minus lepton number, so nucleons may decay in lowest order into antileptons, but not leptons.

[6] See, e.g., Weinberg (1972), chapter 15.

(2) Detailed calculations have been carried out with Nanopoulos (Weinberg and Nanopoulos to be published) to estimate the difference in the branching ratios r, $\bar{r}$ for $X \rightarrow q\ell$ and $X \rightarrow \bar{q}\bar{\ell}$ that arises from the interference of tree graphs with one-loop graphs. In general, a difference between r and $\bar{r}$ could arise from one-loop graphs in which a scalar or vector boson is exchanged between the final fermions, even when all fermion masses are negligible compared with the temperature, provided that CP invariance is violated in the Lagrangian, or is already spontaneously broken at these high temperatures. In various grand unified theories there are relations among the various couplings of Higgs or gauge bosons to fermions, which eliminate most of these contributions to $r - \bar{r}$. However, there will still be a contribution to $r - \bar{r}$ in X_S decay from the exchange of X_S bosons of different species. Since Higgs-boson exchange is naturally weaker than $W^{\pm}$ or Z^0 exchange at ordinary energies, it is possible that the CP-invariance violation is maximal in the coupling of fermions to Higgs bosons, including X_S bosons. In this case, $r - \bar{r}$ is of order $\alpha_H/2\pi \approx 10^{-6}$. With $B_1 - B_2 = 1$ and $N_X/N \approx 10^{-2}$, Eqs. (11) and (9) then give a ratio of baryon number to entropy of order 10^{-9}.

(3) The masses of superheavy gauge bosons were estimated in grand unified gauge theories to be of order 10^{16} GeV, by Georgi, Quinn, and Weinberg (1974). (As shown there, this estimate applies for arbitrary simple grand unified gauge groups, under reasonable general assumptions on the spectrum of fermions. The same assumptions yielded a Z^0–γ mixing angle with $\sin^2\theta \simeq 0.2$.) Presumably the Higgs-boson masses are of the same order. Decay and inverse-decay processes arising from the gauge coupling of vector besons to each other and to Higgs bosons and fermions will bring all these particles into thermal equilibrium at a temperature given by Eq. (7) [with $N \approx 100$, $\alpha_X \approx 10^{-2}$, $m_X \approx 10^{16}$ GeV] as of order 10^{17} GeV. Hence there is no need to invoke gravitational processes at the Planck temperature to establish initial equilibrium distributions, and any preexisting baryon imbalance would have been wiped out at $kT \simeq 10^{17}$ GeV. As the temperature dropped below 10^{16} GeV all superheavy gauge bosons and some of the superheavy Higgs bosons would have disappeared. However, if the lightest superheavy bosons happen to be X_S bosons, then these bosons would have survived as the temperature fell below their mass, because the only decay channels open then would have been two-fermion states, and $\alpha_X \ll \alpha$ for Higgs-fermion couplings. The decay of these scalar bosons when the temperature finally dropped to $kT_D \approx 10^{14}$ GeV $\ll m(X_S)$ would then produce the baryon excess estimated in Note (2).

I am grateful for valuable conversations with J. Ellis, D. Nanopoulos, A. Salam, L. Susskind, F. Wilczek, C. N. Yang, and M. Yoshimura.

REFERENCES

Aharony, A. 1973, in Modern Developments in Thermodynamics (New York: Wiley), pp. 95-114.
Dimopoulos, S., and Susskind, L. 1978, Phys. Rev. D, **18**, 4500.
Dimopoulos, S., and Susskind, L. to be published, Stanford University Report No. ITP-616.
Ellis, J., Gaillard, M. K., and Nanopoulos, D. V. to be published, CERN Report No. Ref. TH-2596.
Georgi, H., Quinn, H., and Weinberg, S. 1974, Phys. Rev. Lett., **33**, 451.
Ignatiev, A. Yu, Krosnikov, N. V., Kuzmin, V. A., and Tavkhelidze, A. N. 1978, Phys. Lett., **76B**, 436.
Steigman, G. to be published, private communication.
Toussaint, B., Treiman, S. B., Wilczek, F., and Zee, A. 1979, Phys. Rev. D, **19**, 1036.
Uehling, E. A., and Uhlenbeck, G. E. 1933, Phys. Rev., **43**, 552.
Weinberg, S. 1964, in Lectures on Particles and Fields, S. Deser and K. Ford, eds. (Englewood Cliffs, NJ: Prentice-Hall), p. 482.
Weinberg, S. 1972, Gravitation and Cosmology - Principles and Applications of the General Theory of Relativity (New York; Wiley).
Weinberg, S., and Nanopoulos, D. V. to be published.
Yoshimura, M. 1978, Phys. Rev. Lett., **41**, 381.

TOPOLOGY OF COSMIC DOMAINS AND STRINGS

T. W. B. Kibble
Blackett Laboratory, Imperial College, London, England

Received: March 11, 1976

ABSTRACT

The possible domain structures which can arise in the universe in a spontaneously broken gauge theory are studied. It is shown that the formation of domain walls, strings or monopoles depends on the homotopy groups of the manifold of degenerate vacua. The subsequent evolution of these structures is investigated. It is argued that while theories generating domain walls can probably be eliminated (because of their unacceptable gravitational effects), a cosmic network of strings may well have been formed and may have had important cosmological effects.

1. INTRODUCTION

Gauge theories with spontaneous symmetry breaking have come to play a central role in elementary particle theory. Kirzhnits (1972), and Kirzhnits and Linde (1972; 1974) suggested that as in ferromagnets and superconductors the full symmetry may be restored above some critical temperature. That this actually happens in a class of theories where the symmetry breaking occurs through the acquisition of a vacuum expectation value by an elementary scalar field has been demonstrated by Weinberg (1974) while Jacobs (1974) and Harrington and Yildiz (1975) have examined models of dynamical symmetry breaking in which the role of the order parameter is played by a composite field operator. (See also Bernard 1974; Dolan and Jackiw 1974; Dashen *et al.* 1975; and Linde 1975).

In the hot big-bang model, the universe must at one time have exceeded the critical temperature so that initially the symmetry was unbroken. It is then natural to enquire whether as it expands and cools it might acquire a domain structure, as in a ferromagnet cooled through its Curie point. Zel'dovich *et al.* (1974; see also Kobzarev *et al.* 1974) have discussed this question, and in particular pointed out the important gravitational effects to be expected of domain walls. Everett (1974) has studied the propagation of waves across a domain boundary.

The aim of this paper is to discuss the topology and scale of the possible cosmic structures that might arise. After reviewing the results of Weinberg and others on phase transitions in a simple class of models in §2, we discuss in §3 the initial formation of 'protodomains' as the universe cools. The possible topological configurations are examined in §4. These include domain walls, strings and monopoles. We show that their occurrence is largely determined by the topology of the manifold M of degenerate vacuum states (specifically by its homotopy groups). (Coleman [1976] has stated the same result in a different context. In the case of monopoles it has been proved by Krive and Chudnovskii 1975.) In §5 we examine the later evolution of these structures. We show that domain walls can be of two main types with very different transmissivity, and that highly reflecting walls may behave very differently from the essentially transparent ones considered by Zel'dovich _et al._ (1974). In all cases however the typical scale of the domain structure will grow with time until it is comparable with the radius of the universe. Hence the argument of Zel'dovich _et al._, to the effect that domain walls cannot have persisted beyond the recombination era because their gravitational effect would have destroyed the isotropy of the 3 K background radiation, applies. If domain walls existed they must have disappeared by then. This in turn is possible only if the universe has a small built-in asymmetry. The exclusion of theories generating domain walls is an interesting example of a restriction on elementary particle theories derived from cosmology.

The general conclusion is that there is a rich variety of possible topological structures which might have appeared in the early history of the universe. Few of these (monopoles excepted) are likely to be stable enough to have survived to the present, but they may nevertheless be of importance in understanding the history of the universe, for example the evolution of galaxies. The conclusions are summarized in more detail in §6.

2. THE PHASE TRANSITION

Although our discussion will be quite general, for illustrative purposes it is convenient to have a specific example in mind. Let us consider an N-component real scalar field ϕ with a Lagrangian invariant under the orthogonal group O(N), and coupled in the usual way to $\frac{1}{2}N(N-1)$ vector fields represented by an antisymmetric matrix B_μ. We can take

$$L = \frac{1}{2}(D_\mu\phi)^2 - \frac{1}{8}g^2(\phi^2 - \eta^2)^2 + \frac{1}{8}\mathrm{Tr}(B_{\mu\nu}B^{\mu\nu}) \qquad (1)$$

with

$$D_\mu\phi = \partial_\mu\phi - eB_\mu\phi$$

$$B_{\mu\nu} = \partial_\nu B_\mu - \partial_\mu B_\nu + e[B_\mu, B_\nu] \,.$$

The coupling constants g and e are not necessarily related, but we shall assume that they are of a similar order of magnitude (and both small).

At zero temperature the O(N) symmetry here is spontaneously broken to O(N-1), with ϕ acquiring a vacuum expectation of order η. In the tree approximation,

$$\langle\phi\rangle^2 = \eta^2 \tag{2}$$

so that the manifold of degenerate vacua is an (N - 1) sphere S^{N-1}.

Let us recall the more general situation. In a model with symmetry group G, the vacuum expectation value $\langle\phi\rangle$ will be restricted to lie on some orbit of G. If H is the isotropy subgroup of G at one point $\langle\phi\rangle$, i.e. the subgroup of transformations leaving $\langle\phi\rangle$ unaltered, then the orbit may be identified with the coset space M = G/H. Physically H is the subgroup of unbroken symmetries, and M is the manifold of degenerate vacua. As we shall see, the topological properties of M (specifically its homotopy groups) largely determine the geometry of possible domain structures.

At a finite temperature T the expectation value of ϕ in a thermal equilibrium state must be found by minimizing the free energy, or equivalently the temperature-dependent effective potential. The leading temperature dependence at high T and small coupling constant comes from the one-loop diagrams (Weinberg 1974; Dolan and Jackiw 1974; Bernard 1974). Including these terms, we have

$$V(\phi) = \frac{1}{8}g^2(\phi^2 - \eta^2)^2 + \frac{1}{48}[(N + 2)g^2 + 6(N - 1)e^2]T^2\phi^2 , \tag{3}$$

as in the Landau-Ginsberg theory of superconductivity. (See for example Schrieffer 1964). The minimum occurs at $\phi = 0$ and so the symmetry is unbroken for T larger than the transition temperature

$$T_c = \eta\left(\frac{N + 2}{12} + \frac{N - 1}{2}\frac{e^2}{g^2}\right)^{-1/2} . \tag{4}$$

This is the normal phase. Below T_c, we have an ordered phase: ϕ acquires a vacuum expectation value, which plays the role of the order parameter, and whose magnitude is determined by

$$\langle\phi\rangle^2 = \eta^2[1 - (T^2/T_c^2)] . \tag{5}$$

Thus the manifold of degenerate equilibrium states for all $T < T_c$ is an $(N - 1)$ sphere, $M = O(N)/O(N - 1) = S^{N-1}$.

In more complicated models there may be several transition temperatures, and as Weinberg (1974) has shown the symmetry may even increase as the the temperature drops. However we shall not consider such cases, but assume a single transition temperature above which the symmetry is unbroken.

3. FORMATION OF PROTODOMAINS

Let us consider a 'hot big-bang' universe and examine what happens as it expands and cools through the transition temperature T_c. In unified models of weak and electromagnetic interactions T_c is of the order of the square root of the Fermi coupling constant, $G_F^{1/2}$, i.e. a few hundred GeV. Thus the transition occurs when the universe is aged between 10^{-10} and 10^{-12} seconds and far above nuclear densities. In other models, however, T_c might be considerably smaller and the transition would occur correspondingly later.

For T near T_c there will be large fluctuations in ϕ. Once T has fallen well below T_c, we may expect ϕ to have settled down with a non-zero expectation value corresponding to some point on M. No point is preferred over any other. As in an isotropic ferromagnet cooled below its Curie point the choice will be determined by whatever small fields happen to be present, arising from random fluctuations. Moreover this choice will be made independently in different regions of space, provided they are far enough apart. (What is far enough we shall discuss shortly.) Thus we can anticipate the formation of an initial domain structure with the expectation value of ϕ, the order parameter, varying from region to region in a more or less random way. Of course for energetic reasons a constant or slowly varying $\langle\phi\rangle$ is preferred and so much of this initially chaotic variation will quickly die away. The interesting question is whether any residue remains - in particular whether normal regions can be 'trapped' like flux tubes in a superconductor.

Because domains are most familiar in the context of ferromagnetism it may be well to point out at the outset a crucial difference between that case and ours. The long-range dipole-dipole interaction between spins ensures that it is energetically favorable for a large ferromagnet to break up into domains with different magnetization directions. However in the models we are considering, the source of the gauge field always involves the derivative of the order parameter $\langle\phi\rangle$, and vanishes for constant $\langle\phi\rangle$. Thus there is no long-range force between differently oriented domains, and the true ground state necessarily has a spatially uniform $\langle\phi\rangle$. The phenomena we are concerned with are non-equilibrium effects.

It is perhaps conceivable that in rather different theories domains might not be neutral with respect to some of the charges associated with gauge fields. However it is hard to see how domain formation in such cases could be energetically allowed. It is important to realize that even though the gauge field acquires a mass (Kislinger and Morley 1975) the effective interaction between non-neutral domains is still long-range. Physically, the situation is the same as for photons in a plasma. Because of the plasmon mass, charge fluctuations are shielded, but an unbalanced net charge will nevertheless yield a long-range Coulomb force. Net charge separation is energetically prohibitive.

Another possibility is that the universe as a whole may be non-neutral, if it is open and expanding. But then its mean charge would define a particular direction in the group space so that the direction of symmetry breaking would be fixed by the initial conditions rather than by random choice, and would be everywhere the same. Thus no domain structure would arise. However, a small degree of non-neutrality would be compatible with domain structure. Initially at least, we shall assume the overall neutrality of the universe, but we shall return to this question later, in §5.

Let us now consider the initial scale of the 'protodomains' formed as the universe cools below T_c. At a temperature $T < T_c$, the difference in free-energy density between states with $\langle\phi\rangle$ equal to zero and to its equilibrium value (i.e. between 'normal' and 'ordered' phases) is easily found from (3)-(5) to be

$$\Delta f = \tfrac{1}{8}g^2(\langle\phi\rangle^2)^2 . \tag{6}$$

It rises rapidly towards its zero-temperature value

$$\Delta f_0 = \tfrac{1}{8}g^2\eta^4$$

as T falls below T_c. The correlation length ξ which determines the scale of fluctuations in ϕ is the inverse of the scalar-meson mass (the 'Higgs scalar'), given approximately by

$$\xi^{-1} = m_S = g|\langle\phi\rangle| . \tag{7}$$

It is of course infinite at the phase transition. As in a superconductor there is a second correlation length, the penetration depth λ given by

$$\lambda^{-1} = m_V = e|\langle\phi\rangle|$$

which is of importance in discussing 'angular' oscillations, i.e. oscillations in the orientation of ϕ. However 'radial' oscillations, in the magnitude of ϕ, are controlled by ξ.

If T is only a little less than T_c, a fluctuation back to $\phi = 0$ remains quite probable. The free energy associated with such a fluctuation with scale ξ is, ignoring factors of order unity,

$$(2\xi)^3 \Delta f \simeq |\langle\phi\rangle| / g .$$

This fluctuation will have high probability so long as the free energy required is substantially less than the thermal energy T. (We use units in which Boltzmann's constant, as well as $\hbar$ and c, is equal to unity.) The two are equal when

$$\frac{1}{T^2} = \frac{1}{T_c^2} + \frac{g^2}{\eta^2} \tag{8}$$

at which time the correlation length ξ is given by

$$\xi^{-1} \simeq g^2 T. \tag{9}$$

For weak coupling this is not too different from $g^2 T_c$, or roughly $g^2\eta$.

Thereafter fluctuations back to $\phi = 0$ rapidly become less likely, so that the distinction between normal and ordered phases is well established as is that between ordered phases corresponding to well separated points on M (i.e. very different orientations of $\langle\phi\rangle$). The correlation length at this time thus determines the initial scale of the protodomains. Beyond this point it continues to fall, but the fluctuations are no longer large enough to disturb the establised long-range order. In fact the scale of the domains over which $\langle\phi\rangle$ is nearly constant will tend to grow with time because of the $(\nabla\phi)^2$ term in the energy.

4. TOPOLOGY OF DOMAIN STRUCTURES

Our next task is to examine the possible geometric configurations of the initial domain structure. It will be convenient to discuss a somewhat idealized model. Let us suppose that space is divided into cells of variable shape and size, with a mean dimension of the order of the correlation length obtained above, and that within each cell $\langle\phi\rangle$ is initially nearly constant, but with random variation from one cell to another. Although in reality $\langle\phi\rangle$ may be expected to vary continuously from point to point, this model seems to be a reasonable idealization of the kind of structure we might expect.

The interesting question is what happens on the boundaries where two or more cells meet. Consider two cells characterized by $\langle\phi\rangle = \phi_1$ and $\langle\phi\rangle = \phi_2$ respectively, with a common boundary surface. If there is a continuous path in the manifold M joining the points ϕ_1 and ϕ_2, then because of the $(\nabla\phi)^2$ term in the energy it will be energetically favorable for the width of the region over which $\langle\phi\rangle$ varies from ϕ_1 to ϕ_2 to expand, leaving a smoothly varying $\langle\phi\rangle$ in place of a sharp boundary. However, if M is disconnected, and ϕ_1 and ϕ_2 belong to different connected components, then no smooth transition is possible. One can pass from one ordered phase to the other only by going through a normal region. Hence a wall of normal phase will be formed between the two cells.

The simplest example of a model with disconnected M is the case N = 1 of the model described in §2. In that case there is only a discrete reflection symmetry, and no gauge fields. The manifold M is the 0-sphere S° comprising two points corresponding to the two possible values of $\langle\phi\rangle$, for example at T = 0, $\langle\phi\rangle = \pm\eta$. In this case there will be two distinct, though internally indistinguishable, ordered phases. Any large enough volume of space will be divided more or less equally between the two with thin walls of normal phase, where $\langle\phi\rangle$ passes through zero, separating them. The width of the domain wall is determined by a balance between Δf and the $(\nabla\phi)^2$ term in the energy, and is in fact the correlation length ξ. (A solution with a plane boundary separating two domains is effectively a 'soliton'. See for example, Scott _et al._ [1973].)

The subsequent development of this complex geometric structure is governed largely by the surface tension σ of the wall which is of order $\xi\Delta f$ or, by (6) and (7),

$$\sigma \simeq \xi\Delta f \simeq g\left|\langle\phi\rangle\right|^3 . \qquad (10)$$

(We again ignore factors of order unity.) Because of surface tension, the area of wall will tend to decrease with time. Thus in any fixed finite volume one of the two ordered phases will eventually tend to predominate. The rate at which this happens and the scale of the structure after a given time we shall discuss later.

Domain walls can form only if M is disconnected, i.e if as in the N = 1 model there is a discrete broken symmetry. There may of course be continuous symmetries that are broken too. If M has two connected components, the structure produced is a kind of emulsion of two (similar) ordered phases. The walls can never terminate and must either form closed surfaces or extend to infinity. If the number of connected components is three or more we have a more complicated structure, an emulsion of several phases.

Let us now consider what happens when three cells meet along an edge. We suppose that the corresponding expectation values ϕ_1, ϕ_2, ϕ_3 belong to the same connected component of M, so that no pair of cells is separated by a domain wall. Then as we go around the edge we will find the expectation value

$\langle\phi\rangle$ following a smooth path in M from ϕ_1 to ϕ_2 to ϕ_3, and back to ϕ_1. Now if this closed path can be continuously deformed in M to a point, i.e. if it is null-homotopic, then the ordered phase can be extended in a continuous manner into the region of the edge. All three domains can then fuse, leaving only a smoothly varying $\langle\phi\rangle$. But if it cannot be so deformed then a region of normal phase will be trapped along the edge, exactly like a vortex filament in a superfluid (see for example Putterman 1974) or a flux tube in a type II superconductor (see Schrieffer 1964). The Nielson-Olesen (1973) string is an example of a solution of this kind, essentially for the model of §2 with N = 2, in which case the manifold M is a circle, S^1. Note that although the gauge field does not seem to play an important topological role, it is vital in making the energy per unit length of the string finite.

We note that the strings have a tension whose order of magnitude is

$$\mu \simeq \xi^2 \Delta f \simeq |\langle\phi\rangle|^2 \,. \tag{11}$$

What is relevant in deciding whether strings can exist is the first homotopy group $\pi_1(M)$, comprising the homotopy classes of maps from S^1 into M, i.e. of closed paths in M (see Hu 1959). If M is simply connected, $\pi_1(M)$ is trivial and strings are impossible. But whenever $\pi_1(M)$ is nontrivial they will occur. Not all edges will of course form strings but we can estimate how many do so if we make two simple assumptions: that the initial choices of $\langle\phi\rangle$ in different cells are uncorrelated, and that where two cells meet and fuse $\langle\phi\rangle$ varies along the shortest path in M between the two original values ϕ_1 and ϕ_2. Then we find for the case where M is a circle that one in four of the original edges will form into strings.

The number of independent generators of $\pi_1(M)$ influences the type of structure that the strings can form. If it is one, as when M is a circle, there is only one basic type of string. In principle n strings might coalesce to form a multiple string around which the phase of $\langle\phi\rangle$ varies by $2n\pi$, though whether this is energetically favored depends on the details of the model (essentially on the ratio e/g, as in the distinction between type I and type II superconductors - see Schrieffer 1964). If $\pi_1(M)$ has more than one generator, then several types of strings can exist. We may then expect to find vertices where three strings join, and space will be filled with a three-dimensional network.

We can go on to discuss what happens at a vertex where four cells meet with order parameters ϕ_1, ϕ_2, ϕ_3, ϕ_4. These points may be regarded as the vertices of a (curvilinear) tetrahedron in M, with edges representing the paths along which $\langle\phi\rangle$ varies across each bounding surface. We assume that each closed circuit, such as ϕ_1-ϕ_2-ϕ_3-ϕ_1, is null-homotopic, so that the faces of the tetrahedron can be filled in without leaving M. If the closed surface so formed can be shrunk to a point in M, then $\langle\phi\rangle$ can be extended continuously into the region around the vertex. But if not, then a normal region will be trapped, as in the monopole solution of t'Hooft (1974). For monopoles to exist we require that the second homotopy group $\pi_2(M)$ be nontrivial, as it is for the N = 3 model. (Compare Krive and Chudnovskii [1975].)

Such objects if they do exist will be localized structures with a scale typical of elementary particles, and regarded as such. They will not be significant on a cosmic scale, and we shall not discuss them further. (The considerations of this paper might perhaps be relevant in estimating the probable density of monopoles. If M is a sphere S^2 then under the same assumptions as before one in eight of the initial vertices should form a monopole. However to obtain an estimate of present density would require a careful study of annihilation mechanisms, which are its principal determinant.)

We have seen that to trap a k-dimensional normal region within the ordered phase requires, for k = 0 or 1, that the homotopy group $\pi_{2-k}(M)$ be nontrivial, i.e. that there exist non-contractible maps of S^{2-k} into M. This criterion applies also for k = 2, although $\pi_0(M)$ is not strictly speaking a homotopy group: it has no group structure but represents merely the number of connected components of M (see for example Hilton and Wylie 1960).

It is perhaps amusing to note that if $\pi_0(M)$, $\pi_1(M)$ and $\pi_2(M)$ are all trivial, we can still obtain a topological characteristic of our particular universe (if it is finite and homeomorphic to S^3) by considering $\pi_3(M)$: namely, the homotopy class of the function $x \mapsto \langle\phi(x)\rangle$, which is with high probability an invariant once we are well past the transition temperature. However it is hard to envisage any possible physical relevance for it.

If we are interested in cosmic structures, we have then two candidates: domain walls when $\pi_0(M)$ is nontrivial and strings when $\pi_1(M)$ is. The next question we must ask is how these structures would evolve, once formed.

5. EVOLUTION OF DOMAIN WALLS AND STRINGS

The motion of a domain wall is controlled primarily by its surface tension (equation [10]). The corresponding surface energy is also the inertial mass of the wall. In other words the wall is 'massless' in the sense usually implied in talking of massless strings: its action integral is simply proportional to the three-volume it sweeps out in space-time. As a result of the equality between areal mass density and surface tension the velocity of waves of short wavelength along the wall is the velocity of light.

The surface is in local equilibrium when its mean curvature is zero, i.e. when the two principal radii of curvature are equal and opposite, $\pm R$ say. (The simplest nontrivial example is the catenoid $(x^2 + y^2)^{1/2} = \cosh z$.) Any deviation from local equilibrium will lead to waves spreading out rapidly from their point of origin. Among the possible damping mechanisms the most obvious is the interaction of the walls with surrounding matter. A wall moving with velocity v through matter of density ρ will experience a retarding force per unit area

$$\alpha v = a\rho\bar{u}v \tag{12}$$

where a is a numerical constant depending on the reflectivity of the wall and $\bar{u}$ a suitably defined mean velocity of the matter particles. The characteristic time for damping out short-wavelength oscillations is then

$$t_d = 2\sigma/\alpha \ . \tag{13}$$

Clearly it is important to estimate the degree of transparency of the walls. If it were possible for the domain structure to survive until near the present, this question would also have a more direct relevance to the possible observability of domains. It was from this point of view that it was tackled by Everett (1974). However, domain walls surviving so long can be ruled out by the argument of Zel'dovich _et al._ (1974).

In the very early stages the wall thickness ξ is comparable with the (relativistic) thermal wavelength T^{-1}. In fact at the time of formation of protodomains, identified in §3, we have $\xi T \simeq g^{-2} > 1$. Thereafter, however, ξ falls rapidly to its limiting value of order η^{-1}, and T decreases steadily though more slowly. Once it has fallen to a small fraction of T_c - say $T \simeq 1$ GeV which occurs at $t \simeq 10^{-6}$ s - we can legitimately treat the domain walls as very thin, and replace them by sharp discontinuities across which the fields are related by boundary conditions of the usual type.

Most of the significant evolution occurs during the radiation era, so we are mainly interested in the transparency of the walls to light. Let us consider an electromagnetic wave striking a domain wall. We suppose, as is now conventional, that the photon belongs to a multiplet of gauge bosons whose other members are very massive. Two very different situations can arise, depending on the precise model of symmetry breaking. If a different component of the field is massless on the far side of the wall then, as Everett (1974) showed, the probability of reflection is large. In that case the constant a in equation (12) will be of order unity and the motion of the walls will be heavily damped by radiation pressure. On the other hand if the electromagnetic field corresponds to the same component on both sides, as in the models considered by Zel'dovich _et al._ (1974), it is clear that the wall will be almost transparent and so little affected by radiation pressure. It is not hard to write down models of both types.

In the latter case, one must examine other damping mechanisms - for example the absorption and emission of radiation - but none seems likely to be very effective. Thus the walls will continue to move with relativistic speeds, and the subsequent evolution will be more or less along the lines described by Zel'dovich et al.

For the moment, however, let us assume that the walls are highly reflecting, so that a is of order unity, and

$$\alpha \simeq \rho \simeq T^4 .$$

Once T is sufficiently below T_c for σ to have reached its asymptotic value, we then have

$$t_d \simeq g\eta^3/T^4 \qquad (14)$$

which grows like t^2.

Initially t_d is roughly of order g/η, and therefore small compared to the length scale $1/g^2\eta$ of the protodomains. Consequently the domain walls will evolve rather rapidly towards local equilibrium. Moreover isolated pockets of one phase surrounded by another will quickly disappear. In the case where M has two components, we are left with an emulsion of two phases each occupying a single connected region (cf Broadbent and Hammersley 1957) and separated by a single, high convoluted wall whose mean curvature is everywhere near zero. (The structure is in some ways rather similar to the matter-antimatter emulsion considered by Omnès [1971].)

Such a configuration is stable against short-wavelength perturbations, but it cannot be completely stable. It is well known that the only stable infinite-area solution of Plateau's problem (to find minimum-area surfaces of given perimeter) is the plane (see Thompson 1942). In general, waves shorter than a typical radius of curvature will be damped but disturbances of longer wavelength will grow exponentially in time. The equation of motion for small deviations from equilibrium is

$$\frac{\partial^2\zeta}{\partial t^2} + \frac{2}{t_d}\frac{\partial\zeta}{\partial t} - \Delta\zeta - \frac{2}{R^2}\zeta = 0 , \qquad (15)$$

where ζ is the normal displacement and Δ the two-dimensional Laplace-Beltrami operator. Thus when $R \gg t_d$, the characteristic time for growth of long-wavelength perturbations is R^2/t_d.

Eventually the surface will jump to a new equilibrium configuration of smaller area which in turn will be unstable to yet longer wavelengths. The length scale ℓ of the emulsion - the volume to wall-area ratio for any large enough volume - will therefore grow with time. Roughly, we may identify the characteristic time for growth of ℓ with that for the long-wavelength perturbations, so that

$$\frac{1}{\ell}\frac{d\ell}{dt} \simeq \frac{t_d}{\ell^2} .$$

Hence ℓ grows like the square root of the elapsed time, as one might expect from the diffusion-like character of the process. Since this is much faster than the slow growth of t_d, the approximation $\ell >> t_d$ remains valid. However over a long period of time the growth of $t_d \propto t^2$ leads to $\ell \propto t^{3/2}$. Clearly therefore t_d will overtake ℓ. It is easy to see that this happens when both are of the same order as the age of the universe; in fact when

$$\ell \simeq t_d \simeq t = t_1 \simeq 1/G\sigma , \tag{16}$$

where G is the Newtonian gravitational constant (which enters via the relation $tT^2 \sim 1/G^{1/2}$). It is interesting to note that the very different mechanism envisaged by Zel'dovich et al. yields a similar result for the time at which the scale of the emulsion becomes comparable with the radius of the universe. With the parameters chosen earlier we have $t_1 \simeq 10^7$ s, but because of its strong η dependence, proportional to η^{-3}, t_1 is very model-dependent.

The degree of inhomogeneity introduced into the universe by the gravitational effect of the domain walls (as shown by Zel'dovich et al. 1974) is of the order $G\sigma\ell \simeq \ell/t_1$. It is clearly unacceptably large unless either there are many domain walls remaining ($\ell/t_1 < 10^{-3}$) or none at all. The first alternative is implausible. Domain walls with the required spacing can be ruled out on observational grounds, at any rate if they are substantially reflecting. The second is impossible if we start from completely random initial choices of the order parameter. The typical scale of the emulsion can never much exceed the distance over which causal corrections are possible, i.e. the present radius of the universe.

There seems to be only one way of accomodating models that generate domain walls, namely, as Zel'dovich et al. suggest, to allow a small initial bias. It would require only a minute non-neutrality in the initial state of the universe with respect to some 'charge' - related perhaps to the preference for matter over antimatter - to ensure a slight but consistent excess in the number of protodomains of one type. This excess would tend to become more pronounced with time and eventually lead to complete dominance of the preferred phase. It would not be hard to realize this case by suitable choice of model parameters. But such initial asymmetry is unaesthetic, and not at all in the spirit of spontaneous symmetry breaking.

In all probability therefore we should rule out models with discrete symmetry breaking, which lead to domain walls - an interesting example of a restriction on elementary particle theories deriving from cosmology.

There is no such objection to cosmic strings. A single domain wall in the universe (with the parameters chosen here) would have a mass far in excess of the accepted upper limit for the mass of the universe. By contrast, the mass of a string might be of order 1 mg m^{-1}, so that the total mass of a string of length equal to the current radius of the universe would be less than that of a minor planet.

The evolution of the network of cosmic strings would in many respects be similar to that of walls, although of course there is only one local equilibrium configuration for a string - a straight line - in contrast to the infinite variety of surfaces of zero mean curvature.

Consider a section of string with radius of curvature R. Because of its 'masslessness', it experiences an initial acceleration 1/R. However there is a damping force arising as before from the interaction with other matter. If the effective (linear) cross section of the string is simply its width ξ then the ratio of retarding force to mass is the same as that found earlier for a wall. Thus we may assume that the damping time t_d is similar. It follows that the string will acquire a limiting velocity t_d/R and the kink will be straightened out in a time of order R^2/t_d.

As the strings move they will sometimes cross. When they do, they may be apt to change partners, leaving a pair of sharply kinked strings which straighten out in time, and so on. It is not entirely clearly how far the existence of the topological invariants of strings studied by Khan (1975) might inhibit this exchange process. Assuming however that the exchange can occur, it is clear that occasionally a small closed loop may form which then shrinks to a point and vanishes. Thus the overall length of string will decrease, or equivalently the length scale ℓ will increase, on a time scale ℓ^2/t_d, exactly as for domain walls. With most choices of parameters, this means that the scale now is comparable with the radius of the universe. We can expect only one string within the visible universe, so that looking for cosmic strings directly would be pointless.

If the manifold of equilibrium states is more complex, so that $\pi_1(M)$ has more than one generator, then as we have seen a network of strings can be formed. Although superficially such a network might seem to have more stability its evolution would probably not in fact be much slower. It seems most unlikely that any structure could have survived until the present on a small enough scale to be directly visible.

Nonetheless the existence of such a network of cosmic strings may have had profound effects on the earlier history of the universe, at a stage where the number of strings was large. For example, strings can produce significantly local inhomogeneities by their interaction with matter. The details of this mechanism and its effectiveness require further study.

6. CONCLUSIONS AND DISCUSSION

It may be well to begin recalling our basic assumptions. We have assumed that the universe is correctly decribed by a spontaneously broken gauge theory exhibiting a phase transition at a critical temperature T_c, above which the symmetry is restored. We have taken for granted the hot big-bang model of the universe, with no maximum temperature, whence it follows that in its very early history the universe was above T_c, in the 'normal' phase. Finally we have generally assumed the overall neutrality of the universe with respect not only to electric charge but also to all the other charges associated with gauge fields.

On this basis we showed that a domain structure can be expected to arise. The topological character of this structure depends on the homotopy groups $\pi_k(M)$ of the manifold M of degenerate vacua. Domain walls can form if $\pi_0(M)$ is nontrivial, i.e. if M is non-connected. If it has n connected components we find an n-phase emulsion. The formation of cosmic strings requires that $\pi_1(M)$ be nontrivial, i.e. that M is not formed of simply connected components. Finally, 'monopoles' can form if $\pi_2(M)$ is nontrivial.

The later evolution of domain walls is governed by their surface tension and their interaction with matter. Different types of domain walls can occur with very different transparency, but in all cases the overall scale of the structure will grow with time. In general we may expect it now to be comparable with the radius of the universe. Domain walls on anything like this scale can be ruled out (Zel'dovich *et al.* 1974) because their gravitational effect would lead to unacceptable anisotropy in the black-body background radiation. The only way of accomodating theories with spontaneously broken discrete symmetries (and hence domain walls) is to relax the requirement of complete neutrality of the universe, so that one of the ordered phases is slightly preferred and eventually comes to occupy all of space. However this is not a very attractive solution, and it may be better to regard this as an argument against such theories.

Networks of strings will evolve in a similar way under the combined effects of tension and interaction with matter. Once again, the scale of the structure will grow with time, probably at a similar rate. One cannot expect to find significant numbers of cosmic strings in the visible universe now, but their presence may have had an important effect on the earlier evolution of the universe.

REFERENCES

Bernard, C. 1974, *Phys. Rev. D*, **9**, 3312.

Broadbent, S. R., and Hammersly, J. M. 1957, *Proc. Camb. Phil. Soc.*, **53**, 629.

Coleman, S. 1976, Int. School on Subnuclear Physics 'Ettore Majorana' Erice 1975, to be published.
Dashen, R. F., Ma, S. K., and Rajaraman, R. 1975, Phys. Rev., **11**, 1499.
Dolan, L., and Jackiw, R. 1974, Phys. Rev. D, **9**, 3320.
Everett, A. E. 1974, Phys. Rev. D, **10**, 3161.
Harrington, B. J., and Yildiz, A. 1975, Phys. Rev. D, **11**, 779, 1705.
Hilton, P. J., and Wylie, S. 1960, Homology Theory (Cambridge: Cambridge University Press), p. 276.
Hu, S.-T. 1959, Homotopy Theory (New York: Academic Press).
Jacobs, L. 1974, Phys. Rev. D, **10**, 3956.
Khan, I. 1975, Trieste preprint IC/75/148.
Kirzhnits, D. A. 1972, Zh. Eksp. Teor. Fiz. Pis'ma Red., **15**, 745 (1972, Sov. Phys.-JETP Lett., **165**, 529).
Kirzhnits, D. A., and Linde, A. D. 1972, Phys. Lett., **42B**, 471.
Kirzhnits, D. A., and Linde, A. D. 1974 Zh. Eksp. Teor. Fiz., **67**, 1263 (1974, Sov. Phys.-JETP, **40**, 628).
Kislinger, M. B., and Morley, P. D. 1975, Chicago Preprint EFI 75/8.
Kobzarev, I. Yu., Okun, L. B., and Zel'dovich, Ya. B. 1974, Phys. Lett., **50B**, 340.
Krive, I. V., and Chudnovskii, E. M. 1975, Zh. Eksp. Teor. Fiz. Pis'ma Red., **21**, 271 (1975 Sov. Phys. - JETP Lett., **21**, 124).
Linde, A. D. 1975, Lebedev Physical Institute preprint no. 123.
Nielsen, H. B., and Olesen, P. 1973, Nucl. Phys. B, **61**, 45.
Omnès, R. 1971, Astron. Astrophys., **10**, 228.
Putterman, S. J. 1974, Superfluid Hydrodynamics (Amsterdam: North Holland).
Schrieffer, J. R. 1964, Theory of Superconductivity (New York: Benjamin).
Scott, A., Chu, F., and McLaughlin, D. 1973, Proc. IEEE, **61**, 1443.
't Hooft, G. 1974, Nucl. Phys. B, **79**, 276.
Thompson, d'A. W. 1942, On Growth and Form (Cambridge: Cambridge University Press), 2nd edn.
Weinberg, S. 1974, Phys. Rev. D, **9**, 3357.
Zel'dovich, Ya. B., Kobzarev, I. Yu., and Okun, L. B. 1974, Zh. Eksp. Teor. Fiz, **67**, 3 (1975, Sov. Phys.-JETP, **40**, 1).

COSMOLOGICAL PRODUCTION OF SUPERHEAVY MAGNETIC MONOPOLES

John P. Preskill
Lyman Laboratory of Physics, Harvard University
Cambridge, Massachusetts

Received: June 21, 1979

Grand unified models of elementary particle interactions contain stable superheavy magnetic monopoles. The density of such monopoles in the early universe is estimated to be unacceptably large. Cosmological monopole production may be suppressed if the phase transition at the grand unification mass scale is strongly first order.

There has been much interest recently in grand unified models of elementary particle interactions (Georgi, Quinn, and Weinberg 1974; Georgi and Glashow 1974), in which a simple gauge group breaks down at a very large mass scale to a group containing the $SU(3)\otimes SU(2)\otimes U(1)$ of the strong, weak, and electromagnetic interactions. Because the unbroken symmetry group contains a U(1) factor, general arguments ('t Hooft 1974; 1976; Polyakov 1974) imply that such models contain topological stable solitons which carry U(1) magnetic charges and have masses of the order of the scale of the symmetry breakdown. In this note, I will argue that an unacceptably large number of such superheavy magnetic monopoles (M) and antimonopoles ($\bar{M}$) might have been produced in the early universe, indicating a possible discrepancy between standard big-bang cosmology and grand unified models.

The masses of the superheavy particles in grand unified models are characterized by a zero-temperature scalar-field[1] expectation value, v_0, which is expected to be of the order of 10^{15} GeV (Goldman and Ross 1979). The M mass is approximately given by $m \approx hv_0$, where h is the M U(1) magnetic charge ('t Hooft 1974; 1976; Polyakov 1974; Bogomol'nyi 1976). The smallest allowed magnetic charge is $h = 2\pi/q$, where q is the minimal U(1) charge of a particle which transforms as a singlet under the unbroken subgroup ('t Hooft 1974; 1976; Polyakov 1974). If symmetry breakdown occurs at many different mass scales, then the M mass is determined by the largest scale at which a U(1) factor appears in the unbroken subgroup, but its classical size at zero temperature is of the order of $(v_{min})^{-1}$, where v_{min} is the smallest mass scale at which the U(1) factor is altered.

If the temperature T is greater than a critical temperature T_c, which is of the order of v_0 (Kirzhnits and Linde

[1] My arguments should apply even if the symmetry breakdown is dynamical; a composite order parameter then replaces the elementary scalar field.

1972; Weinberg 1974; Dolan and Jackiw 1974), the full gauge symmetry is restored, and no M's are present. Suppose that very early in the history of the universe, T exceeds T_c.[2] When the universe cools below T_c, M's can be produced. Because M's, unlike the other superheavy particles in these models, are absolutely stable, their density per comoving volume can be reduced only by annihilation of $M - \bar{M}$ pairs. We will see that the expansion of the universe halts this annihilation process. If M's are produced copiously when $T \approx T_c$, then many M's remain when the temperature is much lower - enough to dominate the mass of the universe by many orders of magnitude.

In a recent paper, Zel'dovich and Khlopov (1979) have considered the cosmological production of M's with a mass of order 10^4 GeV. However, these authors make the implausible assumption that collisions produce a thermal density of M's when $T \lesssim T_c$. I will argue that, if the phase transition at $T = T_c$ is second order (or weakly first order), the production of an appreciable density of M's is a consequence of the large fluctuations near the critical point. If the phase transition is strongly first order, the situation is less clear, and the density of M's may be tolerably small.

We need to estimate both the M density produced initially and the rate at which the density per comoving volume decreases. Since the latter question can be answered more precisely, I consider it first.[3] Below a temperature T_i at which the M production rate is negligible compared with the expansion rate of the universe, the M density is governed by the rate of $M - \bar{M}$ annihilation. If we may ignore $M - \bar{M}$ correlations, we have

$$dn/dt = -Dn^2 - (3\dot{R}/R)n \ , \tag{1}$$

where n is the M density per unit volume (M and $\bar{M}$ densities are assumed equal), R is the scale factor of the universe, and D characterizes the annihilation process. If the expansion is adiabatic ($RT \approx$ const) and the universe is radiation-dominated, the expansion rate is[4]

$$\dot{R}/R = -\dot{T}/T = T^2/Cm_P \ . \tag{2}$$

Here, T is the temperature of the universe, $M_P = G^{-1/2} =$

[2] This assumption is required in calculations of the baryon-to-entropy ratio. Weinberg (1979), and references therein.

[3] This discussion of $M - \bar{M}$ annihilation follows closely that of Zel'dovich and Khlopov (1979).

[4] Units are chosen such that $\hbar = c = k = 1$.

1.2×10^{19} GeV is the Planck mass, and $C = (45/4\pi^3N)^{1/2} = 0.60\ N^{-1/2}$, where N is the effective number of spin degrees of freedom due to particles light compared with the temperature.[5] If D depends on the temperature like a simple power, $D = (A/m^2)(m/T)^p$, Eqs. (1) and (2) can be integrated:

$$r(T) = \{\frac{1}{r(T_i)} + \frac{A}{p-1}\frac{Cm_P}{m}[(\frac{m}{T})^{p-1} - (\frac{m}{T_i})^{p-1}]\}^{-1} , \qquad (3)$$

where $r = n/T^3$. If $p < 1$, the expansion of the universe cuts off the annihilation when $T \ll T_i$, and r(T) approaches a constant value. If $p > 1$, the annihilation persists at low temperatures; for $T \ll T_i$, the density becomes independent of its initial value, and is given by

$$r(T) \approx \frac{p-1}{A}\frac{m}{Cm_P}(\frac{T}{m})^{p-1} , \qquad (4)$$

if $r(T) \ll r(T_i)$.

For T less than the inverse size of the M, the interactions between M and light charged particles, and between M and $\bar{M}$, are dominated by the long-range magnetic coupling. We can estimate scattering cross sections by calculating them classically. $M - \bar{M}$ can annihilate by capturing each other in magnetic Coulomb bound states, and then cascading down. As long as the M mean free path λ is shorter than the capture distance $a_c = h^2/4\pi T$, the annihilation rate is given by the flux of M's diffusing through the dense plasma of charged particles toward an $\bar{M}$. This flux is $\phi = h^2(\tau/m)n$, where τ is the mean time between collisions in which the M is scattered by a large angle. The cross section for large-angle scattering of a thermal relativistic particle with charge q is $\sigma \approx (hq/4\pi)^2T^{-2}$, and the M is itself scattered by a large angle after m/T encounters. Therefore the collision time is $\tau \approx m(BT^2)^{-1}$ where $B = (3/4\pi^2)\zeta(3)\Sigma_i(hq_i/4\pi)^2$. (The sum is over all spin states of relativistic charged particles). We see that D in Eq. (1) is $D = \phi n^{-1} = h^2(BT^2)^{-1}$. At the temperature $T_f = (4\pi/h^2)^2mB^{-2}$ at which $\lambda \approx a_c$, Eq. (4) gives

$$r(T_f) \approx \frac{1}{Bh^2}(\frac{4\pi}{h^2})^2\frac{m}{Cm_P} , \qquad (5)$$

if $r(T_f) \ll r(T_i)$.

When $T < T_f$, M and $\bar{M}$ can capture each other only by emitting radiation, and D in Eq. (3) is $\langle\sigma v\rangle$, the average value of

[5] In a minimal grand unified model, $C \approx 1/20$ near the critical temperature. I take C to be constant, although it varies slowly as particle species freeze out.

the product of the $M - \bar{M}$ capture cross section and the relative velocity. A classical calculation yields $D \approx (h^2/4\pi)^2 m^{-2} (m/T)^{9/10}$. From Eqs. (3) and (5), we see that capture by emission of radiation does not reduce $r(T)$ below $r(T_f)$. If $r(T_i)$ is smaller than the right-hand side of Eq. (5) (about 10^{-10} for $m \approx 10^{16}$ GeV, $h^2/4\pi \approx 75$, and $B \approx 10$) annihilation does not reduce r at all for $T < T_i$, as long as $M - \bar{M}$ correlations are negligible.

I have neglected the size of the M. However, when $T \lesssim v_{min} \approx 250$ GeV, the classical size b of the M grows to $b \approx v_{min}^{-1}$, which is large compared with $(h^2/4\pi T)(T/m)^{3/10}$, the typical impact parameter for which capture can occur by emission of radiation. Hence the capture cross section may be dominated by nonelectromagnetic interactions. If one makes the very optimistic assumption that an $M - \bar{M}$ pair with an impact parameter of order b can capture by emitting light scalar particles, the D in Eq. (1) is $D \approx \langle\sigma v\rangle \approx b^2(T/m)^{1/2}$. For $T_i \approx v_{min}$, Eq. (3) gives us $r(T) \approx (v_{min}/Cm_p)(m/v_{min})^{1/2}$, if $r(T) \ll r(T_i)$. Comparing with Eq. (5), we see that capture by scalar emission cannot reduce r significantly unless $m \gtrsim 10^{18}$ GeV. For smaller values of m, the size of the M can be safely ignored.

I have also neglected correlations between M's which may be produced by the gravitational effects of inhomogeneities in the energy density of the universe. When electrons and positrons freeze out at $T < 1$ MeV, the mean collision time of the M becomes large enough so that such effects are potentially important. When galaxies form, M's and $\bar{M}$'s accumulate rapidly in the cores of galaxies and stars, where the annihilation rate is greatly enhanced by the relatively large number densities.

Experimental limits (Fleischer et al. 1969; Kolm et al. 1971; Alvarez et al. 1971) on the M flux in cosmic rays do not apply if $m \approx 10^{16}$ GeV. Such massive M's are not accelerated to relativistic velocities by galactic magnetic fields, and so do not ionize strongly. Also, they do not bind to matter in Earth's crust, because of the strong pull of Earth's gravitational field. We can, however, obtain a bound on the present value of r from the simple observation that the mass density due to M's must not exceed the limit on the mass density of the universe imposed by the observed Hubble constant and deceleration parameter (Kristian, Sandage, and Westphal 1978). This constraint is $r(2.7\ ^\circ K) \lesssim (10\ eV)/m$, or $r \lesssim 10^{-24}$ for $m \approx 10^{16}$ GeV. To find a bound or r(T) which applies when Eq. (1) is still valid, we note that the standard scenario of helium synthesis (Peebles 1966) requires that the mass of M's does not dominate the universe when $T \approx 1$ MeV. We demand that

$$r(T = 1\ \text{MeV}) \lesssim (1\ \text{MeV})/m\ , \tag{6}$$

or $r \lesssim 10^{-19}$ for $m \approx 10^{16}$ GeV. Since $M - \bar{M}$ annihilation cannot reduce a large initial density below $r \approx 10^{-10}$, we conclude that $r(T_i) \lesssim 10^{-19}$.

Now we must estimate the initial density $r(T_i)$. As the universe cools below T_c, one does not expect $M - \bar{M}$ pairs to be brought into statistical equilibrium by collisions, because $M - \bar{M}$ pair production is expected to be suppressed by a small factor of order $\exp(-a/g^2)$ (Witten 1979). Nevertheless, when $T \lesssim T_c$, the scalar field ϕ undergoes large random fluctuations; the zeros of ϕ can be identified as M's and $\bar{M}$'s. Positions of M's and $\bar{M}$'s are strongly correlated on a scale of the order of the correlation length (Halperin unpublished), which is the inverse of the largest scalar or vector mass. As T decreases, M's and $\bar{M}$'s pair up and annihilate, but pairs which are widely separated can survive.

When T is far enough below T_c, large fluctuations have no significant effect on $M - \bar{M}$ annihilation. The criterion for large fluctuations to be unimportant is $T \lesssim T_G$, where (Ginzburg 1960)

$$\lambda^{1/2} T_G/[4\pi v(T_G)] \approx 1 \ . \tag{7}$$

Here, $v(T) = v_0(1 - T^2/T_c^2)^{1/2}$ is a scalar field expectation value and λ is the largest scalar self-coupling in the theory, normalized so that $m_s(T) = \lambda^{1/2} v(T)$ is the largest scalar mass. The most optimisitic assumption we can make is that M's and $\bar{M}$'s annihilate rapidly enough to remain in statistical equilibrium until $T \approx T_G$. In that case, when $T \approx T_G$, the density of widely separated $M - \bar{M}$ pairs is suppressed by a Boltzmann factor:

$$r(T_G) \approx \exp[-m(T_G)/T_G] \approx \exp(-\lambda^{1/2}/g) \ , \tag{8}$$

where g is the gauge coupling. If $\lambda^{1/2}g^{-1}$ is not large, there are many unpaired monopoles. After closely paired M's and $\bar{M}$'s annihilate, the widely separated M's which remain feel only the long-range magnetic coupling. When typical $M - \bar{M}$ separations are of the order of $a_c \approx h^2/4\pi T$, $M - \bar{M}$ correlations may be ignored, and Eq. (1) applies. Hence the initial M density in Eq. (3) should be of the order of $r(T_i) \approx (4\pi/h^2)^3 \approx 10^{-6}$.

It appears that M's and $\bar{M}$'s are copiously produced when $T \approx T_c$. The rapid expansion of the universe prevents the complete annihilation of M's, so that M's dominate the energy density of the universe before the time of helium synthesis. This conclusion, which is incompatible with standard cosmology, might be avoided if there is scalar self-coupling λ which is sufficiently large that $\exp(-\lambda^{1/2}g^{-1}) \lesssim 10^{-19}$. If $g^2/4\pi \approx 1/45$ at the grand unification scale, the required value of λ is so large that it interferes with the integration of the renormalization-group equations down to ordinary energies. One should note, though, that the suppression factor in Eq. (8) is very sensitive to numerical factors which may occur in the exponent, but cannot be calculated accurately.

The above discussion applies if the phase transition is second order. If the phase transition is strongly first order, $r(T_i)$ is more difficult to estimate; conceivably, M production is severely suppressed. In a first-order phase transition, expanding bubbles of the stable asymmetric vacuum form in the metastable symmetric vacuum when T is of the order of a nucleation temperature T_N (Coleman 1977; Linde 1977). M's can be produced by expanding bubbles or collisions between bubbles, but one might hope that, if the discontinuity in the M mass is large compared with T_N, $r(T_i)$ can be as small as 10^{-19}. A strongly first-order phase transition can be generated by an explicit cubic term in the scalar potential, or by higher-order corrections to the finite-temperature effective potential (Kirzhnits and Linde 1976). Such higher-order corrections are important if the scalar mass matrix has eigenvalues which are sufficiently small. In particular, an interesting possibility is that M production is strongly suppressed when bare scalar masses vanish (Coleman and Weinberg 1973; Gildener and Weinberg 1976).

This possible conflict between big-bang cosmology and grand unified models might be resolved in several less attractive ways. Perhaps there is no grand unification. If the gauge group contains a U(1) factor at arbitrarily large mass scales, there need never be any M's. Perhaps the universe was never so hot that symmetry restoration occurred. Then, however, the baryon excess in the universe is no longer explained (see footnote 2).

Perhaps M production does not occur if the unification scale v_0 is quite close to the Planck mass m_P. If the M mass exceeds m_P, quantum gravity corrections might invalidate the analysis in this note. The standard calculations, however, suggest that v_0 is comfortably below m_P (Georgi, Quinn, and Weinberg 1974; Georgi and Glashow 1974; Goldman and Ross 1979).

Finally, consider the consequences of a lower unification scale. The constraint in Eq. (6) can be satisfied by r as given in Eq. (5) for $m \lesssim 10^{11}$ GeV. However, a grand unification scale as low as $\tilde{10}^{10}$ GeV seems to be ruled out by the observed bound on the lifetime of the proton (Buras, Ellis, Gaillard, and Nanopoulos 1978).

I am grateful to M. Peskin and B. Halperin for many helpful comments and suggestions. I have also benefitted from conversations with P. Ginsparg, A. Guth, E. Purcell, H. Tye, S. Weinberg, and E. Witten. This research was supported in part by the National Science Foundation under Grant No. PHY77-22864.

REFERENCES

Alvarez, L. W., et al. 1971, Phys. Rev. D, **4**, 3260.
Bogomol'nyi, E. B., 1976, Yad. Fiz., **24**, 861 (1976, Sov. J. Nucl. Phys., **24**, 449).

Buras, A., Ellis, J., Gaillard, M. K., and Nanopoulos, D. V. 1978, Nucl Phys., **B135**, 66.
Coleman, S. 1977, Phys. Rev. D, **15**, 2929.
Coleman, S., and Weinberg, E. 1973, Phys. Rev. D, **7**, 1888.
Dolan, L., and Jackiw, R. 1974, Phys. Rev. D, **9**, 3320.
Fleischer, R. L., et al. 1969, Phys. Rev., **184**, 1393, 1398.
Georgi, H., and Glashow, S. L. 1974, Phys. Rev. Lett., **32**, 438.
Georgi, H., Quinn, H. R., and Weinberg, S. 1974, Phys. Rev. Lett., **33**, 451.
Gildener, E., and Weinberg, S. 1976, Phys. Rev. D., **13**, 3333.
Goldman, T. J., and Ross, D. A. 1979, Phys. Lett., **84B**, 208.
Ginzburg, V. L. 1960, Yad. Fiz., **2**, 2031. [Sov. Phys. Solid State, **2**, 1824.]
Halperin, B., private communication.
Kirzhnits, D. A., and Linde, A. D. 1972, Phys. Lett., **42B**, 471.
Kirzhnits, D. A., and Linde, A. D. 1976, Ann. Phys. (N.Y.), **101**, 195.
Kolm, H. H., et al. 1979, Phys. Rev. D, **4**, 1285.
Kristian, J., Sandage, A., and Westphal, J. 1978, Ap.J., **221**, 383.
Linde, A. D. 1977, Phys. Lett., **70B**, 306.
Peebles, P. J. E. 1966, Ap.J., **146**, 542.
Polyakov, A. 1974, Pis'ma Eksp. Teor. Fiz., **20**, 430 (1974, JETP Lett, **20**, 194).
't Hooft, G. 1974, Nucl. Phys., **B79**, 276.
't Hooft. G. 1976, Nucl. Phys., **B105**, 538.
Weinberg, S. 1974, Phys. Rev. D, **9**, 3357.
Weinberg, S. 1979, Phys. Rev. Lett., **42**, 850.
Witten, E. 1979, Harvard University Report No. HUTP-79/A007.
Zel'dovich, Ya. B., and Khlopov, M. Y. 1979, Phys. Lett., **79B**, 239.

INFLATIONARY UNIVERSE: A POSSIBLE SOLUTION TO THE HORIZON AND FLATNESS PROBLEMS

Alan H. Guth[1]
Stanford Linear Accelerator, Stanford University
Stanford, California

Received: August 11, 1980

The standard model of hot big-bang cosmology requires initial conditions which are problematic in two ways: (1) The early universe is assumed to be highly homogeneous, in spite of the fact that separated regions were causally disconnected (horizon problem); and (2) the initial value of the Hubble constant must be fine tuned to extraordinary accuracy to produce a universe as flat (i.e., near critical mass density) as the one we see today (flatness problem). These problems would disappear if, in its early history, the universe supercooled to temperatures 28 or more orders of magnitude below the critical temperature for some phase transition. A huge expansion factor would then result from a period of exponential growth, and the entropy of the universe would be multiplied by a huge factor when the latent heat is released. Such a scenario is completely natural in the context of grand unified models of elementary-particle interactions. In such models, the supercooling is also relevant to the problem of monopole suppression. Unfortunately, the scenario seems to lead to some unacceptable consequences, so modifications must be sought.

I. Introduction
The Horizon and Flatness Problems

The standard model of hot big-bang cosmology relies on the assumption of initial conditions which are very puzzling in two ways which I will explain below. The purpose of this paper is to suggest a modified scenario which avoids both of these puzzles.

By "standard model," I refer to an adiabatically expanding radiation-dominated universe described by a Robertson-Walker metric. Details will be given in Sec. II.

Before explaining the puzzles, I would first like to clarify my notion of "initial conditions." The standard model has a singularity which is conventionally taken to be at time $t = 0$. As $t \to 0$, the temperature $T \to \infty$. Thus, no initial-

[1]Present address: Center for Theoretical Physics, Massachusetts Institute of Technology, Cambridge, Massachusetts 02139.

value problem can be defined at $t = 0$. However, when T is of the order of the Planck mass ($M_p \equiv 1/\sqrt{G} = 1.22\times10^{19}$ GeV)[2] or greater, the equations of the standard model are undoubtedly meaningless, since quantum gravitational effects are expected to become essential. Thus, within the scope of our knowledge, it is sensible to begin the hot big-bang scenario at some temperature T_0 which is comfortably below M_P; let us say $T_0 = 10^{17}$ GeV. At this time one can take the description of the universe as a set of initial conditions, and the equations of motion then described the subsequent evolution. Of course, the equation of state for matter at these temperatures is not really known, but one can make various hypotheses and pursue the consequences.

In the standard model, the initial universe is taken to be homogeneous and isotropic, and filled with a gas of effectively massless particles in thermal equilibrium at temperature T_0. The initial value of the Hubble expansion "constant" H is taken to be H_0, and the model universe is then completely described.

Now I can explain the puzzles. The first is the well-known horizon problem (Rindler 1956; Weinberg 1972; Misner, Thorne, and Wheeler 1973). The initial universe is assumed to be homogeneous, yet it consists of at least $\sim10^{83}$ separate regions which are causally disconnected (i.e., these regions have not yet had time to communicate with each other via light signals).[3] (The precise assumptions which lead to these numbers will be spelled out in Sec. II.) Thus, one must assume that the forces which created these initial conditions were capable of violating causality.

The second puzzle is the flatness problem. This puzzle seems to be much less celebrated than the first, but it has been stressed by Dicke and Peebles (1979). I feel that it is of comparable importance to the first. It is known that the energy density ρ of the universe today is near the critical value ρ_{cr} (corresponding to the borderline between an open and closed universe). One can safely assume that (Weinberg 1972; Misner 1973)

[2] I use units for which $\hbar = c = k$ (Boltzmann constant) = 1. Then 1 m = 5.068×10^{15} GeV^{-1}, 1 kg = 5.610×10^{26} GeV, 1 sec = 1.519×10^{24} GeV^{-1}, and 1° K = 8.617×10^{-14} GeV.

[3] In order to calculate the horizon distance, one must of course follow the light trajectories back to $t = 0$. This violates my contention that the equations are to be trusted only for $T \lesssim T_0$. Thus, the horizon problem could be obviated if the full quantum gravitational theory had a radically different behavior from the naive extrapolation. Indeed, solutions of this sort have been proposed by A. Zee (1980) and F. W. Stecker (1980). However, it is the point of this paper to show that the horizon problem can also be obviated by mechanisms which are more within our grasp, occurring at temperatures below T_0.

$$0.01 < \Omega_p < 10 , \qquad (1.1)$$

where

$$\Omega \equiv \rho/\rho_{cr} = (8\pi/3)G\rho/H^2 , \qquad (1.2)$$

and the subscript p denotes the value at the present time. Although these bounds do not appear at first sight to be remarkably stringent, they, in fact, have powerful implications. The key point is that the condition $\Omega \approx 1$ is unstable. Furthermore, the only time scale which appears in the equations for a radiation-dominated universe is the Planck time, $1/M_P = 5.4\times10^{-44}$ sec. A typical closed universe will reach its maximum size on the order of this time scale, while a typical open universe will dwindle to a value of ρ much less than ρ_{cr}. A universe can survive $\sim10^{10}$ years only by extreme fine tuning of the initial values of ρ and H, so that ρ is very near ρ_{cr}. For the initial conditions taken at $T_0 = 10^{17}$ GeV, the value of H_0 must be fine tuned to an accuracy of one part in 10^{55}. In the standard model this incredibly precise initial relationship must be assumed without explanation. (For any reader who is not convinced that there is a real problem here, variations of this argument are given in the Appendix.)

The reader should not assume that these incredible numbers are due merely to the rather large value I have taken for T_0. If I had chosen a modest value such as $T_0 = 1$ MeV, I would still have concluded that the "initial" universe consisted of at least $\sim10^{22}$ causally disconnected regions, and the the initial value of H_0 was fine tuned to one part in 10^{15}. These numbers are much smaller than the previous set, but they are still very impressive.

Of course, any problem involving the initial conditions can always be put off until we understand the physics of $T \gtrsim M_P$. However, it is the purpose of this paper to show that these puzzles might be obviated by a scenario for the behavior of the universe at temperatures well below M_P.

The paper is organized as follows. The assumptions and basic equations of the standard model are summarized in Sec. II. In Sec. III, I describe the inflationary universe scenario, showing how it can eliminate the horizon and flatness problems. The scenario is discussed in the context of grand unified models in Sec. IV, and comments are made concerning magnetic monopole suppression. In Sec. V I discuss briefly the key undersirable feature of the scenario: the inhomogeneities produced by the random nucleation of bubbles. Some vague ideas which might alleviate these difficulties are mentioned in Sec. VI.

II. THE STANDARD MODEL OF THE VERY EARLY UNIVERSE

In this section I will summarize the basic equations of the standard model, and I will spell out the assumptions which lead to the statements made in the Introduction.

The universe is assumed to be homogeneous and isotropic, and therefore described by the Robertson-Walker metric (Weinberg 1972; Misner, Thorne, and Wheeler 1973):

$$d\tau^2 = dt^2 - R^2(t)\left[\frac{dr^2}{1-kr^2} + r^2(d\theta^2 + \sin^2\theta\, d\phi^2)\right] , \quad (2.1)$$

where $k = +1$, -1, or 0 for a closed, open, or flat universe, respectively. It should be emphasized that any value of k is possible, but by convention r and R(t) are rescaled so that k takes on one of the three discrete values. The evolution of R(t) is governed by the Einstein equations

$$\ddot{R} = -\frac{4\pi}{3} G(\rho + 3p)R , \quad (2.2a)$$

$$H^2 + \frac{k}{R^2} = \frac{8\pi}{3} G\rho , \quad (2.2b)$$

where $H \equiv \dot{R}/R$ is the Hubble "constant" (the dot denotes the derivative with respect to t). Conservation of energy is expressed by

$$\frac{d}{dt}(\rho R^3) = -p \frac{d}{dt}(R^3) , \quad (2.3)$$

where p denotes the pressure. In the standard model one also assumes that the expansion is adiabatic, in which case

$$\frac{d}{dt}(sR^3) = 0 , \quad (2.4)$$

where s is the entropy density.

To determine the evolution of the universe, the above equations must be supplemented by an equation of state for matter. It is now standard to describe matter by means of a field theory, and at high temperatures this means that the equation of state is to a good approximation that of an ideal quantum gas of massless particles. Let $N_b(T)$ denote the number of bosonic spin degrees of freedom which are effectively massless at temperature T (e.g., the photon contributes two units to N_b); and let $N_f(T)$ denote the corresponding number for fermions (e.g., electrons and positrons together contribute four units). Provided that T is not near any mass thresholds, the thermodynamic functions are given by

$$\rho = 3p = \frac{\pi^2}{30} \mathbf{N}(T)T^4 , \qquad (2.5)$$

$$s = \frac{2\pi^2}{45} \mathbf{N}(T)T^3 , \qquad (2.6)$$

$$n = \frac{\zeta(3)}{\pi^2} \mathbf{N}'(T)T^3 , \qquad (2.7)$$

where

$$\mathbf{N}(T) = N_b(T) + \frac{7}{8} N_f(T) , \qquad (2.8)$$

$$\mathbf{N}'(T) = N_b(T) + \frac{3}{4} N_f(T) . \qquad (2.9)$$

Here n denotes the particle number density, and $\zeta(3)$ = 1.202 06 ... is the Riemann zeta function.

The evolution of the universe is then found by rewriting (2.2b) solely in terms of the temperature. Again assuming that T is not near any mass thresholds, one finds

$$\left(\frac{\dot{T}}{T}\right)^2 + \varepsilon(T)T^2 = \frac{4\pi^3}{45} G\mathbf{N}(T)T^4 , \qquad (2.10)$$

where

$$\varepsilon(T) = \frac{k}{R^2T^2} = k\left[\frac{2\pi^2}{45} \frac{\mathbf{N}(T)}{S}\right]^{2/3} , \qquad (2.11)$$

where $S \equiv R^3 s$ denotes the total entropy in a volume specified by the radius of curvature R.

Since S is conserved, its value in the early universe can be determined (or at least bounded) by current observations. Taking $\rho < 10\, \rho_{cr}$ today, it follows that today

$$\left|\frac{k}{R^2}\right| < 9H^2 . \qquad (2.12)$$

From now on I will take $k = \pm 1$; the special case $k = 0$ is still included as the limit $R \to \infty$. Then today $R > \frac{1}{3} H^{-1}$ $\sim 3\times 10^9$ years. Taking the present photon temperature T_γ as 2.7° K, one then finds that the photon contribution to S is bounded by

$$S_\gamma > 3\times 10^{85} . \qquad (2.13)$$

Assuming that there are three species of massless neutrinos (e, μ, and τ), all of which decouple at a time when the other effectively massless particles are the electrons and photons, then $S_\nu = 21/22\ S_\gamma$. Thus,

$$S > 10^{86} \tag{2.14}$$

and

$$|\varepsilon| < 10^{-58}\mathcal{N}^{2/3} . \tag{2.15}$$

But then

$$\left|\frac{\rho - \rho_{cr}}{\rho}\right| = \frac{45}{4\pi^3}\frac{M_P^2}{\mathcal{N}T^2}|\varepsilon| < 3\times 10^{-59}\mathcal{N}^{-1/3}(M_P/T)^2 . \tag{2.16}$$

Taking $T = 10^{17}$ GeV and $\mathcal{N} \sim 10^2$ (typical of grand unified models), one finds $|\rho - \rho_{cr}|/\rho < 10^{-55}$. This is the flatness problem.

The εT^2 term can now be deleted from (2.10), which is then solved (for temperatures higher than all particle masses) to give

$$T^2 = \frac{M_P}{2\gamma t} , \tag{2.17}$$

where $\gamma^2 = (4\pi^3/45)\mathcal{N}$. (For the minimal SU_5 grand unified model, $N_b = 82$, $N_f = 90$, and $\gamma = 21.05$.) Conservation of entropy implies RT = constant, so $R \propto t^{1/2}$. A light pulse beginning at $t = 0$ will have traveled by time t a physical distance

$$\ell(t) = R(t) \int_0^t dt' R^{-1}(t') = 2t , \tag{2.18}$$

and this gives the physical horizon distance. This horizon distance is to be compared with the radius $L(t)$ of the region at time t which will evolve into our currently observed region of the universe. Again using conservation of entropy,

$$L(t) = [s_p/s(t)]^{1/3}\ L_p , \tag{2.19}$$

where s_p is the present entropy density and $L_p \sim 10^{10}$ years is the radius of the currently observed region of the universe. One is interested in the ratio of volumes, so

$$\frac{\ell^3}{L^3} = \frac{11}{43}\left(\frac{45}{4\pi^3}\right)^{3/2} \mathbf{N}^{-1/2}\left(\frac{M_P}{L_p T_p T_\gamma}\right)^3$$

$$= 4\times 10^{-89}\mathbf{N}^{-1/2}(M_P/T)^3 \ . \tag{2.20}$$

Taking $\mathbf{N} \sim 10^2$ and $T_0 = 10^{17}$ GeV, one finds $\ell_o{}^3/L_o{}^3 = 10^{-83}$. This is the horizon problem.

III. THE INFLATIONARY UNIVERSE

In this section I will describe a scenario which is capable of avoiding the horizon and flatness problems.

From Sec. II one can see that both problems could disappear if the assumption of adiabaticity were grossly incorrect. Suppose instead that

$$S_p = Z^3 S_0 \ , \tag{3.1}$$

where S_p and S_0 denote the present and initial values of R^3s, and Z is some large factor.

Let us look fist at the flatness problem. Given (3.1), the right-hand side (RHS) of (2.16) is multiplied by a factor of Z^2. The "initial" value (at $T_0 = 10^{17}$ GeV) of $|\rho-\rho_{cr}|/\rho$ could be of order unity, and the flatness problem would be obviated, if

$$Z > 3\times 10^{27} \ . \tag{3.2}$$

Now consider the horizon problem. The RHS of (2.19) is multiplied by Z^{-1}, which means that the length scale of the early universe, at any given temperature, was smaller by a factor of Z than had been previously thought. If Z is sufficiently large, then the initial region which evolved into our observed region of the universe would have been smaller than the horizon distance at that time. To see how large Z must be, note that the RHS of (2.20) is multiplied by Z^3. Thus, if

$$Z > 5\times 10^{27} \ , \tag{3.3}$$

then the horizon problem disappears. (It should be noted that the horizon will still exist; it will simply be moved out to distances which have not be observed.)

It is not surprising that the RHS's of (3.2) and (3.3) are approximately equal, since they both correspond roughly to S_0 of order unity.

I will now describe a scenario, which I call the inflationary universe, which is capable of such a large entropy production.

Suppose the equation of state for matter (with all chemical potentials set equal to zero) exhibits a first-order phase transition at some critical temperature T_c. Then as the universe cools through the temperature T_c, one would expect bubbles of the low-temperature phase to nucleate and grow. However, suppose the nucleation rate for this phase transition is rather low. The universe will continue to cool as it expands, and it will then supercool in the high-temperature phase. Suppose that this supercooling continues down to some temperature T_s, many orders of magnitude below T_c. When the phase transition finally takes place at temperature T_s, the latent heat is released. However, this latent heat is characteristic of the energy scale T_c, which is huge relative to T_s. The universe is then reheated to some temperature T_r which is comparable to T_c. The entropy density is then increased by a factor of roughly $(T_r/T_s)^3$ (assuming that the number **N** of degrees of freedom for the two phases are comparable), while the value of R remains unchanged, Thus,

$$Z \approx T_r/T_s \ . \tag{3.4}$$

If the universe supercools by 28 or more orders of magnitude below some critical temperature, the horizon and flatness problems disappear.

In order for this scenario to work, it is necessary for the universe to be essentially devoid of any strictly conserved quantities. Let n denote the density of some strictly conserved quantity, and let $r \equiv n/s$ denote the ratio of this conserved quantity to entropy. Then $r_p = Z^{-3} r_0 < 10^{-84} r_0$. Thus, only an absurdly large value for the initial ratio would lead to a measurable value for the present ratio. Thus, if baryon number were exactly conserved, the inflationary model would be untenable. However, in the context of grand unified models, baryon number is not exactly conserved. The net baryon number of the universe is believed to be created by CP-violating interactions at a temperature of 10^{13}-10^{14} GeV (Yoshimura 1978; 1979a; 1979b; Dimopoulos and Susskind 1978; 1979; Ignatiev, Krashikov, Kuzmin, and Tavkhelidze 1978; Toussaint, Treiman, Wilczek, and Zee (1979); Weinberg 1979; Nanopoulos and Weinberg 1979; Ellis, Gaillard, and Nanopoulos 1979a; 1979b; Honda and Yoshimura 1979; Toussaint and Wilczek 1979; Barr, Segre, and Weldon 1979; Sakharov 1979; Ignatiev, Krashikov, Kuzmin, and Shaposhnikov 1979; Kolb and Wolfram 1980a; 1980b; Fry, Olive, and Turner 1980a; 1980b; Treiman and Wilczek 1980; Senjanović and Stecker to be published). Thus, provided that T_c lies in this range or higher, there is no problem. The baryon production would take place after the reheating. (However, strong constraints are imposed on the entropy which can be generated in any phase transition with $T_c << 10^{14}$ GeV, in particular, the Weinberg-Salam phase transition.)[4]

Let us examine the properties of the supercooling universe in more detail. Note that the energy density $\rho(T)$, given the standard model by (2.5), must now be modified. As $T \to 0$, the system is cooling not toward the true vacuum, but rather toward some metastable false vacuum with an energy density ρ_0 which is necessarily higher than that of the true vacuum. Thus, to a good approximation (ignoring mass thresholds)

$$\rho(T) = \frac{\pi^2}{30} \mathcal{N}(T) T^4 + \rho_0 \ . \tag{3.5}$$

Perhaps a few words should be said concerning the zero point of energy. Classical general relativity couples to an energy-momentum tensor of matter, $T_{\mu\nu}$, which is necessarily (covariantly) conserved. When matter is described by a field theory, the form of $T_{\mu\nu}$ is determined by the conservation requirement up to the possible modification

$$T_{\mu\nu} \to T_{\mu\nu} + \lambda g_{\mu\nu} \ , \tag{3.6}$$

for any constant λ. (λ cannot depend on the values of the fields, nor can it depend on the temperature or the phase.) The freedom to introduce the modification (3.6) is identical to the freedom to introduce a cosmological constant into Einstein's equations. One can always choose to write Einstein's equations without an explicit cosmologial term; the cosmological constant Λ is then defined by

$$\langle 0 | T_{\mu\nu} | 0 \rangle = \Lambda g_{\mu\nu} \ , \tag{3.7}$$

where $|0\rangle$ denotes the true vacuum. Λ is identified as the energy density of the vacuum, and, in principle, there is no reason for it to vanish. Empirically Λ is known to be very small ($|\Lambda| < 10^{-46}$ GeV^4) so I will take its value to be

[4]The Weinberg-Salam phase transition has also been investigated by a number of authors: Witten to be published; Sher 1980; Steinhardt 1980; and Guth and Weinberg 1980.

zero.[5] It is typically of $O(T_c^4)$.

Using (3.5), Eq. (2.10) becomes

$$\left(\frac{\dot{T}}{T}\right)^2 = \frac{4\pi^2}{45} G\mathbf{N}(T)T^4 - \varepsilon(T)T^2 + \frac{8\pi}{3} G\rho_0 \, . \qquad (3.8)$$

This equation has two types of solutions, depending on the parameters. If $\varepsilon > \varepsilon_0$, where

$$\varepsilon_0 = \frac{8\pi^2\sqrt{30}}{45} G\sqrt{\mathbf{N}\,\rho_0} \, , \qquad (3.9)$$

then the expansion of the universe is halted at a temperature T_{min} given by

$$T_{min}^{\;4} = \frac{30\,\rho_0}{\pi^2} \left\{\frac{\varepsilon}{\varepsilon_0} + \left[\left(\frac{\varepsilon}{\varepsilon_0}\right)^2 - 1\right]^{1/2}\right\}^2 , \qquad (3.10)$$

and then the universe contracts again. Note that T_{min} is of $O(T_c)$, so this is not the desired scenario. The case of interest is $\varepsilon < \varepsilon_0$, in which case the expansion of the universe is unchecked. [Note that $\varepsilon_0 \sim \sqrt{\mathbf{N}} T_c^{\,2}/M_P^{\,2}$ is presumably a very small number. Thus $0 < \varepsilon < \varepsilon_0$ (a closed universe) seems unlikely, but $\varepsilon < 0$ (an open universe) is quite plausible.] Once the temperature is low enough for the ρ_0 term to dominate over the other two terms on the RHS of (3.8), one has

$$T(t) \approx \text{const} \times e^{-\chi t} \, , \qquad (3.11)$$

where

$$\chi^2 = \frac{8\pi}{3} G\rho_0 \, . \qquad (3.12)$$

Since RT = const, one has[6]

$$R(t) = \text{const} \times e^{\chi t} \, . \qquad (3.13)$$

[5]The reason Λ is so small is of course one of the deep mysteries of physics. The value of Λ is not determined by the particle theory alone, but must be fixed by whatever theory couples particles to quantum gravity. This appears to be a separate problem from the ones discussed in this paper, and I merely use the empirical fact that $\Lambda \approx 0$.

[6]The effects of a false vacuum energy density on the evolution of the early universe have also been considered by Kolb and Wolfram (1979) and Bludman (1979).

The universe is expanding exponentially, in a false vacuum state of energy density ρ_0. The Hubble constant is given by $H = \dot{R}/R = \chi$. (More precisely, H approaches χ monotonically from above. This behavior differs markedly from the standard model, in which H falls as t^{-1}.)

The false vacuum state is Lorentz invariant, so $T_{\mu\nu} = \rho_0 g_{\mu\nu}$. It follows that $p = -\rho_0$, the pressure is negative. This negative pressure allows for the conservation of energy, Eq. (2.3). From the second-order Einstein equation (2.2a), it can be seen that the negative pressure is also the driving force behind the exponential expansion.

The Lorentz invariance of the false vacuum has one other consequence: The metric described by (3.13) (with $k = 0$) does not single out a comoving frame. The metric is invariant under an O(4,1) group of transformations, in contrast to the usual Robertson-Walker invariance of O(4).[7] It is known as the de Sitter metric, and it is discussed in the standard literature (Weinberg 1972).

Now consider the process of bubble formation in a Robertson-Walker universe. The bubbles form randomly, so there is a certain nucleation rate $\lambda(t)$, which is the probability per (physical) volume per time that a bubble will form in any region which is still in the high-temperature phase. I will idealize the situation slightly and assume that the bubbles start at a point and expand at the speed of light. Furthermore, I neglect k in the metric, so $d\tau^2 = dt^2 - R^2(t)d\vec{x}^2$.

I want to calculate p(t), the probability that any given point remains in the high-temperature phase at time t. Note that the distribution of bubbles is totally uncorrelated except for the exclusion principle that bubbles do not form inside of bubbles. This exclusion principle causes no problem because one can imagine fictitious bubbles which form inside the real bubbles with the same nucleation rate $\lambda(t)$. With all bubbles expanding at the speed of light, the fictitious bubbles will be forever inside the real bubbles and will have no effect on p(t). The distribution of all bubbles, real and fictitious, is then totally uncorrelated.

p(t) is the probability that there are no bubbles which engulf a given point in space. But the number of bubbles which engulf a given point is a Poisson distributed variable, so $p(t) = \exp[-\bar{N}(t)]$, where $\bar{N}(t)$ is the expectation value of the number of bubbles engulfing the point. Thus (Guth and Tye 1980a; 1980b)

$$p(t) = \exp\left[-\int_0^t dt_1 \lambda(t_1) R^3(t_1) V(t,t_1)\right] , \qquad (3.14)$$

where

[7] More precisely, the usual invariance is O(4) if $k = 1$, O(3,1) if $k = -1$, and the group of rotations and translations in three dimensions if $k = 0$.

$$V(t,t_1) = \frac{4\pi}{3}\left[\int_{t_1}^{t} \frac{dt_2}{R(t_2)}\right]^3 \tag{3.15}$$

is the coordinate volume at time t of a bubble which formed at time t_1.

I will now assume that the nucleation rate is sufficiently slow so that no significant nucleation takes place until $T << T_c$, when exponential growth has set in. I will further assume that by this time $\lambda(t)$ is given approximately by the zero-temperature nucleation rate λ_0. One then has

$$p(t) = \exp\left[-\frac{t}{\tau} + O(1)\right] , \tag{3.16}$$

where

$$\tau = \frac{3\chi^3}{4\pi\lambda_0} , \tag{3.17}$$

and O(1) refers to terms which approach a constant $\chi t \to \infty$. During one of these time constants, the universe will expand by a factor

$$Z_\tau = \exp(\chi\tau) = \exp\left(\frac{3\chi^4}{4\pi\lambda_0}\right) . \tag{3.18}$$

If the phase transition is associated with the expectation value of a Higgs field, then λ_0 can be calculated using the method of Coleman and Callan (Coleman 1977; 1979; Callan and Coleman 1977). The key point is that nucleation is a tunneling process, so that λ_0 is typically very small. The Coleman-Callan method gives an answer of the form

$$\lambda_0 = A\rho_0 \exp(-B) , \tag{3.19}$$

where B is a barrier penetration term and A is a dimensionless coefficient of order unity. Since Z_τ is then an exponential of an exponential, one can very easily (Guth and Weinberg 1980; see footnote 4)[8] obtain values as large as $\log_{10} Z \approx 28$, or even $\log_{10} Z \approx 10^{10}$.

Thus, if the universe reaches a state of exponential growth, it is quite plausible for it to expand and supercool

[8]E. J. Weinberg and I are preparing a manuscript on the possible cosmological implications of the phase transition in the SU_5 grand unified model.

by a huge number of orders of magnitude before a significant fraction of the universe undergoes the phase transition.

So far I have assumed that the early universe can be described from the beginning by a Robertson-Walker metric. If this assumption were really necessary, then it would be senseless to talk about "solving" the horizon problem; perfect homogeneity was assumed at the outset. Thus, I must now argue that the assumption can probably be dropped.

Suppose instead that the initial metric, and the distribution of particles, was rather chaotic. One would then expect that statistical effects would tend to thermalize the particle distribution on a local scale (Ellis and Steigman 1980; Ellis, Gaillard, and Nanopoulos 1980). It has also been shown (in idealized circumstances) that anisotropies in the metric are damped out on the time scale of $\sim 10^3$ Planck times (Hu and Parker 1978). The damping of inhomogeneities in the metric has also been studied (Weinberg 1972), and it is reasonable to expect such damping to occur. Thus, assuming that at least some region of the universe started at temperatures high compared to T_c, one would expect that, by the time temperature in one of these regions falls to T_c, it will be locally homogeneous, isotropic, and in thermal equilibrium. By locally, I am talking about a length scale ξ which is of course less than the horizon distance. It will then be possible to describe this local region of the universe by a Robertson-Walker metric, which will be accurate at distance scales small compared to ξ. When the temperature of such a region falls below T_c, the inflationary scenario will take place. The end result will be a huge region of space which is homogeneous, isotropic, and of nearly critical mass density. If Z is sufficiently large, this region can be bigger than (or much bigger than) our observed region of the universe.

IV. GRAND UNIFIED MODELS AND MAGNETIC MONOPOLE PRODUCTION

In this section I will discuss the inflationary model in the context of grand unified models of elementary-particle interactions.[9,10]

[9]The simplest grand unified model is the SU(5) model of H. Georgi and S. L. Glashow (1974). See also Georgi, Quinn, and Weinberg (1974) and Buras, Ellis, Gaillard, and Nanopoulos (1978).

[10]Other grand unified models include the SO(10) model: Georgi (1975), Fritzsch and Minkowski (1975), and Georgi and Nanopoulos (1979a; 1979b). The E(6) model: Gürsey, Ramond, and Sikivie (1975) and Gürsey and Serdaroglu (1978). The E(7) model: Gürsey and Sikivie (1976; 1977) and Ramond (1976). For some general properties of grand unified models, see Gell-Mann, Ramond, and Slansky (1978). For a review, see Langacker (1980).

A grand unified model begins with a simple gauge group G which is a valid symmetry at the highest energies. As the energy is lowered, the theory undergoes a hierarchy of spontaneous symmetry breaking into successive subgroups: $G \to H_n \to \cdots \to H_0$, where $H_1 = SU_3 \times SU_2 \times U_1$ [QCD (quantum chromodynamics) × Weinberg-Salam] and $H_0 = SU_3 \times U_1^{EM}$. In the Georgi-Glashow model (see footnote 9), which is the simplest model of this type, $G = SU_5$ and $n = 1$. The symmetry breaking of $SU_5 \to SU_3 \times SU_2 \times U_1$ occurs at an energy scale $M_x \sim 10^{14}$ GeV.

At high temperatures, it was suggested by Kirzhnits and Linde (1972) that the Higgs fields of any spontaneously broken gauge theory would lose their expectation values, resulting in a high-temperature phase in which the full gauge symmetry is restored. A formalism for treating such problems was developed (Weinberg 1974; Dolan and Jackiw 1974; Kirzhnits and Linde 1976; Linde 1979)[11] by Weinberg and Dolan and Jackiw. In the range of parameters for which the tree potential is valid, the phase structure of the SU_5 model was analyzed by Tye and me (Guth and Tye 1980a; 1980b)[12]. We found that the SU_5 symmetry is restored at $T > \sim 10^{14}$ GeV and that for most values of the parameters there is an intermediate-temperature phase with gauge symmetry $SU_4 \times U_1$, which disappears at $T \sim 10^{13}$ GeV. Thus, grand unified models tend to provide phase transitions which could lead to an inflationary scenario of the universe.

Grand unified models have another feature with important cosmological consequences: They contain very heavy magnetic monopoles in their particle spectrum. These monopoles are of the type discovered by 't Hooft and Polyakov ('t Hooft 1974; Polyakov 1974a; 1974b),[13] and will be present in any model satisfying the above description.[14] These monopoles typically have masses of order $M_x/\alpha \sim 10^{16}$ GeV, where $\alpha = g^2/4\pi$ is the grand unified fine structure constant. Since the monopoles are really topologically stable knots in the Higgs field expectation value, they do not exist in the high-temperature phase of the theory. They therefore come into existence during the course of a phase transition, and the dynamics of the phase transition is then intimately related to the monopole production rate.

The problem of monopole production and the subsequent annihilation of monopoles, in the context of a second-order or

[11] ε-expansion techniques are employed by Ginsparg (to be published).

[12] In the case that the Higgs quartic couplings are comparable to g^4 or smaller (g = gauge coupling) the phase structure has been studied by Daniel and Vayonakis (1980) and by Suranyi (1980).

[13] For a review, see Goddard and Olive (1978).

[14] If $\Pi_1(G)$ and $\Pi_2(G)$ are both trivial, then $\Pi_2(G/H_0) = \Pi_1(H_0)$. In our case $\Pi_1(H_0)$ is the group of integers. For a general review of topology written for physicists, see Mermin (1979).

weakly first-order phase transition, was analyzed by Zeldovich and Khlopov (1978) and by Preskill (1979). In Preskill's analysis, which was more specifically geared toward grand unified models, it was found that relic monopoles would exceed present bounds by roughly 14 orders of magnitude. Since it seems difficult to modify the estimated annihilation rate, one must find a scenario which suppresses the production of these monopoles.

Kibble (1976) has pointed out that monopoles are produced in the course of the phase transition by the process of bubble coalescence. The orientation of the Higgs field inside one bubble will have no correlation with that of another bubble not in contact. When the bubbles coalesce to fill the space, it will be impossible for the uncorrelated Higgs fields to align uniformly. One expects to find topological knots, and these knots are the monopoles. The number of monopoles so produced is then comparable to the number of bubbles, to within a few orders of magnitude.

Kibble's production mechanism can be used to set a "horizon bound" on monopole production which is valid if the phase transition does not significantly disturb the evolution of the universe.[15] At the time of bubble coalescence T_{coal} the size ℓ of the bubbles cannot exceed the horizon distance at that time. So

$$\ell < 2t_{coal} = \frac{M_P}{\gamma T_{coal}^{\;2}} . \tag{4.1}$$

By Kibble's argument, the density n_M of monopoles then obeys

$$n_M \gtrsim \ell^{-3} > \frac{\gamma^3 T_{coal}^{\;6}}{M_P^{\;3}} . \tag{4.2}$$

By considering the contribution to the mass density of the present universe which could come from 10^{16} GeV monopoles, Prekill (1979) concludes that

$$n_M/n_\gamma < 10^{-24} , \tag{4.3}$$

where n_γ is the density of photons. This ratio changes very little from the time of the phase transition, so with (2.7) one concludes

$$T_{coal} < [10^{-24} \frac{2\zeta(3)}{\pi^2}]^{1/3} \gamma^{-1} M_P \approx 10^{10} \text{ GeV} . \tag{4.4}$$

[15]This argument was first shown to me by John Preskill. It is also described by Einhorn et al. (1980), except that they make no distinction between T_{coal} and T_c.

If $T_c \sim 10^{14}$ GeV, this bound implies that the universe must supercool by at least about four orders of magnitude before the phase transition is completed.

The problem of monopole production in a strongly first-order phase transition with supercooling was treated in more detail by Tye and me (Guth and Tye 1980a; 1980b).[16] We showed how to explicitly calculate the bubble density in terms of the nucleation rate, and we considered the effects of the latent heat released in the phase transition. Our conclusion was that (4.4) should be replaced by

$$T_{coal} < 2\times 10^{11} \text{ GeV} , \qquad (4.5)$$

where T_{coal} refers to the temperature just before the release of the latent heat.

Tye and I omitted the crucial effects of the mass density ρ_0 of the false vacuum. However, our work has one clear implication: If the nucleation rate is sufficiently large to avoid exponential growth, then far too many monopoles would be produced. Thus, the monopole problem seems to also force one into the inflationary scenario.[17]

[16] The problem of monopole production was also examined by Einhorn, Stein, and Toussaint (1980), who focused on second-order transitions. The structure of SU(5) monopoles has been studied by Dokos and Tomaras (1980); and by Daniel, Lazarides, and Shafi (1980). The problem of suppression of the cosmological production of monopoles is discussed by Lazarides and Shafi (1980), and Lazarides, Magg, and Shafi (1980); the suppression discussed here relies on a novel confinement mechanism, and also on the same kind of supercooling as in Guth and Tye (1980a; 1980b). See also Fry and Schramm (1980).

[17] An alternative solution to the monopole problem has been proposed by Langacker and Pi (1980). By modifying the Higgs structure, they have constructed a model in which the high-temperature SU_5 phase undergoes a phase transition to an SU_3 phase at $T \sim 10^{14}$ GeV. Another phase transition occurs at $T \sim 10^3$ GeV, and below this temperature the symmetry is $SU_3 \times U_1^{EM}$. Monopoles cannot exist until $T < 10^3$ GeV, but their production is negligible at these low temperatures. The suppression of monopoles due to the breaking of U_1^{EM} symmetry at high temperatures was also suggested by Tye, (1980).

V. PROBLEMS OF THE INFLATIONARY SCENARIO[18]

As I mentioned earlier, the inflationary scenario seems to lead to some unacceptable consequences. It is hoped that some variation can be found which avoids these undesirable features but maintains the desirable ones. The problems of the model will be discussed in more detail elsewhere (see footnote 18), but for completeness I will give a brief description here.

The central problem is the difficulty in finding a smooth ending to the period of exponential expansion. Let us assume that $\lambda(t)$ approaches a constant as $t \to \infty$ and $T \to 0$. To achieve the desired expansion factor $Z > 10^{28}$, one needs $\lambda_0/\chi^4 < 10^{-2}$ [see (3.18)], which means that the nucleation rate is slow compared to the expansion rate of the universe. (Explicit calculations show that λ_0/χ^4 is typically much smaller than this value [Guth and Weinberg 1980; see footnotes 4 and 8]). The randomness of the bubble formation process then leads to gross inhomogeneities.

To understand the effects of this randomness, the reader should bear in mind the following facts.

(i) All of the latent heat released as a bubble expands is transferred initially to the walls of the bubble (Coleman 1977; 1979; Callan and Coleman 1977). This energy can be thermalized only when the bubble walls undergo many collisions.

(ii) The de Sitter metric does not single out a comoving frame. The O(4,1) invariance of the de Sitter metric is maintained even after the formation of one bubble. The memory of the original Robertson-Walker comoving frame is maintained by the probability distribution of bubbles, but the local comoving frame can be reestablished only after enough bubbles have collided.

(iii) The size of the largest bubbles will exceed that of the smallest bubbles by roughly a factor of Z; the range of bubble sizes is immense. The surface energy density grows with the size of the bubble, so the energy in the walls of the largest bubbles can be thermalized only by colliding with other large bubbles.

(iv) As time goes on, an arbitrarily large fraction of the space will be in the new phase [see (3.16)]. However, one can ask a more subtle question about the region of space which is in the new phase: Is the region composed of finite separated clusters, or do these clusters join together to form an infinite region? The latter possibility is called "percolation." It can be shown[19] that the system percolates for large values

[18]This section represents the work of Weinberg, Kesten, and myself. Weinberg and I are preparing a manuscript on this subject.

[19]The proof of this statement was outlined by Kesten (Department of Mathematics, Cornell University), with details completed by me.

of λ_0/χ^4, but that for sufficiently small values it does not. The critical value of λ_0/χ^4 has not been determined, but presumably an inflationary universe would have a value of λ_0/χ^4 below critical. Thus, no matter how long one waits, the region of space in the new phase will consist of finite clusters, each totally surrounded by a region in the old phase.

(v) Each cluster will contain only a few of the largest bubbles. Thus, the collisions discussed in (iii) cannot occur.

The above statements do not quite prove that the scenario is impossible, but these consequences are at best very unattractive. Thus, it seems that the scenario will become viable only if some modification can be found which avoids these inhomogeneities. Some vauge possibilities will be mentioned in the next section.

Note that the above arguments seem to rule out the possibility that the universe was ever trapped in a false vacuum state, unless $\lambda_0/\chi^4 \gtrsim 1$. Such a large value of λ_0/χ^4 does not seem likely, but it is possible (see footnote 8).

VI. CONCLUSION

I have tried to convince the reader that the standard model of the very early universe requires the assumption of initial conditions which are very implausible for two reasons:

(i) The horizon problem. Causally disconnected regions are assumed to be nearly identical; in particular, they are simultaneously at the same temperature.

(ii) The flatness problem. For a fixed initial temperature, the initial value of the Hubble "constant" must be fine tuned to extraordinary accuracy to produce a universe which is as flat as the one we observe.

Both of these problems would disappear if the universe supercooled by 28 or more orders of magnitude below the critical temperature for some phase transition. (Under such circumstances, the universe would be growing exponentially in time.) However, the random formation of bubbles of the new phase seems to lead to a much too inhomogeneous universe.

The inhomogeneity problem would be solved if one could avoid the assumption that the nucleation rate $\lambda(t)$ approaches a small constant λ_0 as the temperature $T \to 0$. If, instead, the nucleation rate rose sharply at some T_1, then bubbles of an approximately uniform size would suddenly fill space as T fell to T_1. Of course, the full advantage of the inflationary scenario is achieved only if $T_1 \lesssim 10^{-28}\ T_c$.

Recently Witten (private communication) has suggested that the above chain of events may in fact occur if the parameters of the SU_5 Higgs field potential are chosen to obey the Coleman-Weinberg condition (Coleman and Weinberg 1973; Ellis, Gaillard, Nanopoulos, and Sachrajda 1979; Ellis, Gaillard, Peterman, and Sachrajda 1980) (i.e., that $\partial^2V/\partial\phi^2 = 0$ at $\phi = 0$). Witten (to be published) has studied this possibility in detail for the case of the Weinberg-Salam phase transition. Here he finds that thermal tunneling is totally ineffective, but instead the phase transition is driven when the temperature of the QCD chiral-symmetry-breaking phase transition is reached. For the SU_5 case, one can hope that a much larger amount of supercooling will be found; however, it is difficult to see how 28 orders of magnitude could arise.

Another physical effect which has so far been left out of the analysis is the production of particles due to the changing gravitational metric (Parker 1977; Lukash, Novikov, Starobinsky, and Zeldovich 1976). This effect may become important in an exponentially expanding universe at low temperatures.

In conclusion, the inflationary scenario seems like a natural and simple way to eliminate both the horizon and the flatness problems. I am publishing this paper in the hope that it will highlight the existence of these problems and encourage others to find some way to avoid the undesirable features of the inflationary scenario.

ACKNOWLEDGEMENTS

I would like to express my thanks for the advice and encouragement I received from Sidney Coleman and Leonard Susskind, and for the invaluable help I received from my collaborators Henry Tye and Erick Weinberg. I would also like to acknowledge very useful conversations with Michael Aizenman, Beilok Hu, Harry Kesten, Paul Langacker, Gordon Lasher, So-Young Pi, John Preskill, and Edward Witten. This work was supported by the Department of Energy under Contract No. DE-AC03-76SF00515.

APPENDIX: REMARKS ON THE FLATNESS PROBLEM

This appendix is added in hope that some skeptics can be convinced that the flatness problem is real. Some physicists would rebut the argument given in Sec. I by insisting that the equations might make sense all the way back to $t = 0$. Then if one fixes the value of H corresponding to some arbitrary temperature T_a, one always finds that when the equations are extrapolated backward in time, $\Omega \to 1$ as $t \to 0$. Thus, they would argue, it is natural for Ω to be very nearly equal to 1 at early times. For physicists who take this point of view, the flatness problem must be restated in other terms. Since H_0 and T_0 have no significance, the model universe must be specified by its conserved quantities. In fact, the model universe is completely specified by the dimensionless constant $\varepsilon \equiv k/R^2T^2$, where k and R are parameters of the Robertson-Walker metric, Eq. (2.1). For our universe, one must take $|\varepsilon| < 3\times 10^{-57}$. The problem then is the to explain why $|\varepsilon|$ should have such a startingly small value.

Some physicists also take the point of view that $\varepsilon \equiv 0$ is plausible enough, so to them there is no problem. To these physicists I point out that the universe is certainly not described exactly by a Robertson-Walker metric. Thus it is difficult to imagine any physical principle which would require a parameter of that metric to be exactly equal to zero.

In the end, I must admit that questions of plausibility are not logically determinable and depend somewhat on intuition. Thus I am sure that some physicists will remain convinced that there really is no flatness problem. However, I am also sure that many physicists agree with me that the flatness of the universe is a peculiar situation which at some point will admit a physical explanation.

REFERENCES

Barr, S., Segre, G., and Weldon, H. A. 1979, Phys. Rev. D, **20**, 2494.

Bludman, S. A. 1979, University of Pennsylvania Report No. UPR-0143T (unpublished).

Bludman, S. A., and Ruderman, M. A. 1977, Phys. Rev. Lett., **38**, 255.

Buras, A. J., Ellis, J., Gaillard, M. K., and Nanopoulos, D. V. 1978, Nucl. Phys., **B135**, 66.

Callan, C. G., and Coleman, S. 1977, Phys. Rev. D, **16**, 1762.

Coleman, S. 1977, Phys. Rev. D, **15**, 2929.

Coleman, S. 1979, in The Ways of Subnuclear Physics, proceedings of the International School of Subnuclear Physics, Ettore Majorana, Erice, 1977, A. Zichichi, ed. (New York: Plenum).

Coleman, S., and Weinberg, E. J. 1973, Phys. Rev. D, **7**, 1888.

Daniel, M., Lazarides, G., and Shafi, Q. 1980, Nucl. Phys., **B170**, 156.

Daniel, M., and Vayonakis, C. E. 1980, CERN Report No. TH. 2860 (unpublished).

Dicke, R. H., and Peebles, P. J. E. 1979, General Relativity: An Einstein Centenary Survey, S. W. Hawking and W. Israel, eds. (London: Cambridge University Press).

Dimopoulos, S., and Susskind, L. 1978, Phys. Rev. D, **18**, 4500.

Dimopoulos, S., and Susskind, L. 1979, Phys. Lett., **81B**, 416.

Dokos, C. P., and Tomaras, T. N. 1980, Phys. Rev. D, **21**, 2940.

Dolan, L., and Jackiw, R. 1974, Phys. Rev. D., **9**, 3320.

Einhorn, M. B., Stein, D. L., and Toussaint, D. 1980, Phys. Rev. ., **21**, 3295.

Ellis, J., Gaillard, M. K., and Nanopoulos, D. V., 1979a, Phys. Lett., **80B**, 360.

Ellis, J., Gaillard, M. K., and Nanopoulos, D. V., 1979b, Phys. Lett., **82B**, 464.

Ellis, J., Gaillard, M. K., and Nanopoulos, D. V., 1980, Phys. Lett., **90B**, 253.

Ellis, J., Gaillard, M. K., Nanopoulos, D. V., and Sachrajda, C. 1979, Phys. Lett., **83B**, 339.

Ellis, J., Gaillard, M. K., Peterman, A., and Sachrajada, C. 1980, Nucl. Phys., **B164**, 253.

Ellis, J., and Steigman, G. 1980, Phys. Lett., **89B**, 186.

Fritzsch, H., and Minkowski, P. 1975, Ann. Phys. (N.Y.), **93**, 193.
Fry, J. N., Olive, K. A., and Turner, M. S. 1980a, Phys. Rev. D., **22**, 2953.
Fry, J. N., Olive, K. A., and Turner, M. S. 1980b, Phys. Rev. D., **22**, 2977.
Fry, J. N., and Schramm, D. N. 1980, Phys. Rev. Lett., **44**, 1361.
Gell-Mann, M., Ramond, P., and Slansky, R. 1978, Rev. Mod. Phys., **50**, 721.
Georgi, H. 1975, in Particles and Fields - 1975, proceedings of the 1975 meeting of the Division of Particles and Fields of the American Physical Society, Carl Carlson, ed., (New York: American Institute of Physics).
Georgi, H., and Nanopoulos, D. V. 1979a, Phys. Lett., **82B**, 392.
Georgi, H., and Nanopoulos, D. V. 1979b, Nucl. Phys., **B155**, 52.
Georgi, H., and Glashow, S. L. 1974, Phys. Rev. Lett., **32**, 438.
Georgi, H., Quinn, H. R., and Weinberg, S. 1974, Phys. Rev. Lett., **33**, 451.
Ginsparg, P., Nucl. Phys. B, to be published.
Goddard, P., and Olive, D. I. 1978, Rep. Prog. Phys., **41**, 1357.
Gürsey, F., Ramond, P., and Sikivie, P. 1975, Phys. Lett., **60B**, 177.
Gürsey, F., and Serdaroglu, M. 1978, Lett. Nuovo Cimento, **21**, 28.
Gürsey, F., and Sikivie, P. 1976, Phys. Rev. Lett., **36**, 775.
Gürsey, F., and Sikivie, P. 1977, Phys. Rev. D, **16**, 816.
Guth, A. H., and Tye, S.-H. 1980a, Phys. Rev. Lett., **44**, 631.
Guth, A. H., and Tye, S.-H. 1980b, Phys. Rev. Lett., **44**, 963.
Guth, A. H., and Weinberg, E. J. 1980, Phys. Rev. Lett., **45**, 1131.
Honda, M., and Yoshimura, M. 1979, Prog. Theor. Phys., **62**, 1704.
Hu, B. L., and Parker, L. 1978, Phys. Rev. D, **17**, 933.
Ignatiev, A., Krashikov, N. V., Kuzmin, V. A., and Tavkhelidze, A. N. 1978, Phys. Lett., **76B**, 436.
Ignatiev, A., Krashikov, N. V., Kuzmin, V. A., and Shaposhnikov, M. E. 1979, Phys. Lett., **87B**, 114.
Kibble, T. W. B. 1976, J. Phys. A, **9**, 1387.
Kirzhnits, D. A., and Linde, A. D. 1972, Phys. Lett., **42B**, 471.
Kirzhnits, D. A., and Linde. A. D. 1976, Ann. Phys. (N.Y., **101**, 195.
Kolb, E. W., and Wolfram, S. 1980a, Phys. Lett., **91B**, 217.
Kolb, E. W., and Wolfram, S. 1980b, Nucl. Phys., **B172**, 224.
Kolb, E. W., and Wolfram, S. (1979), CAL TECH Report No. 79-0984 (unpublished).
Langacker, P. 1980, Report No. SLAC-PUB-2544 (unpublished).
Langacker, P., and Pi, S.-Y. 1980, Phys. Rev. Lett., **45**, 1.

Lazarides, G., Magg, M., and Shafi, Q. 1980, CERN Report No. TH. 2856 (unpublished).
Lazarides, G., and Shafi, Q. 1980, Phys. Lett., **94B**, 149.
Linde, A. D. 1979, Rep. Prog. Phys., **42**, 389.
Lukash, V. N., Novikov, I. D., Starobinsky, A A., Zeldovich, Ya. B. 1976, Nuovo Cimento, **35B**, 293.
Mermin, N. D. 1979, Rev. Mod. Phys., **51**, 591.
Misner, C. W., Thorne, K. S., and Wheeler, J. A. 1973, Gravitation (San Francisco: Freeman).
Nanopoulos, D. V., and Weinberg, S. 1979, Phys. Rev. D, **20**, 2484.
Parker, L. 1977, in Asymptotic Structure of Spacetime, F. Esposito and L. Witten, eds. (New York: Plenum).
Polyakov, A. M. 1974, Pis'ma Zh. Teor. Fiz., **20**, 430. [Sov. Phys. -- JETP Lett., **20**, 194.]
Preskill, J. P., 1979, Phys. Rev. Lett., **43**, 1365.
Rindler, W. 1956, M.N.R.A.S., **116**, 663.
Ramond, P. 1976, Nucl. Phys., **B110**, 214.
Sakharov, A. D. 1979, Zh. Eksp. Teor. Fiz., **76**, 1172. [Sov. Phys. -- JETP, **49**, 594.]
Senjanović, G., and Stecker, F. W., Phys. Lett. B, to be published.
Sher, M. A. 1980, Phys. Rev. D, **22**, 2989.
Stecker, F. W. 1980, Ap.J. (Lett.), **235**, L1.
Steinhardt, P. J. 1980, Harvard report (unpublished).
Suranyi, P. 1980, University of Cincinnati Report No. 80-0506 (unpublished).
't Hooft, G. 1974, Nucl. Phys., **B79**, 276.
Toussaint, D., Treiman, S. B., Wilczek, F., and Zee, A. 1979, Phys. Rev. D, **19**, 1036.
Toussaint, D., amd Wilczek, F. 1979, Phys. Lett., **81B**, 238.
Treiman, S. B., and Wilczek, F. 1980, Phys. Lett., **95B**, 222.
Tye, S.-H. 1980, talk given at the 1980 Guangzhou Conference on Theoretical Physics, Canton (unpublished).
Weinberg, S. 1972, Gravitation and Cosmology (New York: Wiley).
Weinberg, S. 1974, Phys. Rev. D, **9**, 3357.
Weinberg, S. 1979, Phys. Rev. Lett., **42**, 850.
Witten, E., Nucl. Phys. B, to be published.
Witten, E., private communication.
Yoshimura, M. 1978, Phys. Rev. Lett., **41**, 281.
Yoshimura, M. 1979a, Phys. Rev. Lett., **42**, 746(E).
Yoshimura, M. 1979b, Phys. Lett., **88B**, 294.
Zee, A. 1980, Phys. Rev. Lett., **44**, 703.
Zeldovich, Y. B., and Khlopov, M. Y. 1978, Phys. Lett., **79B**, 239.

COSMOLOGY FOR GRAND UNIFIED THEORIES WITH RADIATIVELY INDUCED SYMMETRY BREAKING

Andreas Albrecht and Paul J. Steinhardt
Department of Physics, University of Pennsylvania
Philadelphia, Pennsylvania

Received: January 25, 1982

The treatment of first-order transitions for standard grand unified theories is shown to break down for models with radiatively induced spontaneous symmetry breaking. It is argued that proper analysis of these transitions which would take place in the early history of the universe can lead to an explanation of the cosmological homogeneity, flatness, and monopole puzzles.

Hot big-bang cosmology depends upon special conditions for the early universe to explain the high degree of homogeneity (the "homogeneity puzzle") (Weinberg 1972) and the nearly critical mass density (the "flatness puzzle") (Dicke and Peebles 1979) found in the universe today. In addition, it has been shown that in typical grand unified theories (GUT's) phase transitions should occur in the early history of the universe which lead to many more magnetic monopoles being produced and surviving to the present epoch than are consistent with experiment (the "monopole puzzle") (Preskill 1979).

In this paper we will argue that first-order phase transitions in a special class of GUT's - models in which the GUT symmetry is broken by radiatively induced corrections to the tree approximation to the effective potential - can lead to a solution to these and other cosmological puzzles. (Models with radiatively induced symmetry breaking, a mechanism discovered by Coleman and Weinberg [1973], will be referred to as CW models.) In particular, we will present results for the standard GUT with a finite-temperature effective (scalar) potential:

$$V_T(\phi) = (2A - B)\sigma^2\phi^2 - A\phi^4 + B\phi^4 \ln(\phi^2/\sigma^2) + \tag{1}$$

$$+ 18\,(T^4/\pi^2) \int_0^\infty dx\, x^2 \ln\{1 - \exp[-(x^2 + 25g^2\phi^2/8T^2)^{1/2}]\} ,$$

where the adjoint Higgs field, Φ, has been reexpressed as $\phi(1,1,1,-3/2,-3/2)$ (the fundamental Higgs field will be irrelevant for this discussion); g is the gauge coupling constant; σ is chosen to be 4.5×10^{14} GeV; $B = 5625g^4/1024\pi^2$; and A is a free parameter. Equation (1) includes the one-loop quantum and thermal corrections to the effective potential. For a CW model, the coefficient of the quadratic term, $2A - B$, is set equal to zero and the Higgs mass is $m_{CW} = 2.7\times10^{14}$ GeV. We will also present results for non-CW models in which 2A - B is small and therefore the Higgs mass, m_H, is such that $\Delta_H \equiv (m_H^2 - m_{CW}^2)/m_{CW}^2$ is small.

As for more general GUT models, the process of the first-order phase transition from the SU(5) symmetric phase to the SU(3)⊗U(2)⊗(1) symmetry-breaking phase for the CW model can be understood by studying the shape of the effective potential as a function of the scalar field for various values of the temperature, as shown in Figure 1. For temperatures above the critical temperture (T_{GUT}) for the transition, the symmetric phase ($\phi = 0$) is the global stable minimum of the effective potential. At $T = T_{GUT}$, the symmetric phase and the symmetry-breaking phase have equal energy densities. As the temperature drops below T_{GUT}, the symmetric phase becomes metastable - it has a higher-energy density than stable symmetry-breaking phase but a potential barrier prevents it from becoming unstable.

The decay of a metastable phase to a stable phase has been compared to a classical nucleation process. At $T < T_{GUT}$ there is a rate per unit time per unit volume, $\Gamma(t)$, for producing finite-sized fluctuations containing stable phase - bubbles - within the metastable system. Once produced, the bubbles grow, coalesce and convert the system to the stable phase. For cases where the barrier is sufficiently large, Coleman (1977) and Linde (1977) have found methods for computing $\Gamma(t)$ using a steepest descent (SD) approximation and found it to be of the form $S \exp[-F_f(T)/kT]$. S has the dimensions of $(\text{length})^{-4}$ and k is Boltzmann's constant. $F_f(T)$ is the free energy associated with the bubble computed by SD approximation to be the dominant path across the potential barrier. For this SD bubble fluctuation, the value of $\langle\phi\rangle$ varies from ϕ_f (on the stable-phase side of the barrier) in the center of the bubble to $\phi = 0$ far from the bubble center. As T decreases, $\Gamma(t)$ increases and ϕ_f/σ decreases, where $\phi = \sigma$ corresponds to the symmetry-breaking minimum.

The maintenance of a system in a metastable phase during a first-order phase transition as $T \lesssim T_{GUT}$ continues to decrease is known as supercooling. For phase transitions in the early universe, Guth and Tye (1980), taking into account the expansion of the universe, found the expression for the fractional volume [p(T)] of the universe which at temperature T has decayed to the stable phase during supercooling. When p(T), which depends upon $\Gamma(t)$, is of order unity, the decay is said to be terminated.

Guth (1981) recently suggested that supercooling of first-order phase transitions of typical GUT models can lead to a solution of the cosmological puzzles. His idea was tha the energy density of the universe, ρ, during supercooling is dominated (once $T \lesssim T_{GUT}/10$) by the $\rho_0 \approx T_{GUT}^4$, the difference in energy density between the metastable and stable phases. Then, ρ_0 can act as a cosmological constant in Einstein's equation for standard cosmology described by a Robertson-Walker metric:

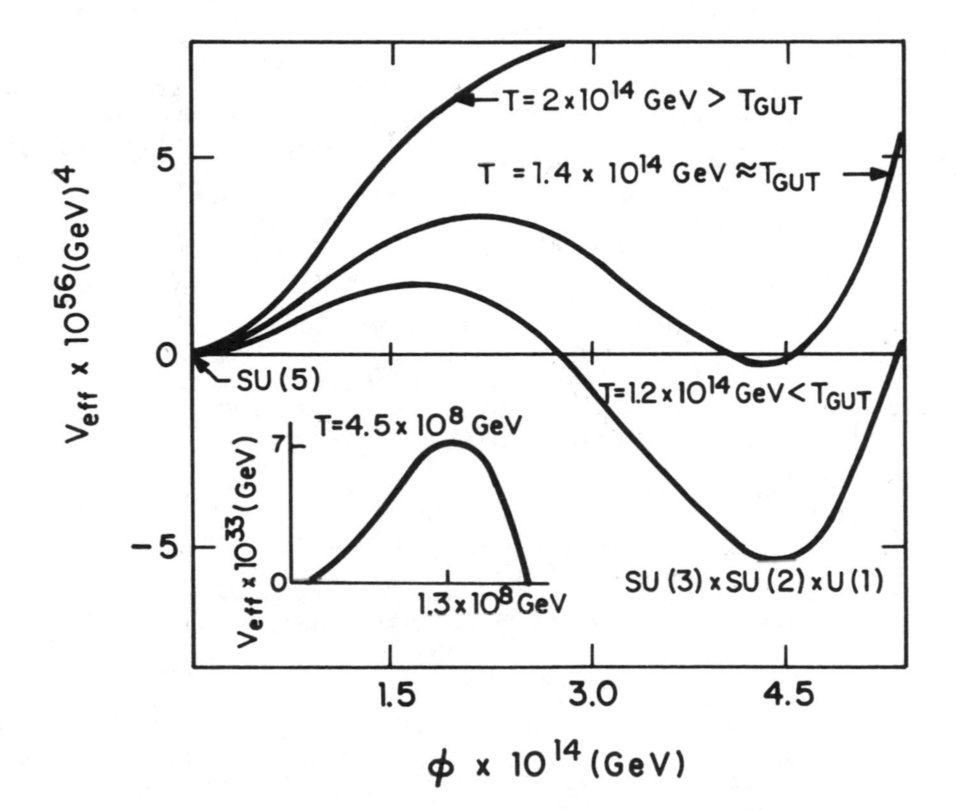

Figure 1. Effective potential vs. ϕ for various values of T.

$$\left(\frac{\dot{R}}{R}\right)^2 = \frac{8\pi}{3M_P^{\,2}}\rho \approx \frac{8\pi}{3M_P^{\,2}} T_{GUT}^{\;\;4} \equiv t_{exp}^{\;\;-2} , \tag{2}$$

where M_P is the Planck mass. The result is exponential growth of the scale factor, $R(t) \sim R_0 \times \exp(t/t_{exp})$. If the growth is continued long enough, each nearly homogeneous region of the universe experiences an expansion which, Guth showed, could explain the cosmological homogeneity, flatness, and monopole puzzles. The problem with Guth's scenario is that when this high degree of expansion can be arranged in typical GUT's (by adjusting free parameters) the rate of expansion of the universe dominates the rate of production and growth of bubbles; the bubbles never coalesce to complete the transition. Since our own universe exhibits the symmetry breaking of the stable phase, it would have to lie within a single, rare bubble, in which case it is difficult to understand how the high entropy found in our universe could have been generated (Steinhardt, to be published).

We claim that CW models and near-CW models in which $\Delta_H < 7\times10^{-6}$ possess special properties that result in completion of the transition to the symmetry-breaking phase along with tremendous expansion. Initially, the analysis of the supercooling for $T < T_{GUT}$ proceeds as in more general GUT models. However, as has been pointed out previously (Steinhardt, to be published), two important features must be taken into account. Firstly, the GUT fine-structure constant, $\alpha\left(= \frac{1}{45}\right.$ at $\left.T = T_{GUT}\right)$, increases as a function of temperature (Sher 1981) until at $T_a \approx 10^6$ GeV, it is of order unity and the one-loop approximation to $V_T(\phi)$ is no longer valid. Secondly, the prefactor in the expression for $\Gamma(t)$, S, is given by T^4 for CW models since, near $\phi = 0$, the only parameter with the dimensions of length that affects the barrier and, thus, the decay, is T^{-1} (Steinhardt, to be published).

When these features are taken into account in the standard SD analysis, they combine to yield the peculiar behavior for $F_f(T)/kT$ and $p(T)$ that is illustrated in Fig. 2. The curves of $F_f(T)/kT$ have been terminated at temperatures, T_{term}, for which the fractional volume of stable phase, $p(T)$, is of order unity. For $\Delta_H > 7.0\times10^{-6}$, $F_f(T_{term})/kT_{term}$ is large, but there is insufficient expansion to solve the cosmological puzzles.

For smaller values of Δ_H, including CW models where $\Delta_H = 0$, $F_f(T_{term})/kT_{term}$ is of order unity or less, values which are too small to trust the SD approximation. The value of the temperature for which the SD approximation fails, T_{SD}, is a function of Δ_H and is roughly 10^8 GeV for $\Delta_H = 0$. T_{SD} is larger than T_α in all cases though, so that one can discuss the breakdown of the SD approximation while still considering only the one-loop approximation to the effective potential, Eq. (1).

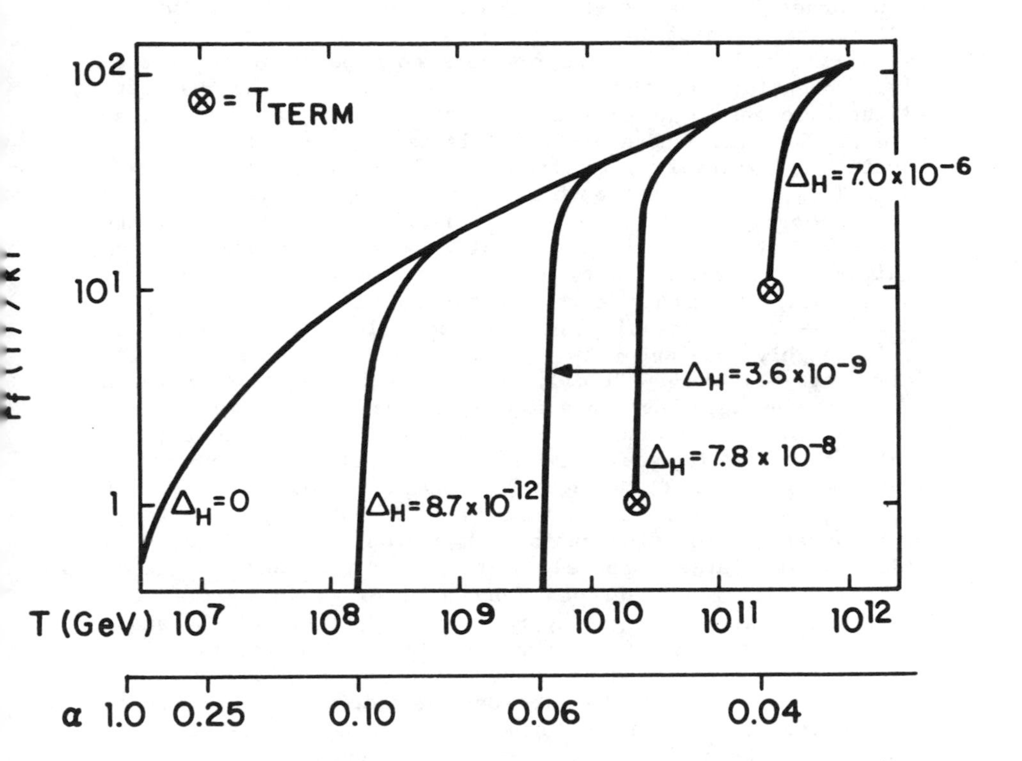

Figure 2. $F_f(T)/kT$ vs. T. Curves terminate for $p(T) \approx 1$. $\Delta_H = (M_H^2 - M_{CW}^2)/M_{CW}^2$.

For the CW and near-CW models, the fact that T reaches T_{SD} before the completion of the transition means that many other types of fluctuations (i.e., paths across the barrier) besides the SD bubble become important. The effect of the barrier becomes negligible. The system (the universe) can be thought of as balancing at a point of unstable (not metastable) equilibrium near $\phi = 0$. Thermal fluctuations drive different regions of the universe away from the SU(5) symmetric phase but towards different symmetry-breaking minima. Since T is the ony dimensional parameter relevant for the fluctuations, the average size of a fluctuation region should be of order T^{-1} and the (roughly constant) value of $\langle\phi\rangle$ within a fluctuation region is of order T.

Even though $\langle\phi\rangle \sim T$ corresponds to a point of the effective potential on the stable-phase side of the barrier, slightly to the right of the barrier in Fig. 1, a crucial feature of CW models and near-CW models is that the effective potential is extremely flat from $\phi = 0$ up to a value of $\phi \sim T_{GUT}$. Thus, even though each fluctuation region has a value of ϕ that corresponds to a point of classical instability for the effective potential, the motion of ϕ towards the stable-phase minimum is characterized by a time constant τ that can be very large. Since ϕ within each fluctuation is of order $T_{SD} \ll T_{GUT}$, the energy density within each fluctuation region is still roughly constant $\sim T_{GUT}^4$. As a result, as concluded independently by Linde (1981),[1] exponential expansion in which ϕ has a value much less than T_{GUT} continues for a time τ.

We have determined an estimate of τ by considering the evolution according to the classical field equations of a state with $\langle\phi\rangle = T$ throughout space (presumably a similar method to what was used by Linde [see footnote 1], for a range of temperatures for fluctuation production. We found τ for CW models to be large compared to t_{exp}. This means that each fluctuation region undergoes many e foldings in spatial expansion before $\langle\phi\rangle$ changes appreciably. Multiplying the scale factor after time τ by the average size of the initial fluctuation region (the size of an SD bubble was used for $T > T_{SD}$) we obtain the size of the fluctuation region, R_U, at time t, after which $\langle\phi\rangle \sim \sigma$ and the exponential expansion ceases. In Table I are shown the results for this computation for a range of temperatures. Column 2 shows R_U(flat) computed with use of the ordinary Klein-Gordon equation. Column 3 shows R_U (exp) derived by using the same equation with an extra drag term $(3\dot{R}\dot{\phi}/R)$ included to account for the time dependence of the scale factor (Albrecht and Steinhardt, to be published). Column 4 shows R_U for $\Delta_H = 3.6\times10^{-11}$, where the time dependence for the scale factor has been included. For this value of Δ_H, the barrier disappears at $T = 3.7\times10^8$ GeV and the maximum value that R_U can achieve is $10^{-0.2}$ cm. Since the

[1] Differences between our results and those of Linde will be discussed in a later paper. (Albrecht and Steinhardt, to be published.)

observed universe, according to the standard model, had a radius of ~1 cm for $T = 10^{14}$ GeV, the choice of Δ_H must be tuned to a value less than 3.6×10^{-11} in order for the observed universe to fit inside a single fluctuation region. Similarly, we have shown that if our scalar field couples to the curvature through a term $bC\phi^2$ (C = curvature), $|b|$ must be $\lesssim 10^{-2}$. Please note, we have treated this calculation as if it were in flat space; curvature effects will be discussed in future publications (Albrecht and Steinhardt, to be published).

Table 1

FLUCTUATION RADIUS VS. T

T (GeV)	R_U(flat) (cm)	R_U(exp) (cm)	$R_U(\Delta_H = 3.6\times10^{-11})$ (cm)
4.5×10^6	10^{81}	$>> 10^{470}$	...
4.5×10^7	$10^{-4.6}$	10^{470}	...
1×10^8	10^{-13}	10^{98}	...
3×10^8	10^{-19}	$10^{-4.4}$	...
4.5×10^8	10^{-20}	10^{-13}	$10^{-7.2}$
4.5×10^9	10^{-23}	10^{-23}	10^{-23}

The result is that the size of a fluctuation region once $T < T_{SD}$ and the SD approximation breaks down is much greater than the size of our present observed universe (10^{28} cm). If one considers the "observed universe" as lying within such a fluctuation region of the "total universe" the special conditions of hot big-bang cosmology can be satisfied. Because the observed universe would be only a small portion of the total universe resulting from the extreme expansion of a small homogeneous region, the homogeneity puzzle is solved. The exponentially large value of scale factor accounts for the flatness puzzle (Guth 1981). Because $\langle\phi\rangle \approx$ const over a fluctuation region, there is no reason to expect to find any monopoles (beyond a small thermally produced number) in the observed universe. Even though there is a discrete symmetry ($\phi \to -\phi$) in the theory, the distance between domain walls (separating regions with $\langle\phi\rangle$ of opposite sign) produced in the transition should be greater than 10^{28} cm, and hence, unobservable. The potential energy stored in the scalar field is eventually converted to thermal energy (Albrecht, Steinhardt, Turner, and Wilczek to be published), thus producing a sizeable entropy density inside each region. Thus, it appears that all the fundamental cosmological puzzles are solved.

A more complete discussion of these results (Albrecht and Steinhardt, to be published) and an analysis of how a fluctuation evolves to thermal equilibrium and produces baryon asymmetry (Albrecht, Steinhardt, Turner, and Wilczek, to be published) will appear in forthcoming publications.

The authors thank A. Guth and E. Witten for useful discussions and G. Segre for his advice and support during the preparation of the manuscript. This work has been supported in part by the U.S. Department of Energy under Contract No. EY-76-C-02-3071.

REFERENCES

Albrecht, A., and Steinhardt, P., to be published.

Albrecht, A., Steinhardt, P., Turner, M., and Wilczek, F., to be published.

Coleman, S. 1977, Phys. Rev. D, **15**, 2929.

Coleman, S., and Weinberg, E. 1973, Phys. Rev. D, **7**, 788.

Dicke, R.H., and Peebles, P. J. E. 1979, in General Relativity: An Einstein Centenary Survey, eds. S. W. Hawking and W. Israel (Cambridge: Cambridge University Press).

Guth, A., and Tye, S. H. H. 1980, Phys. Rev. Lett., **44**, 631 963(E).

Guth, A. 1981, Phys. Rev. D, **23**, 347.

Linde, A. 1977, Phys. Lett., **70B**, 306.

Linde, A. 1981, Lebedev Institute Report No. 229.

Preskill, J. 1979, Phys. Rev. Lett., **43**, 1365.

Sher, M. 1981, Phys. Rev. D., **24**, 1699.

Steinhardt, P., in Proceedings of the 1981 Banff Summer Institute, to be published.

Weinberg, S. 1972, Gravitation and Cosmology (New York: Wiley).